Springer Series in

MATERIALS SCIENCE 115

Springer Series in

MATERIALS SCIENCE

Editors: R. Hull R. M. Osgood, Jr. J. Parisi H. Warlimont

The Springer Series in Materials Science covers the complete spectrum of materials physics, including fundamental principles, physical properties, materials theory and design. Recognizing the increasing importance of materials science in future device technologies, the book titles in this series reflect the state-of-the-art in understanding and controlling the structure and properties of all important classes of materials.

99 **Self-Organized Morphology in Nanostructured Materials**
Editors: K. Al-Shamery, and J. Parisi

100 **Self Healing Materials**
An Alternative Approach to 20 Centuries of Materials Science
Editor: S. van der Zwaag

101 **New Organic Nanostructures for Next Generation Devices**
Editors: K. Al-Shamery, H.-G. Rubahn, and H. Sitter

102 **Photonic Crystal Fibers**
Properties and Applications
By F. Poli, A. Cucinotta, and S. Selleri

103 **Polarons in Advanced Materials**
Editor: A.S. Alexandrov

104 **Transparent Conductive Zinc Oxide**
Basics and Applications in Thin Film Solar Cells
Editors: K. Ellmer, A. Klein, and B. Rech

105 **Dilute III-V Nitride Semiconductors and Material Systems**
Physics and Technology
Editor: A. Erol

106 **Into The Nano Era**
Moore's Law Beyond Planar Silicon CMOS
Editor: H.R. Huff

107 **Organic Semiconductors in Sensor Applications**
Editors: D.A. Bernards, R.M. Ownes, and G.G. Malliaras

108 **Evolution of Thin-Film Morphology**
Modeling and Simulations
By M. Pelliccione and T.-M. Lu

109 **Reactive Sputter Deposition**
Editors: D. Depla and S. Mahieu

110 **The Physics of Organic Superconductors and Conductors**
Editor: A. Lebed

111 **Molecular Catalysts for Energy Conversion**
Editors: T. Okada and M. Kaneko

112 **Atomistic and Continuum Modeling of Nanocrystalline Materials**
Deformation Mechanisms and Scale Transition
By M. Cherkaoui and L. Capolungo

113 **Crystallography and the World of Symmetry**
By S.K. Chatterjee

114 **Piezoelectricity**
Evolution and Future of a Technology
Editors: W. Heywang, K. Lubitz, and W. Wersing

115 **Lithium Niobate**
Defects, Photorefraction and Ferroelectric Switching
By T. Volk and M. Wöhlecke

116 **Einstein Relation in Compound Semiconductors and Their Nanostructures**
By K.P. Ghatak, S. Bhattacharya, and D. De

117 **From Bulk to Nano**
The Many Sides of Magnetism
By C.G. Stefanita

Volumes 50–98 are listed at the end of the book.

Tatyana Volk
Manfred Wöhlecke

Lithium Niobate

Defects, Photorefraction and Ferroelectric Switching

With 102 Figures

Springer

Professor Dr. Tatyana Volk
Russian Academy of Sciences, Institute for Crystallography
Leninskii prospekt 59, 119333 Moscow, Russia
E-mail: volk@ns.crys.ras.ru

Apl. Professor Dr. Manfred Wöhlecke
Universität Osnabrück, Fachbereich Physik
Barbarastr. 7, 49069 Osnabrück, Germany
E-mail: manfred.woehlecke@uos.de

Series Editors:

Professor Robert Hull
University of Virginia
Dept. of Materials Science and Engineering
Thornton Hall
Charlottesville, VA 22903-2442, USA

Professor R. M. Osgood, Jr.
Microelectronics Science Laboratory
Department of Electrical Engineering
Columbia University
Seeley W. Mudd Building
New York, NY 10027, USA

Professor Jürgen Parisi
Universität Oldenburg, Fachbereich Physik
Abt. Energie- und Halbleiterforschung
Carl-von-Ossietzky-Strasse 9–11
26129 Oldenburg, Germany

Professor Hans Warlimont
Institut für Festkörper-
und Werkstofforschung,
Helmholtzstrasse 20
01069 Dresden, Germany

Springer Series in Materials Science ISSN 0933-033X

ISBN 978-3-540-70765-3   ISBN 978-3-540-70766-0 (eBook)

Library of Congress Control Number: 2008931476

Typesetting: Data prepared by SPi using a Springer TEX macro package
Cover concept: eStudio Calamar Steinen
Cover production: WMX Design GmbH, Heidelberg

SPIN: 11930808   57/856/SPi
Printed on acid-free paper

9 8 7 6 5 4 3 2 1
springer.com

Preface

Lithium niobate, $LiNbO_3$, is an oxide ferroelectric with various kinds of pronounced physical properties. This versatility has promoted its career in science and devices. It has been particularly fruitful in the optical regime, where many effects have been found in $LiNbO_3$ and devices introduced using it as a host. One of the few big drawbacks, namely the low level laser damage threshold based on photorefraction due to extrinsic defects was discovered very early. A relatively new topic, not involved so far in any general description, is a fundamental dependence of the optical properties of $LiNbO_3$ on intrinsic defects. Their importance has been realised out due to the development of varies growth techniques in the recent past. The progress in the growth and studies of $LiNbO_3$ crystals with different composition, particularly almost stoichiometric ones, has revealed a significant and sometimes decisive role of the intrinsic defects. For example, the photoinduced charge transport, and therefore the photorefractive properties governing the recording of the phase gratings in $LiNbO_3$, are strongly controlled by the content of intrinsic defects. The recently found impact of intrinsic defects on the coercive field in $LiNbO_3$ is of fundamental importance for the creation of periodically poled structures (PPLN) aimed at the optical-frequency conversion in the quasi-phase matching (QPM) mode of operation. As a consequence of these results, an idea of the intrinsic defects in $LiNbO_3$ has been developed during the last decade and involves microscopic studies on defects, photorefraction and ferroelectric switching using spectroscopic and structure methods.

This monograph is written for researchers as well as for the graduate student. An extensive bibliography is provided, to allow a study of all subjects in greater depth and detail.

We are very grateful and much indebted to Natalya Rubinina for growing all the optical-damage-resistant $LiNbO_3$ crystals for our studies, especially for growing for the first time Zn-doped and In-doped lithium niobate. We acknowledge the support of several projects by International Foundations INTAS (projects 94 1080, 96 0599) and ESF (Oxide Crystals Network), by the Russian Foundation for Basic Research (RFBR, projects 96-02-18851,

00-02-16624), by the German Science Foundation DFG (SFB 225, Graduate College 695) as well as the physics department of the university of Osnabrück and the Institute of Crystallography of the Russian Academy of Sciences during our collaboration for more than a decade.

Moscow, Osnabrück,
July 2008

Tatyana Volk
Manfred Wöhlecke

Contents

List of Greek Symbols

α_{a}, α_{c}	Thermal expansion coefficient
$\alpha(\omega)$	Absorption coefficient
β_{ijk}	Photovoltaic tensor
Γ	Coupling gain factor
η	Diffraction efficiency
$\epsilon_{\mathrm{ij}}^{\mathrm{T}}$, $\epsilon_{\mathrm{ij}}^{\mathrm{S}}$	Dielectric constant (constant tension, stress)
χ	Dielectric susceptibility
λ_{g}, λ_{rec}	Gating and recording wavelength
Λ	Grating period
μ_{n}, μ_{p}	Mobility of electrons and holes
σ_{d}, σ_{ph}	Dark, photo conductivity
τ_{di}	Dielectric relaxation time
τ_{rec}, τ_{er}	Recording and erasure time
τ_{s}	Dark-storage time

List of other Symbols

a_{H}, $\mathrm{c_H}$	Lattice constant (hexagonal setting)
d_{ij}^{T}, d_{ij}^{S}	Nonlinear electro-optical coefficient (const. tension, stress)
D_{n}, D_{p}	Diffusion coefficient of electrons and holes
E_{a}	activation energy
E_{c}, E_{ext}	Coercive field, external field
E_{sc}	Space-charge field
E_{pv}, E_{dif}	Photovoltaic, diffusion field
g_{ij}	Quadratic electro-optic coefficient
$g_{\perp}$, $g_{\parallel}$	g-factor
$G(P, T)$	Elastic Gibbs function
j_{drift}, j_{pv}, j_{dif}	Drift, photovoltaic, Diffusion current
k_{B}	Boltzmann constant
k_{G}	Glass constant
K_{eff}	distribution coefficient
L_{c}, l_{d}	coherence, Debye screening length
N_{A}	Density of neutral acceptor
N_{D}, N_{D}^{+}	Density of neutral and ionized donor
n_{o}, n_{e}	Ordinary and extraordinary index of refraction
P_{s}	Spontaneous polarization
r_{i}, r_{d}	Ion radius, domian radius
r_{ij}^{T}, $\mathrm{r_{ij}^{S}}$	Electro-optical coefficient (constant tension, stress)
t_{p}	Pulse duration
S	Photorefractive sensitivity

T_c	Curie temperature
U	polaron stabilization energy
U_i	energy levels
v_{sw}	sideways domain velocity
w, H	domain width, height
W_E	Depolarization energy
W_w	Domain wall energy

1 Introduction

Summary. Lithium niobate and barium niobate take a key role position in the research on ferroelectricity and its applications. During the last three decades scientific activities grew up in a parallel manner. After a short introduction of review reports, basic properties (structural, mechanical, thermal, electrical, and optical) of congruent and near-stoichiometric crystals are presented. Attempts to grow lithium niobate of various compositions are discussed.

Lithium niobate ($LiNbO_3$, LN) and lithium tantalate ($LiTaO_3$, LT) belong like barium niobate ($BaTiO_3$, BT) to the ABO_3-type ferroelectrics with oxygen octahedra. Ferroelectricity, the spontaneous electric polarization in a material, was reported for all three compounds in 1949 and even earlier. As in the early days of other compounds many physical investigations were hampered by the limited quality of the material with respect to purity, homogeneity, and size of crystals. Many features and properties caused or related to ferroelectricity have been found for the first time in $BaTiO_3$. This is the reason for the frequent use of BT to introduce new effects in the later sections. $BaTiO_3$ is not only a precursor material of the lithium niobate family, but still a competitor with respect to the fundamental science and applications, as may be seen from Fig. 1.1 on p. 2, which shows the publications devoted to LN and BT between 1965 and 2007, according to the WebOfScience database. There were clearly more activities in BT up to about 1970. This is probably due to a delay of about 15 years before Ballman succeeded in growing large $LiNbO_3$ single crystals using the Czochralski technique [1].

Before going into details, we mention early and recent reviews on $LiNbO_3$ and $LiTaO_3$. With respect to Fig. 1.1, it is not surprising that the old classical books on ferroelectricity do not cover LN or LT [2, 3]. One of the most well-known books on principles and applications of ferroelectrics was written by Lines and Glass [4], covering many properties of $LiNbO_3$. More than a decade later, a similar comprehensive overview by Xu followed [5]. In 1999 Kuzminov presented a book completely devoted to $LiNbO_3$ [6]. A very often cited review on "Chemistry and Physics of Lithium Niobate" was written by Räuber [7],

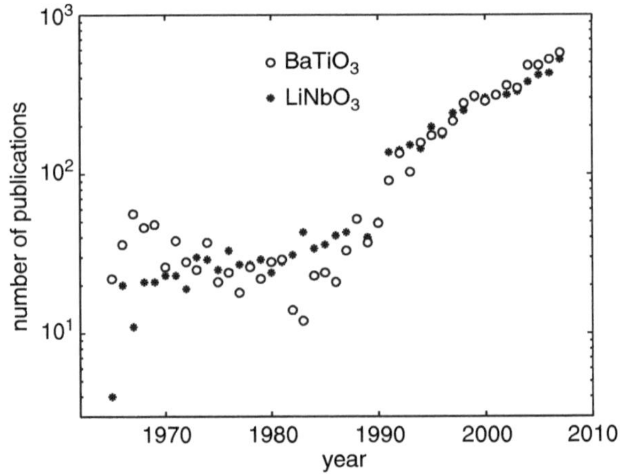

Fig. 1.1. Number of publications for $BaTiO_3$ and $LiNbO_3$ between 1965 and 2007 according to the WebOfScience searching for barium titanate and lithium niobate

an extensive compilation of physical properties and the crystal structure can be found in [8]. A comprehensive description of the photovoltaic effect is found in [9]. In 1989, data at full length are published by the Institution of Electrical Engineers [10] and updated in 2002 [11]. Finally, we mention the two volumes of photorefractive materials and their applications published in the 1989s [12, 13] and the three new editions from 2006 and 2007 [14–16].

The structure of $LiNbO_3$ at room temperature (RT) belongs to the rhombohedral (trigonal) space group *R3c*, with point group 3m. Above the phase transition temperature the crystal transforms to the centrosymmetric space group R3m. The setting of the crystallographic axes for the trigonal symmetry is not unambiguous. There are three choices of axes for $LiNbO_3$, namely, rhombohedral, hexagonal, and orthohexagonal cells. Former two are convenient for crystallographic aims and structure determination. For most applications, the orthohexagonal setting is preferred and the tensor components of properties are given with respect to these axes. In this setting, all axes are mutually orthogonal. Their directions according to the "Standards of Piezoelectric Crystals", Proc. IRE 46, 764, 1958 are settled in the following way. The threefold axis is z, the axis y lies in the mirror plane, and the axis x is orthogonal to both of them. Both z and y are polar (piezoelectric); by convention their positive ends correspond to appearance of the *negative* charge under a uniaxial compression. Additionally, the z-axis is pyroelectric; by convention its positive end corresponds to the appearance of a *positive* charge on cooling the crystal. The x-axis in this setting is non-polar. Experimentally, it is very easy to find the y-axis in usual congruent crystals grown by the Czochralski technique along the polar axis z (so-called 0°-crystals), because in these crystals the mirror planes "outcrop" on the cylindrical boule surface as three very pronounced facets aligned along the growth axis. If the facets

Table 1.1. Properties of congruent and near-stoichiometric $LiNbO_3$ and $LiTaO_3$ (*Part I*: structure, mechanical, thermal, electrical)

Property	LN	NSLN	LT	NSLT	More in
Melting point (°C)	1,255	<1,200	1,650		[17]
Mohs hardness	≈5		≈5.5		[17]
Crystal structure		Trigonal			
Space and point group		R3c	3m		
(RT) Hexagonal setting					
Lattice constant a_H (pm)	515.0	514.7	515.4	515.1	[18]
Lattice constant c_H (pm)	1,386.4	1,385.6	1,378.1	1,377.3	
Density (g cm^{-1})	4.647	4.635	7.456	7.458	[19]
Thermal expansion					
α_a 10^{-6} (K^{-1}) at 300 K	14.1	14.1	12		[17,20]
α_c 10^{-6} (K^{-1}) at 300 K	4.1	6.0	4.2		
Specific heat					
at RT (kJ kg^{-1} K^{-1})	0.628	0.651	0.676		[21]
Thermal conductivity					
at RT (W m^{-1} K^{-1})	3.92	5.97	4.6		[21]
Curie temp. (°C)	1,140	1,206	605	695	[17,22]
Dielectric constant					
ϵ^T_{11}; ϵ^T_{33}	84; 30		54; 42		[23]
ϵ^S_{11}; ϵ^S_{33}	44; 29		42; 41		
Spontaneous polarization					
P_s (μC cm^{-2})	71	62	60	55	[22,24]

are mentally connected to the boule axis, then these imaginary lines oriented mutually by exactly 120° will coincide with y.

$LiNbO_3$ and $LiTaO_3$ crystals are colorless, chemically stable and insoluble in water and organic solvents, and have high melting points. They are well known for their low acoustic losses and are thus excellent materials for surface acoustic wave (SAW) devices, causing a commercial growth of several tons every year. A selection of general properties not related to optics is given in Table 1.1. Although sometimes several digits for a value are given, a scatter of more than 10% is often found in the literature. Therefore, the values should be regarded as examples for the order of the magnitude of the respective property. Further, important applications are based on the electro-optic, nonlinear optical, piezoelectric, and pyroelectric properties with high coefficients advantageous for specific effects and devices. A selection of related optical properties is presented in Table 1.2. Usually LN and LT crystals grown from a melt have a congruently melting composition which has an Li deficiency of about 1.5% compared to the stoichiometric one, see Fig. 1.2.

Table 1.2. Properties of congruent and near-stoichiometric $LiNbO_3$ and $LiTaO_3$ (*Part II*: optical and electrooptical, for an extensive reference list see [17])

Prop.	LN	NSLN	LT	NSLT	More in
Transparent (μm)					
	0.32–5	0.3–5	0.28–5	0.26–5	Sect. 6.1.2
Refractive index n_o; n_e at 633 nm					Sect. 6.2
	2.286; 2.203	2.288; 2.190	2.179; 2.179	2.177; 2.174	[26–28]
EO coefficient r^T_{33}; r^T_{31}; r^T_{22} (pm V^{-1}) at 633 nm					Sect. 6.4
	32; 10; 6.8	38; 10.4;	30; 8; 0		[17]
EO coefficient r^S_{33}; r^S_{31}; r^S_{22} (pm V^{-1}) at 633 nm					Sect. 6.4
	31; 8.6; 3.4	; ; 4.5	30; 7; 0		[17]
NLO coefficient d_{33}; d_{31}; d_{22} (pm V^{-1}) at 1.06 μm					[17]
	34; 6; 3	42; 5; 2.5	14; 0.9;		

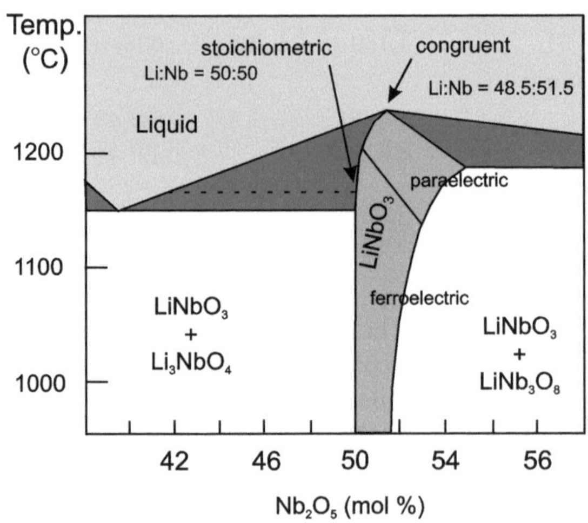

Fig. 1.2. Schematic phase diagram of the Li_2O–Nb_2O_5 pseudobinary system near the congruent and stoichiometric composition of $LiNbO_3$, redrawn from [29]

As seen from the phase diagram of the Li_2O–Nb_2O_5 system (Fig. 1.2), the lithium metaniobate $LiNbO_3$ is a material with a variable composition having a large solid solution range of about 6 mol.% at $T > 1{,}150\,°C$. The liquidus–solidus curve reveals a diffuse maximum at approximately 48.45% Li_2O, whereas for 50% Li_2O (the stoichiometric composition) no singularity exists within the homogeneity range in the phase diagram. To characterize the cation ratio [Li]/[Nb] in $LiNbO_3$ two presentations are used R_1 = [Li]/[Nb] or R_2 = [Li]/([Li] + [Nb]), where the square brackets denote concentration in mol.%. So, the congruently melting composition corresponds to $R_1 = 0.969$, or, which is the same, to $R_2 = 0.4845$. The deviations from $R_1 = 1$ exist only on the Li-deficient side of the equilibrium on the phase diagram. It should be mentioned that in the literature certain scatter is observed concerning the exact position of the congruent point. One of possible reasons is that the accuracy of the determination of the melt composition to some extent depends on the oxygen amount in the feedstock. For example, the initial Nb_2O_5 used in the charge may be oxygen-deficient owing to coexistence of several polymorphous modifications [25].

At the congruent composition the melt and the growing crystal are identical with respect to the composition, so these crystals show the highest uniformity of properties. In other cases, $R_1 = 1$ among them, the composition of the melt and the crystal are slightly varying during the growth and the crystal becomes compositionally non-uniform, particularly along the growth axis. Several physical properties, like the phase transition temperature (Fig. 6.2), and those related to optics (birefringence, the phase-matching temperature, and UV band edge) strongly depend on the [Li]/[Nb] ratio and will be discussed in Chap. 6. To provide the uniformity of these properties, the value of [Li]/[Nb] should be held fixed within 10^{-3}. A high sensitivity of these properties to a compositional non-uniformity is the reason for the of preference of congruent $LiNbO_3$ in optical applications. More recently, it was found that the ratio [Li]/[Nb] controls the linewidths of several spectra, and the coercive field of ferroelectric switching, etc. As shown in Chap. 6, these compositional dependencies are often used as tests for determining [Li]/[Nb] in a crystal. The effect of the ratio metal/oxygen is far less studied, mainly due to the fact that, as discussed in detail in Sect. 2, $LiNbO_3$ crystals, at least congruently melting ones, tend to grow without oxygen vacancies and in as-grown crystals the anion sublattice is fully occupied.

Although large samples of congruent material with excellent optical homogeneity were produced, several methods have been invented to grow stoichiometric crystals (NSLN for $LiNbO_3$ and NSLT for $LiTaO_3$). Stoichiometric crystals have much less lattice defects and thus a more perfect lattice, reducing the density of local field distortions and a weaker phonon coupling, all requirements for all kinds of resonances with small linewidths. Of course electrical and dielectric properties like the coercive field should be influenced in a favorable manner, too, while no improvements are expected for SAW applications.

The commercial congruent $LiNbO_3$ crystals are grown by the usual Czochralski technique by pulling from a melt. This procedure is described in details in several monographs cited below, so we will not dwell on it. Relatively, new methods permitting to grow near-stoichiometric $LiNbO_3$ and $LiTaO_3$ crystals are worthy of a more detailed description. Growth using the Czochralski method with off-congruent melts failed to produce stoichiometric samples, because the composition changed along the growth axis due to a varying concentration ratio [Li]/[Nb]. In 1992, three methods for growing near-stoichiometric LN have been reported. We start with the vapor transport equilibration (VTE). Here, a crystal and a powder with an Li content higher than the crystal are placed in a Pt crucible without any direct contact. Heating to more than 1,000°C for some days increases the Li in the crystal by diffusion. Details like the preparation of the starting material and the conditions are found in the pioneering work of Bordui et al. [30, 31]. Usually plates with a thickness less than 1 mm where used, but recently strategies to apply VTE to plates up to 3 mm thick were reported [32].

Besides this post-growth procedure two more direct growth methods have been developed. Malovichko et al. [34,35] reported on samples being very close to the stoichiometric composition. They were grown from a melt containing up to 10% K_2O, but only a negligible amount of potassium was found in the crystals. Compositions with 49.95 mol.% or even more were obtained and checked experimentally. Boules with a diameter of 20 mm and excellent optical properties where grown by Polgár et al. [36]. They used a mixture containing K_2O, too, but called the method top-seeded solution growth (TSSG), because the mixture has to be considered as a high temperature solution (or flux) rather than a potassium-doped melt with low segregation coefficient.

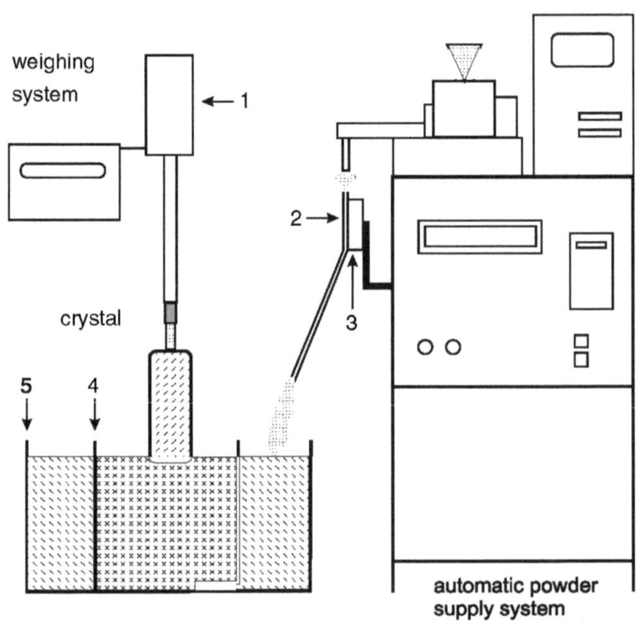

Fig. 1.3. The double crucible Czochralski method after Kitamura et al. [33] (**1**: load cell, **2**: Pt tube, **3**: piezo-vibrator, **4**: inner crucible, **5**: outer crucible)

The third procedure, suitable for growing large boules, is called double crucible Czochralski method (DCCZ) and based on the coexistence of near-stoichiometric $LiNbO_3$ with an Li-rich melt, see the dotted horizontal line in Fig. 1.2. A schematic scheme of the apparatus invented by Kitamura et al. [33] is shown in Fig. 1.3. The crystal is pulled out of an Li-rich melt (about 58–60 Li_2O mol.%) in the inner part of two platinum or iridium crucibles placed into each other. The outer one contains a stoichiometric melt, which can flow into the inner one through a hole at the bottom. Stoichiometric $LiNbO_3$ powder is continuously filled into crucible (5) depending on the increase of the weight of the growing sample, without disturbing the growth process in (4).

Until the beginning of the 1990s, mainly congruently melting samples were grown and investigated; this means compositional specifications referred to the melt or reliable information on the composition was missing. A first comprehensive study on methods to characterize the composition and homogeneity of LN was published in 1996 [37]. Besides non-optical methods like chemical analysis, empirical equations linking the Li content in the melt to that of the crystal, measurements of the Curie temperature, the density, the linewidths of NMR and EPR signals and diffraction methods, non-destructive optical methods have been presented and compared. Most of the latter methods are quantitative and fast featuring a high absolute accuracy of 0.1 mol.% which is increased to about 0.01 mol.% for relative studies. Later, two optical methods have been extended to LT [38, 39]. The situation becomes quite complex when damage-resistent dopants like Mg, Zn, In, or Sc are involved, because such crystals contain several percent of these dopants which may strongly influence the [Li]/[Nb] ratio.

2

Point Defects in $LiNbO_3$

Summary. In a survey of recent publications, the relationship between photorefraction and defect structure is outlined. We start with a description of the situation in undoped lithium niobate at moderate temperature. Changes in spectral characteristics (optical absorption, EPR) caused by reduction treatments and their interpretation using the polaron model are discussed then. The characteristics are used as tests for Nb antisites, i.e., Nb on an Li site. Hydrogen is the first candidate to form an extrinsic defect influencing via the protonic conductivity the charge transport (photorefraction, thermal fixation).

In a next step the photorefractive impurities like the transition metals Fe, Cu, Mn, and Ni are discussed in detail, with special emphasis on iron. Some interesting features of Cr and lanthanide ions supplement this topic. Optical-damage-resistant impurities (Mg, Zn, Sc, In, Hf, and Zr) are antagonists of Nb antisites and strongly reduce the photorefraction, when doping above certain threshold concentrations. These dopants increase the phase transition temperature, the density and shift the bandedge absorption towards the UV. Many properties differ below and above the threshold concentrations, which strongly depend on the crystal composition. Near stoichiometric samples have thresholds at much lower concentrations then congruent ones. A table with energies of the OH stretch mode in lithium niobate closes this chapter.

The origin of the *intrinsic* (structure) and *extrinsic* (impurity) point defects in the $LiNbO_3$ crystal lattice is a subject of several monographs and reviews listed in Sect. 1. Later, we summarize briefly the fundamentals formulated in this literature, dwelling more specifically on recent publications. The discussion is carried out mainly in view of the relationship between photorefraction and defect structure. This is the reason for choosing certain defects in the subsequent sections. We start with a presentation of the ionic radii r_i of the elements discussed in this chapter, see Table 2.1. We use the system proposed by Shannon and Prewitt [40], so the presented data correspond to the coordination number 6 in accordance with the $LiNbO_3$ crystal structure. As known, several definitions of r_i exist, the discrepancy between them being within 15–20 pm.

Table 2.1. Ionic radii

Ion	Li^+	Nb^{5+}	Fe^{2+}	Fe^{3+}	Cr^{3+}	Mg^{2+}	Zn^{2+}	
r_i (pm)	76	64	78	64.5	61.5	72	74	
Ion	Sc^{3+}	In^{3+}	Hf^{4+}	Zr^{4+}	Nd^{3+}	Er^{3+}	Na^+	K^+
r_i (pm)	74.5	81	71	79	98	89	102	138

2.1 Intrinsic Defect Structure of $LiNbO_3$

As shown in Chap. 1, $LiNbO_3$ belongs to the pseudoilmenite structure formed by distorted niobium–oxygen octahedra. Chains of the distorted oxygen octahedra having common faces, are aligned along the polar axis c. In the nonpolar centrosymmetric phase at $T > T_c$ the Li^+ ions are localized within the oxygen planes, whereas the Nb^{5+} ions are in the center of the oxygen octahedra (between the oxygen planes). In the ferroelectric phase at $T < T_c$, the Li^+ ions are shifted with respect to the oxygen plane by 44 pm, and the Nb^{5+} ions by 26–27 pm with respect to the center of the octahedron. As a result, in the polar phase structure the octahedral interstitials are one-third filled by Li ions, one-third by Nb ions and one-third are empty, see Fig. 2.1. The alternation of cation sites may be schematically depicted as a chain $\cdots$Li–Nb–□$\cdots$Li–Nb–□$\cdots$, where □ denotes the *empty* octahedron (referred sometimes to as *vacancy*). The Li octahedron is larger, than the Nb one, because the distances between a Li ion and the closest oxygen ions are 206.8 and 211.2 pm, whereas those for an Nb ion are 188.9 and 211.2 pm, respectively [41]. A larger size of the Li octahedron may qualitatively explain a predominant incorporation of impurity ions onto Li sites as discussed later. As seen from the phase diagram (Fig. 1.2), the $LiNbO_3$ (methaniobate)-structure is stable within Li_2O concentrations from 47 to 50 mol.%. The congruently melting (Li-deficient) composition corresponds to the oxide concentrations 48.45 mol.% Li_2O and 51.55 mol.% Nb_2O_5 with the ratio [Li]/[Nb] = 0.94. Correspondingly, the congruent $LiNbO_3$ (denoted often as CLN) contains about 6% empty Li sites in the lattice. There is a long story of growing defect free stoichiometric $LiNbO_3$ (SLN), which has been briefly discussed in Chap. 1. However, in spite of the development of new growth methods and attempts of practical applications of SLN, the commercial CLN exhibit the best optical quality among all compositions of $LiNbO_3$. Now, we only mention that even for ratios $[Li_2O]/[Nb_2O_5]$ >1 in the melt, the crystal is Li-deficient to some extent anyhow.

At the earliest stages of the discussions of the CLN defect structure, it was reasonable to assume that the most probable defect is the oxygen vacancy, which is characteristic for all oxides. However, structure measurements [41–44] in agreement with model calculations [45] have shown a low probability for the existence of oxygen vacancies in *as-grown* crystals. Therefore $LiNbO_3$ is to some extent a unique oxide containing no oxygen vacancies. Further experimental studies have shown that with decreasing Li_2O content, i.e., increasing

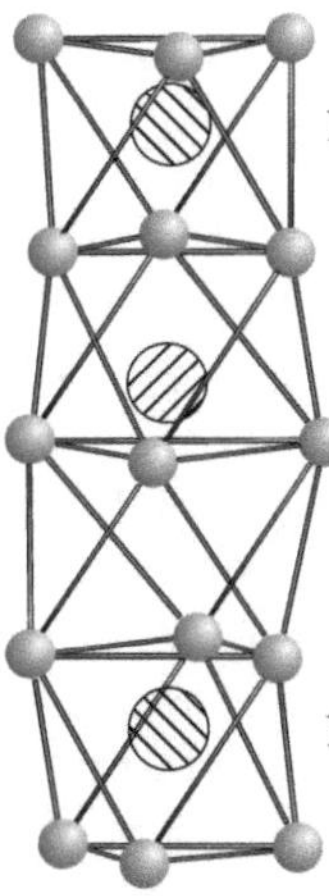

Fig. 2.1. The ideal lithium niobate with the chain ···Li–Nb–□···Li–Nb–□···, where □ denotes the *empty* octahedron

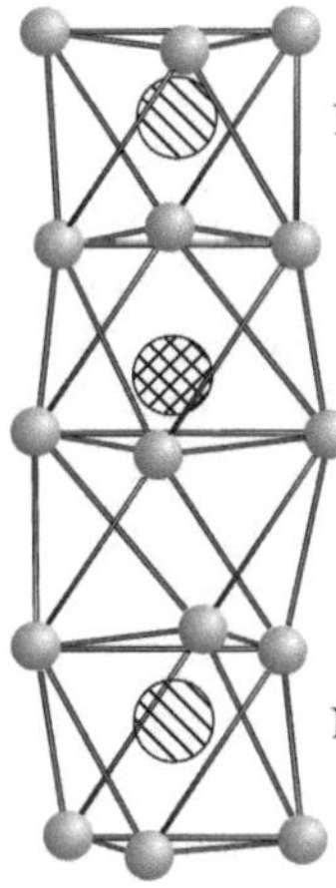

Fig. 2.2. Lithium niobate with an antisite niobium, i.e., Nb on an Li site

concentration of V_{Li}, the density of $LiNbO_3$ increases [46]. To overcome this paradox, some authors assumed that Nb ions incorporate partially on the Li sites [46, 47]. This is very probable, because an Nb^{5+} ion has a smaller ion radius than Li^+ (64 and 76 pm, respectively, in the sixfold coordination [40]). Therefore, the Li-deficient crystal may be formally regarded as one with a Nb surplus, or, in other words, a decreasing Li content is accompanied by an increasing content of the heavier Nb.

The existence of such a stacking fault, Nb on an Li site, (*Nb antisite*, Fig. 2.2) was repeatedly proved by structure studies [41–44], which formed the basis for defect models in $LiNbO_3$.

The shortage of Li during the growth may be described by several possible processes of Li_2O outdiffusion [48]:

$$\langle\langle LiNbO_3\rangle\rangle \rightarrow Li_2O + 2\,V'_{Li} + V_O^{2\bullet}, \tag{2.1}$$
$$\langle\langle LiNbO_3\rangle\rangle \rightarrow 3\,Li_2O + 4\,V'_{Li} + Nb_{Li}^{4\bullet}, \tag{2.2}$$
$$\langle\langle LiNbO_3\rangle\rangle \rightarrow 3\,Li_2O + 4\,V_{Nb}^{5\prime} + 5\,Nb_{Li}^{4\bullet} \tag{2.3}$$

(Note that according to the Kroeger–Vink notation [49], the prime and the bullet, respectively, represent the positive and negative charge states of a defect with respect to the host; however, many authors when writing the compensation equations use the plus sign and the minus sign instead of a prime and bullet.) As mentioned earlier, the formation of oxygen vacancies is energetically unfavorable, so the compensation of the available ${V_{Li}}'$ by ${V_O}^{2\bullet}$ via Reaction 2.1 should be omitted. The compensation within the cation sublattice should occur via Reaction 2.2 or 2.3.

The latter may be modified in view of the compensation of Nb_{Li} by V_{Nb} and presented in the form

$$\langle\langle LiNbO_3\rangle\rangle \rightarrow 3\,Li_2O + 4\,\left(Nb_{Li}^{4\bullet} - V_{Nb}^{5\prime}\right) + Nb_{Li}^{4\bullet}. \tag{2.4}$$

The calculated energies of the defect formation via mechanisms (2.2), (2.3), or (2.4) are 4.56, 15.17, and 10.67 eV, respectively [48]. Therefore, the most energetically favorable one is mechanism (2.2), thus the compensation of Nb antisites by Li vacancies. However, this mechanism was in contradiction with the results of the pioneering work of Abrahams and Marsh [41], according to which Nb ions occupy 4.9% of the Li sites in the lattice, i.e., the *empty* Li sites are mostly populated by Nb. This gave rise to a stronger appropriateness of mechanism (2.4) for conclusions drawn from their experiments [41]. To substantiate mechanism (2.4), i.e., a compensation of $Nb_{Li}^{4\bullet}$ by Nb vacancies, despite its lower energetic benefit, Smyth [50] proposed to regard the stacking fault ($Nb_{Li}^{4\bullet}$–$V_{Nb}^{5\prime}$) as an ilmenite-type point defect (with the cation sequence ···AB–BA–AB···) within the pseudo-ilmenite host, characterized by the cation sequence ···AB–AB–AB···. The lattice energy calculated per unit cell of this ilmenite-type is only 0.1 eV higher than the corresponding energy for perfect $LiNbO_3$. These calculations supported the so-called *Nb site vacancy model* of compensation, see (2.4) and [47, 48]. The formula for congruent (Li-deficient) $LiNbO_3$ in this model may be described by

$$\{[Li]_{1-5x}[Nb_{Li}]_5\}\{[Nb_{Nb}]_{1-4x}[V_{Nb}]_{4x}\}O_3 \tag{2.5}$$

(the groups in the braces correspond to the population of the Li and Nb sites).

However, further structure results led to different conclusions. According to precise data of X-ray and neutron diffraction [42] only 1% of the Li sites are occupied by Nb, whereas about 4% of the Li sites are empty (Li vacancies). This existence of a high concentration of Li vacancies in congruent $LiNbO_3$ was confirmed by other structure reports [43,44] and NMR studies [51]. These data

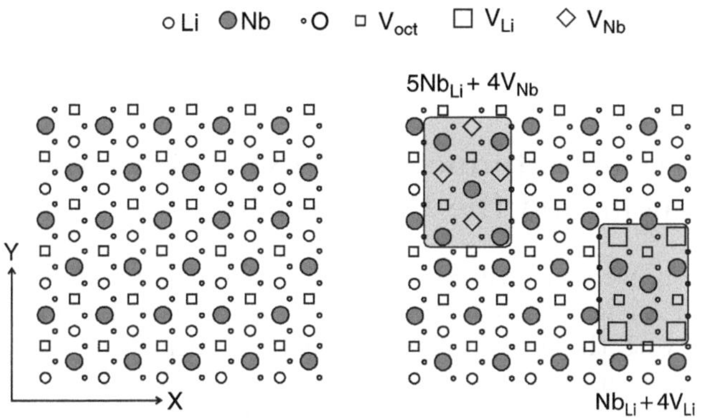

Fig. 2.3. The ideal lithium niobate structure in a projection onto the (001) plane (*left*) and the intrinsic defect Nb_{Li} with two alternatives of its compensation by Li or Nb vacancies (*right*)

supported the *Li site vacancy model*, see Reaction 2.2 and [42, 46] in which an Nb antisite is compensated by four V'_{Li}, for an illustration see Fig. 2.3. In this model a congruent $LiNbO_3$ is described as

$$\{[Li]_{1-5x}[Nb_{Li}]_x[V_{Li}]_{4x}\}[Nb_{Nb}]O_3 \tag{2.6}$$

(here the group in the braces corresponds to the population of the Li sites).

Nowadays, the Li site vacancy model (2.6) is commonly accepted as valid and considerations of all defect reconstructions in $LiNbO_3$ are discussed in its framework. Nevertheless, Wilkinson et al. [43] suggested that the degree of filling the Li sites by Nb ions is accidental and depends, for example, on the growth conditions, so one may expect in as-grown Li-deficient crystals the coexistence of all three cation stacking faults Nb_{Li}, V_{Li}, and V_{Nb}.

The defect scenario discussed in this section characterizes the $LiNbO_3$ structure at moderate temperatures and cannot be applied to high temperatures. For example, neutron diffraction studies led to the conclusion that at high temperatures the Li position is split, so at 600 K about 9% Li atoms are missing on the regular site, being distributed between Li and vacant positions, i.e., the empty octahedron [52, 53]; this fraction increases with rising temperature. A synchrotron X-ray diffraction analysis of stoichiometric $LiNbO_3$ evidenced a small fraction of Li atoms on vacant sites even at room temperature [54]. It should be noted that the properties of the relative crystal $LiTaO_3$ is very often discussed in terms of identical structure defects Ta_{Li} and missing oxygen vacancies. Actually, the structure and optical properties of $LiNbO_3$ and $LiTaO_3$ are very similar. However, so far there is no direct proof of the existence of such stacking faults in $LiTaO_3$, so considerations based on this defect model are speculative in contrast to $LiNbO_3$. Summarizing this section, we emphasize once again that the cornerstone of the $LiNbO_3$ defect structure model is the existence of Nb_{Li} defects and missing oxygen vacancies, established by structure investigations. The discussion of many properties of these crystals is based on this model, as will be shown later.

2.2 Point Defects Induced in $LiNbO_3$ by a Reduction Treatment and the Polaron Model

In this section, we start with a description of the spectral characteristics of $LiNbO_3$ after a chemical reduction. The scope of these experimental data in combination with the intrinsic defect structure outlined in the earlier section are the foundation of the currently accepted polaron model in $LiNbO_3$, for references see for example [55–58]. This model was initially developed and applied to reduced crystals. Later it was extended to the interpretation of the short-living electronic states arising in $LiNbO_3$ under intensive photoexcitation. At present short-living excited states in $LiNbO_3$ and relative crystals are often discussed in terms of the small polaron model, sometimes rather speculatively. Let us follow the historical development of the polaron concept in $LiNbO_3$. Recall that the small bound polaron state is caused by a coupling of a photoinduced electron with the host phonons; the formed local polarization (or deformation) of the host serves as a potential well for the electron producing it. So, polarons are formed due to the Coulomb interaction and phonon coupling of the photoexcited charge carriers with the host ions. One should distinguish between large and small polarons. The former means that the disturbed host region exceeds significantly the lattice constant. In the latter case, the electron coupling with the host is higher, due to which the polaron radius is within the single lattice site. The transport of large polarons may be regarded as a band movement of a free electron with an effective mass, whereas a movement of small polarons occurs by hopping. Both large and small polarons may be related to electrons and holes. The polaron formation is possible if the polaron energy exceeds the electron bandwidth. Provided that the host contains charged defects, the polaron energy is decreased and polarons are localized at defect sites. This is the case of bound small polarons, unlike nonlocalized free polarons in a defect-free lattice. The transport of bound polarons as well as of free polarons occurs via hopping. The essentials, one may find in several reports [59–62]. With respect to $LiNbO_3$ the polaron concept was first applied at an early stage of studies of the photovoltaic effect and photorefraction to account for the origin of a broad absorption band within 1.5–3 eV induced in $LiNbO_3$ by X-ray irradiation or two-photon excitation at low temperatures [63–65]. The appearance of this band was accompanied by a rise of the EPR signal with a g-value characteristic for trapped holes ($2.002 < g < 2.047$). On this basis the induced center was identified as a capture of a photoexcited hole by one of the equivalent O^{2-} ions surrounding a cation defect, or, in other words as an O^- trapped hole, i.e., a small hole polaron.

The bound hole polaron exhibits a broad band in the visible with a peak at about 500 nm [65]. A coexistence of several bands attributed to different centers in this spectral range (the hole polaron O^-, Fe^{2+} and the bipolaron discussed later) extremely hinders the interpretation. According to [66], the optical absorption of O^- is due to a hopping of a hole between equivalent adjacent O^{2-} ions surrounding a cation fault, in the case of $LiNbO_3$ an Li vacancy, thus binding the hole polaron to a single Li vacancy

In addition to a broad band in the visible, Arizmendi et al. [67] detected in the low temperature X-ray-induced spectral bands at 3.2 and 1.6 eV, characteristic for all $LiNbO_3$ crystals independent of the composition. Simultaneously, X-ray irradiation and two-photon excitation induced in $LiNbO_3$ a characteristic 10-line EPR spectrum, see Fig. 2.4 and [63], testifying the existence of an axial center with $g_\perp = 1.72$ and $g_\parallel = 1.9$. This center was interpreted as an electron self-trapped by the Nb_{Li}^{5+} ion (a small polaron Nb_{Li}^{4+})

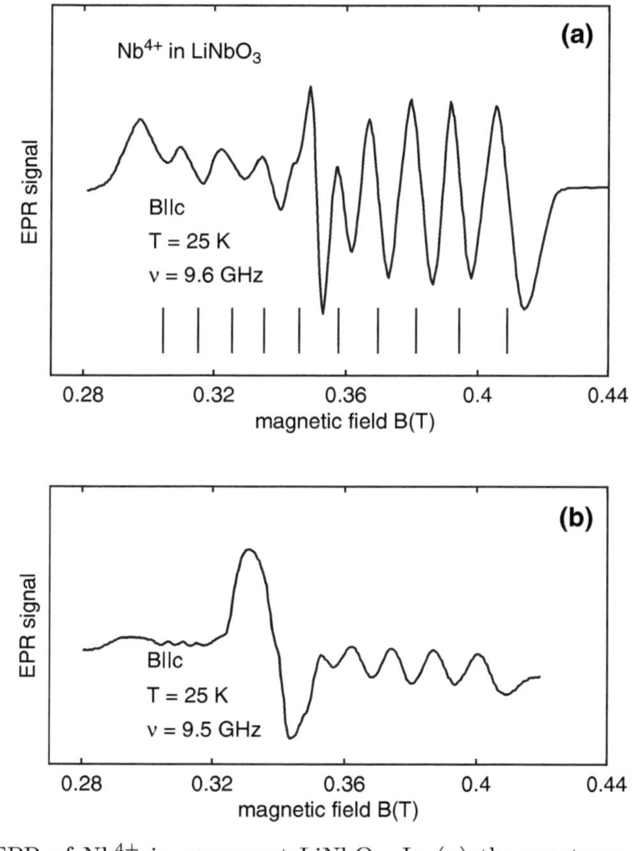

Fig. 2.4. EPR of Nb^{4+} in congruent $LiNbO_3$. In (**a**) the spectrum of a reduced and illuminated sample including the expected 10-line hyperfine structure of ^{93}Nb is shown. The additional strong line is due to Ti^{3+}. In (**b**) the spectrum after X-ray irradiation at low temperature is shown, the wide absorption close to 0.34 T belongs to O^-, data taken from [55]

responsible for the band 1.6 eV observed by Arizmendi et al. [67]. This conclusion formed a base for the further development of the polaron model as applied to the trapping phenomena in $LiNbO_3$.

We now switch to the chemical reduction of $LiNbO_3$. This process is performed by annealing a crystal either in vacuum, in a hydrogen atmosphere, or in an inert gas. The desirable annealing temperature is within 400–600°C, because as mentioned in the earlier section, at higher temperatures the cation sublattice seems to be disturbed [52, 53] and the reduction process is accompanied by the formation of oxygen vacancies [69]. To avoid an Li loss during reduction it is better to embed the crystal into an Li salt (e.g., Li carbonate). The spectrum of a reduced congruent (Li-deficient) undoped $LiNbO_3$ crystal reveals a stable broad dichroic absorption band with a maximum at about 2.5 eV, see Fig. 2.5. The origin of this band was the subject of a long discussion, one may find an appropriate bibliography in [70]. At those times the most likely was to relate it to a formation of oxygen vacancies according to (2.7)

$$O_O \leftrightarrow \frac{1}{2}O_2 + V_O \,, \tag{2.7}$$

whose thermal ionization (2.8)

$$V_O \leftrightarrow V_O^{\bullet} + e^- \tag{2.8}$$

would provide an increased conductivity in reduced $LiNbO_3$. However, a structure proof of missing oxygen vacancies in $LiNbO_3$ undermined this mechanism. There are additional experimental arguments disproving the role of oxygen vacancies in the reduction process. First, after a weak reduction oxygen sites

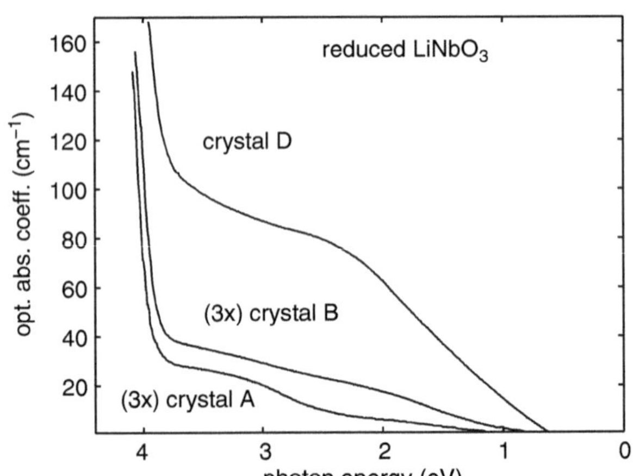

Fig. 2.5. Optical absorption of three reduced crystals with different [Li]/[Nb] ratios (crystal D with $[Li]/[Nb]_{cystal} < 0.92$, crystal B with $[Li]/[Nb]_{cystal} = 0.976$, and crystal A with $[Li]/[Nb]_{cystal} = 1$, respectively). The reduction temperature is 900°C in all cases, redrawn from [68]

remain almost fully occupied. Second, the density of the crystals increases after reduction, which is incompatible with the creation of V_O [71]. Finally, the intensity of this band strongly decreases in Li-enriched crystals (Fig. 2.5), which may scarcely be interpreted in terms of a formation of oxygen vacancies [68]. We should note that the role of the anion sublattice in the reduction of $LiNbO_3$ has not been completely ruled out yet. Although at present, the spectroscopic properties of reduced $LiNbO_3$ are discussed dominantly in terms of the polaron model discussed later, there is a scope of data not lying within its framework. These data will be listed later.

After finding in CLN an Nb_{Li} structure defect an alternative model (2.9) for the reduction process was proposed [48, 71]

$$3O_O + 2V'_{Li} + Nb_{Nb} \leftrightarrow \frac{3}{2}O_2 + Nb_{Li}^{4\bullet} + 6e^- \,. \tag{2.9}$$

According to this model a loss of oxygen from an $LiNbO_3$ surface molecule, is accompanied by an incorporation of the left Nb^{5+} ion onto the empty Li site and ultimately to a total redistribution of V_{Li} and Nb_{Li}; the electrons are trapped by Nb. The oxygen leaves the crystal surface, so no oxygen vacancies appear at the end; the released electrons are captured by traps. The reduced crystal is diamagnetic, so the absorption band itself cannot be assigned to the electron capture by Nb_{Li} forming a paramagnetic Nb_{Li}^{4+}. As was assumed by Schirmer et al. [72] and supported by model calculations [48, 70] the electrons are captured by the pre-existing or newly formed neighboring complexes Nb_{Nb}–Nb_{Li} to form a diamagnetic bipolaron, the stable bonded electron pair $(Nb_{Nb}\text{–}Nb_{Li})^{2-}$. The formation of a bipolaron is energetically favorable [45, 48] and qualitatively means a self-stabilization of an electron pair. In the framework of this model, the broad band at 2.5 eV (Fig. 2.5) is attributed to the transfer of one of the paired electrons from the bipolaron and further capturing by an isolated Nb_{Li}, or, in other words to a dissociation of the bipolaron. The dissociation may be performed thermally or optically.

In line with this consideration, a thermal annealing of a reduced $LiNbO_3$ at $T > 600°C$ resulted in a reversible shift of the band to 1.6 eV [73], which was interpreted as a thermal dissociation of bipolarons and a formation of isolated small polarons. From the temperature dependence of the *turnover* between the bands of the corresponding bipolarons (2.5 eV) and polarons (1.6 eV), the bipolaron dissociation energy was determined to be of about 0.3 eV [73].

The experiments on photoinduced changes of this spectrum at low temperatures seemed to support this interpretation [74, 75]. If a reduced crystal is illuminated by blue–green light at $T \leqslant 80\,K$, then the band at 2.5 eV is pumped over to the metastable band at 1.6 eV, see (Fig. 2.6) and [67, 74, 75]. When heating to $T > 100\,K$, this new band is bleached and the initial spectrum is restored. The appearance of the band at 1.6 eV is accompanied by the rise of a characteristic 10-line EPR pattern (Fig. 2.4), which is thermally annealed simultaneously with the decay of the band at 1.6 eV with the same activation energy [70]. The photoinduced EPR pattern was identical to that

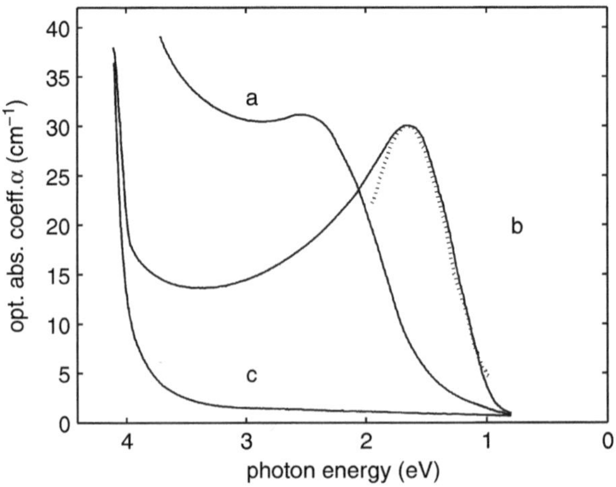

Fig. 2.6. Optical absorption α of congruent reduced $LiNbO_3$. (**a**) The bipolaron band at 2.5 eV before illumination. (**b**) After illumination at $T < 100\,K$ with the $Nb_{Li}^{3\bullet}$ polaron. The *dotted part* is a fit with (2.11)

observed under X-ray irradiation of $LiNbO_3$ at low temperatures [63], which, recall, was attributed to a formation of the small polaron Nb_{Li}^{4+}. The scope of these data allowed to regard the reconstruction of the absorption spectrum under illumination as a proof of the optical dissociation of the bipolaron following the scheme

$$[Nb_{Li}\text{–}Nb_{Nb}]^{2-} \xrightarrow{(h\nu=2.5\,eV)} [Nb_{Li}^{4\bullet}] + Nb_{Nb}^{-} \,. \tag{2.10}$$

The polaron band at 1.6 eV is ascribed to an optical transfer of an electron from Nb_{Li}^{4+} to one of the surrounding Nb_{Nb} ions [55, 70]. The validity of the interpretation of the band at 1.6 eV as a Nb_{Li}^{4+} small polaron was supported by certain estimations. The absorption of the small bound polaron is described with an expression like [55, 58, 59]

$$\alpha(\omega) = \frac{D}{\hbar\omega} \exp\left(-\frac{(4U + \epsilon - \hbar\omega)^2}{8\,\hbar\omega_0}\right), \tag{2.11}$$

where $\omega = 2\pi c/\lambda$, $\hbar\omega_0$ is the phonon energy, U the polaron stabilization energy, ϵ the energy corresponding to the optical transfer of a small polaron between nonequivalent sites, and D a measure for the amplitude. The photoinduced absorption band at 1.6 eV was fitted [55] by the function (2.11) using a typical value of $\hbar\omega_0 \sim 0.1\,eV$ and a polaron stabilization energy $U \sim 0.3\,eV$ according to the estimates from the thermal dissociation of bipolarons [73]. The fitted curve (dotted in Fig. 2.6) is in agreement with the experimental absorption band. Another proof was provided by measurements of the temperature dependence of the conductivity, which found an activation energy of 0.49 eV [70]. The ratio of the optical energy of 1.6 eV to the thermal activation energy is $\approx$4:1, so being in a satisfactory agreement with the ratio required for the small polaron model [60, 61]. Recent experiments in reduced $LiNbO_3$, enlarging the spectral range of studies in the polaron states to 5 µm [57] found one more absorption band at 0.38 eV. These authors assigned it to an optical transition between Nb_{Li} and Nb_{Nb} polaron levels.

This whole consideration concerned the Li-deficient (congruent) $LiNbO_3$. One may see (see Fig. 2.5) an essential decrease of this band with increasing [Li], i.e., with a decrease of $[Nb_{Li}]$. With respect to the reduction equation (2.9) this dependence is qualitatively clear. However, no such experiments have been performed in stoichiometric crystals (totally free of Nb_{Li}), because at those times, when intensive studies of the $LiNbO_3$ defect structure were performed, the growth of the stoichiometric crystals had not been elaborated, yet. At the same time, Sweeney et al. [76] observed a fundamental dissimilarity of the absorption spectrum in reduced highly doped (*above-threshold*) $LiNbO_3$:Mg from the spectra in reduced undoped (or low-doped) crystals. After a rather hard reduction treatment in highly doped $LiNbO_3$:Mg no absorption band appeared in the visible area, whereas a stable broad band arose at about 0.95 eV, which was assigned in [76] to an electron captured by the trap $[Mg_{Li}]^+$. For this band, some authors [77, 78] proposed another explanation, starting from the assumption that the threshold concentration corresponds to the disappearance of Nb antisites from the lattice due to their substitution by Mg ions, for details see the Sect. 2.5. So, in the framework of this substitution scheme, highly doped $LiNbO_3$:Mg crystals were taken as a model for an $LiNbO_3$ free of Nb_{Li}, and distinctive features of their spectroscopic characteristics after reduction were attributed to missing Nb antisites. The band at 0.95 eV was ascribed to a free small polaron Nb_{Nb}^{4+} resulting from a capture of electrons released under reduction, by a regular Nb_{Nb}^{5+} ion and coupling to an unknown lattice perturbation [77,78]. In this presentation, the band at 0.95 eV characterizes an electron hopping between equivalent sites Nb_{Nb}. A low temperature illumination does not affect this absorption spectrum and induces no EPR lines, thus, in other words, no events happen which are characteristic for the congruent crystals and related to Nb antisites.

We now list the data that are inconsistent with the above explanation of the reducing process in $LiNbO_3$ crystals. The interpretation of the band in the visible is not unambiguous. Formerly some authors [67, 68, 79] assigned it, at least partially to a contribution from oxygen vacancies. Some authors [65, 74, 75] found the photoinduced transformations of this band at low temperatures (Fig. 2.6), attributed the observed effects to a photoexcitation of charged oxygen vacancies. The existence of V_O in $LiNbO_3$ may be concluded from the experiments on reducing $LiNbO_3$ in an inert atmosphere under external field [69] or from optical absorption spectra of Ar-implanted $LiNbO_3$ [80]. The detailed in situ studies of the optical absorption spectra under reduction/oxidation of congruent $LiNbO_3$ established a removal of oxygen, that is the formation of V_O and their contribution to the absorption band in visible, if the reduction treatment was performed at $T > 600°C$.

According to the polaron model, all the spectral characteristics of reduced crystals are often used as a *test* for Nb antisites. The appearance of a band at 0.95 eV after reduction is an evidence of missing Nb_{Li}, whereas a photoinduced EPR structure arising in a reduced crystal (Fig. 2.4) combined with a red-shift of the absorption band (1.6 eV, see Fig. 2.6) produces evidence for the presence of Nb_{Li}. The main assumption underlying these tests is that a moderate reduction treatment does not affect or affects very slightly the initial concentration of Nb antisites. As shown later, in terms of these tests the defect structure in the optical-damage-resistant $LiNbO_3$ crystals was analyzed. At the same time, we should mention again that there is a lack of a careful study of these effects in undoped stoichiometric crystals.

According to Fig. 2.5, in near-stoichiometric crystals (NSLN for $LiNbO_3$ and NSLT for $LiTaO_3$) the absorption band in the visible may be also induced by a reduction treatment, although it is weaker than in Li-deficient ones. Its existence is supported by recent data of reduction experiments in NSLN grown by the Czochralski technique from a K_2O-supplemented melt [81] and in NSLN crystals produced by the VTE-technique [82]. No investigations of this band similar to those in the congruent crystals (the photoinduced effects, EPR measurements, etc., described earlier), have been performed in reduced near-stoichiometric ones, which could help in the interpretation of the observed optical absorption. As in stoichiometric crystals, an interpretation of this band in the framework of the polaron model is not appropriate, so it may be related either to oxygen vacancies or to unintentional admixtures. We emphasize again that the formation of oxygen vacancies is provoked by using too high temperatures for the reduction. Recently, the reduction conditions aimed to the studies of bipolarons in reduced crystals were refined in [83]. When annealing congruent $LiNbO_3$ samples in vacuum at $T \approx 600\,°C$ for several hours, a stable noticeable absorption band in the visible appears and steeply grows with T (Fig. 2.7).

Experiments on the photoinduced dissociation of bipolarons by Reaction 2.10 formed a basis for an IR hologram recording in reduced $LiNbO_3$ in the two-color gating regime [84]. This subject will be discussed in more detail in Sect. 5.1 devoted to a nonvolatile optical readout of the photorefractive gratings, so we now restrict ourselves to an outline of the principle of this recording. The absorption band corresponding to the small polaron provides a possible photoexcitation of free electrons at 1.6 eV ($\lambda \approx 800\,nm$). This can be used for recording a grating by a spatially modulated IR light of this spectral range. In this case, a two-step photoexcitation process (the bipolaron photodissociation and subsequent small polaron photoexcitation) should be involved

$$[Nb_{Li}–Nb_{Nb}]^{2-\,(h\nu=2.5\,eV)} \rightarrow \left[Nb_{Li}^{4\bullet}\right]^{-\,(h\nu=1.6\,eV)} \rightarrow Nb_{Li}^{4\bullet} + e^- \,. \qquad (2.12)$$

Actually, in accordance with this scheme in a reduced $LiNbO_3$ a recording grating was achieved with $\lambda = 852\,nm$ (from a Ti–sapphire laser) under simultaneously pumping with spatially uniform visible light $\lambda = 488\,nm$ [84,85].

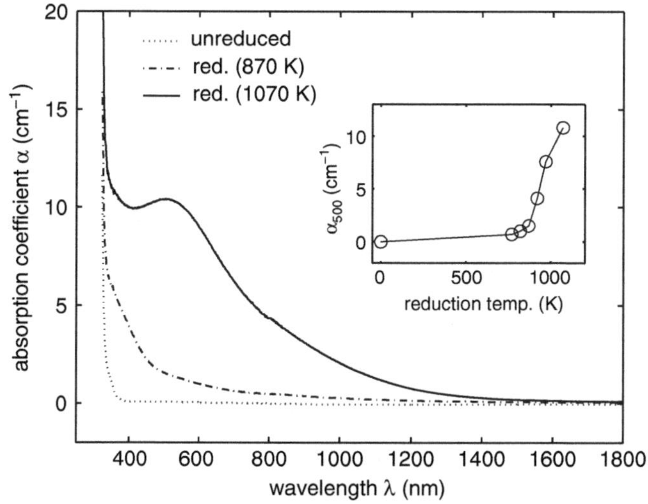

Fig. 2.7. Steady-state absorption of unreduced and reduced $LiNbO_3$. The absorption at $\lambda = 500\,\text{nm}$ as a function of the reduction temperature is shown in the inset, redrawn from [83]

This process was interpreted in terms of the two-step scheme (2.12). Thus, in reduced $LiNbO_3$ crystals a small polaron Nb_{Li}^{4+} serves as an intermediate short-living center for the IR recording. These practically interesting results stimulated wide studies of the absorption band at 1.6 eV in reduced $LiNbO_3$ of different composition. As mentioned earlier, in undoped $LiNbO_3$ this band is stable at $T < 80\,\text{K}$ and short-living at room temperature, i.e., about μs in $LiNbO_3$:Fe [86]. It was found, that its decay time (i.e., the lifetime of the small polaron Nb_{Li}^{4+}) very strongly depends on the crystal composition, for example, it increases up to 10–100 ms in the crystals close to the stoichiometric composition [84, 87] and even up to several seconds in Li-enriched $LiNbO_3$:Mg crystals [88]. Doping of $LiNbO_3$ with some rare earth impurities, such as Pr [89–91] or Tb [92, 93], also seems to increase this time. The dependence of the lifetime on the crystal composition might be accounted for by a kinetic compromise between the electron photoexcitation and the capturing $Nb_{Li}^{4+} \leftrightarrow Nb_{Li}^{5+} + e^-$. The role of the rare earths is still a mystery.

The polaron model was further applied to the interpretation of the short-living photoexcited states in $LiNbO_3$. Due to the development of ultrashort optical pulses, recently a great progress was achieved in this topic since the pioneering investigations of the photoexcitation of $LiNbO_3$ under intensive light pulses [94, 95]. This subject will be outlined in the Sect. 5.1 devoted to the two-color holographic recording. We now dwell on the recent experiments [58, 86, 96, 97] aimed to analyze the short-living photoinduced coloration in $LiNbO_3$. A novelty of these experiments is their particular concentration on the formation of small polarons Nb_{Li}^{4+} and thus a discussion of

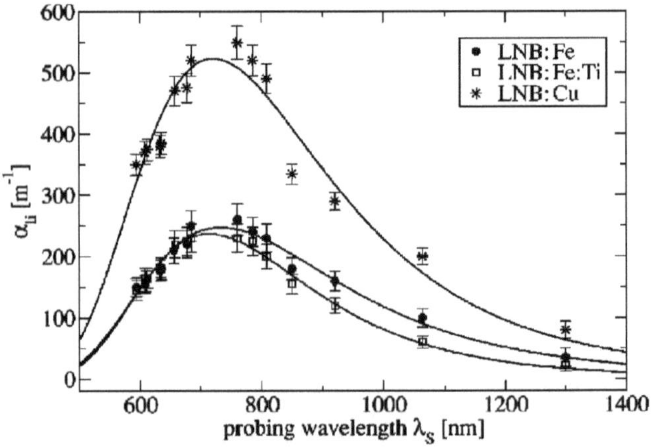

Fig. 2.8. Maximum of the light-induced absorption on the probing beam. *Solid lines* represent fits using (2.11), taken from [58]. Reprinted figure with permission from P. Herth, D. Schaniel, T. Woike, T. Granzow, M. Imlau, E. Krätzig, Physical Review B, 71, 125128 (2005). Copyright 2005 by the American Physical Society

the results in terms of the polaron model. In these works the photoinduced optical absorption $\Delta\alpha$ was studied using the pump–probe technique. This means that $\Delta\alpha$ was pumped by a short light pulse with a wavelength in the visible, i.e., $\lambda_p = 532$ nm and pulse duration in the range of nanoseconds [58, 86, 96]. An even shorter wavelength $\lambda_p = 388$ nm and pulse duration (240 fs) were recently used [97]. The probing was performed with a delayed short pulse or even CW-light at a suitable wavelength. The spectral dependencies of $\Delta\alpha(\lambda)$ obtained in several compositions ($LiNbO_3$:Fe, $LiNbO_3$:Fe:Ti, $LiNbO_3$:Cu), are characterized by two ranges, namely, by a broad band in the visible and a broad band in the near IR with the maximum at about 740 nm (1.67 eV), see Fig. 2.8. The former is trivial, its spectral features depend on the composition and are related to a nonequilibrium overfilling of the capture levels. The band at 740 nm does not depend on the composition, thus it is common for all $LiNbO_3$ crystals. As the spectral position of this band is close to the photoinduced band in reduced $LiNbO_3$ (Fig. 2.6), so the authors analyzed its relation to the Nb_{Li}^{4+} center and fitted the bandshape using the function (2.11), the fits are shown by the solid curves in Fig. 2.8. The curves in all crystals are in good agreement with the experimental data and the estimated values of the polaron stabilization energy are in the range of 0.3 eV and thus close to the value of U obtained for the photoinduced IR band in reduced crystals. Herth et al. [58] interpreted the observed photoinduced IR band as a bound small polaron Nb_{Li}^{4+}. Spectral studies of the photoinduced absorption in the near IR led Beyer et al. [97] to the same conclusion. Their usage of ultrashort pumping pulses permitted them to resolve the formation time of this center, which is according to their estimates within 400 fs. The decay of the band follows a single stretched exponential law $A_0 \exp[-(t/\tau)^\beta]$ with $\beta =$ constant for a given temperature and τ in the range of μs at room temperature.

The kinetics via stretched exponentials do not reply to the requirements of the two-center model [98] according to which the relaxation of a photoinduced absorption should occur by a single-exponent. Recall that the stretched exponential (referred to sometimes as the Kohlrausch law) may be regarded as a superposition of an infinity number of single-exponents, so τ characterizes a mean value of different lifetimes. To account for the observed kinetics Berben et al. [86] suggested that the relaxation occurs via hopping of charge carriers over shallow levels up to a capture by a deep trap. The probability P of this process depends on the distance r between sites ($P \propto \exp(-r)$), in other words in the discussed case on the concentration of Nb_{Li}. The numerical simulations gave a satisfactory agreement with the experimentally observed decay kinetics.

Probing $\Delta\alpha(\lambda)$ in the spectral range 1,300–1,500 nm detected no evidences for the presence of free small polarons [97]. The scope of these data is of fundamental importance for the charge transport scheme in $LiNbO_3$, because, as will be shown later, the defect Nb_{Li} is regarded as the most probable shallow (secondary) electron trap, whose participation in the charge transport scheme can explain the observed peculiarities of the photoconductivity, the photovoltaic currents, the photorefraction and others, especially under an intensive irradiation. Therefore, the data presented [58,86,96,97] independently of their interpretation evidently support the assumption that Nb_{Li} acts as a shallow electron trap.

Recently, an interest to the formation of photoexcited bound hole polarons O^- in $LiNbO_3$, which was formerly attracted for interpretation of the broad absorption band in the visible resulted from X-ray irradiation or the two-photon excitation at low temperatures [63, 64] (for a bibliography on bound hole polarons in oxides see the recent review of Schirmer [66]) was resumed. A similar band with weakly pronounced maxima at about 500 and 400 nm was repeatedly reported in $LiNbO_3$ after X-, γ-, UV-, and electron-irradiation, one may find some references in [58, 96]. The data on its shape and relaxation kinetics are very scattered, the characteristic times depend on the composition and irradiation conditions. As mentioned earlier, the coexistence of absorption bands attributed to various centers (Fe^{2+}-like ions, O^-, ionized oxygen vacancies, bipolarons in reduced crystals) in this spectral range extremely hampers the interpretation. When analyzing the spectra and relaxation kinetics of the band in the blue–green spectral range induced in $LiNbO_3$ and $LiNbO_3$:Fe by irradiation with 532 nm, pulse durations of the order of nanoseconds, intensities within 0.5 GW cm^{-2}, the authors of [58, 96] concluded that a contribution from hole polarons was detected. Their conclusion was expanded to the reduced $LiNbO_3$ [83] and a general model of the photoexcitation with account for participation of bound small polarons, bipolarons, and Fe^{2+}/Fe^{3+} was proposed.

2.3 Hydrogen in the $LiNbO_3$ Lattice

The bibliography on OH^- ions in oxides on the whole one may find in a recent review [99] and on hydrogen in $LiNbO_3$ in the review [100]. We shortly summarize the available material. In synthetic oxides the incorporation mechanism of hydrogen is not well understood, but it is widely accepted that it enters the lattice from the ambient atmosphere during or after the growing process. Among the possible hydrogen-based defects the hydroxyl ion OH^- attracts the attention by its presence in a wide range of materials. In oxide crystals, hydroxyl ions occupy oxygen sites. Its surplus positive charge with respect to the lattice ($[OH^-_O]^\bullet$) compensates some other kinds of intrinsic or extrinsic defects, allowing one to probe the defect structure by studying the spectroscopic properties of the hydroxyl ions affected by their surroundings. A relatively low concentration of OH^- permits to regard them as isolated diatomic molecules in the lattice. As the hydrogen impurity is much lighter than the host atoms, so it gives rise to a localized vibrational mode with a frequency much higher than the vibrational frequencies of the host. The vibrational modes of OH^- are studied by IR absorption spectroscopy and Raman scattering and can be interpreted in the framework of an anharmonic oscillator model. The stretch mode frequency of the hydroxyl ion in various oxide crystals is in the range between 3,200 and 3,700 cm^{-1}, the typical halfwidth of the OH^- band being between 5 and 50 cm^{-1}.

An absorption band in $LiNbO_3$ have been reported as early as 1968 and assigned to hydrogen incorporated into the crystal [101], later polarization characteristics and the fine structure were analyzed [102]. A large number of experimental studies on the OH stretch mode in $LiNbO_3$ (see [99, 100] and references therein) may be shortly summarized as follows. In congruent $LiNbO_3$, the usually observed broad IR band has a full width at half-maximum (FWHM) of about 32 cm^{-1} , whereas in $LiTaO_3$ the FWHM is about 50 cm^{-1}. In $LiNbO_3$ strongly doped with hydrogen, the band is sufficiently broadened [103]. The absorption in congruent $LiNbO_3$ consists of a strong peak at 3,482 cm^{-1} (2.87 μm) and a weaker one at 3,467 cm^{-1} (2.88 μm), see Fig. 2.9.

Raman scattering measurements in congruent $LiNbO_3$ crystals have shown that the local symmetry of the OH-related defect is C_1 [104]. As well as in other oxides, the band is strongly dichroic, namely the optical absorption for the ordinary light polarization (σ) significantly exceeds that for extraordinary (π) one, i.e., the hydrogen stretching vibration occurs in the plane perpendicular to the polar *c*-axis. For a given composition of $LiNbO_3$ the IR band does not depend on temperature from room temperature down to liquid helium, which is unlike in other ABO_3 compounds. At the same time, the shape of the IR band in $LiNbO_3$ and its peak position noticeably depend on the [Li]/[Nb ratio [102, 105]. When increasing the Li content, the peak position is shifted and the band is narrowed (Fig. 2.9) owing to the disappearance of intrinsic defects, thus a more homogeneous local environment is achieved. Similar narrowing of the OH-band was found in near-stoichiometric $LiTaO_3$ [39,106]. The

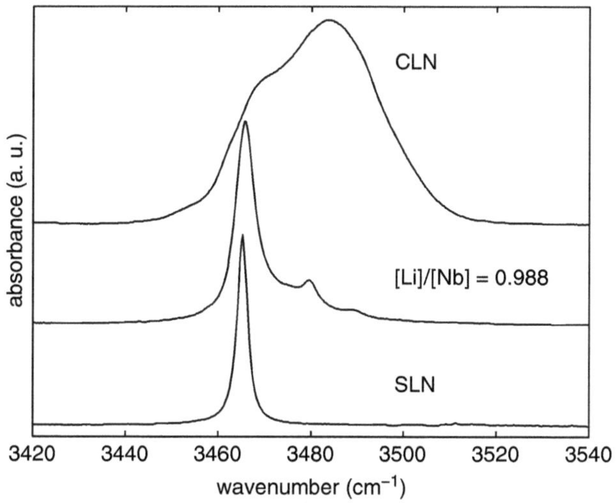

Fig. 2.9. OH absorption bands in $LiNbO_3$ of various composition, redrawn from [36]

dependence of the IR on other factors, such as doping with optical-damage-resistant impurities Mg, Zn, In, Sc, and even on the ferroelectric switching will be discussed in the appropriate sections.

The model proposed for the OH absorption in $LiNbO_3$ will be introduced in brief later [105, 107]. $LiNbO_3$ belongs to the pseudo-ilmenite group composed of six planar rows of oxygen atoms in a distorted close-packed hexagonal structure. The chains of distorted facet-linked oxygen octahedra are oriented along the polar *c*-axis, the oxygen planes being perpendicular. The OH stretch mode is due to protons coupling to one of the six nearest O^{2-} ions. The distortion of the oxygen octahedra results in different lengths for the available O–O bonds, i.e., in six nonequivalent sites for the hydroxyl ions. The model assumes that the longest (336 pm) O–O bonds, i.e., positions 5 and 6 in Fig. 2.10, are not occupied by OH, so the OH absorption band should be formed by four components corresponding to four different OH-bonds. Actually, a careful deconvolution of the band shapes of the OH spectra for different [Li]/[Nb] ratios gives 4 or 5 lines. From a deconvolution of the band in nearly stoichiometric $LiNbO_3$ the four Lorentzian components 3,465, 3,472, 3,479, and 3,488 cm^{-1} were obtained [108], formerly only two being well resolved. So, following this model in $LiNbO_3$ the hydroxyl ions (or, which is the same, protons) are statistically distributed over four nonequivalent positions. The total area under the band is proportional to the whole OH concentration [109] and the intensities of the individual (deconvoluted) lines are assumed to be proportional to the amount of OH ions on the corresponding O sites [108]. As shown in [108], this distribution is not in equilibrium, because in stoichiometric $LiNbO_3$ at elevated temperatures the intensities of the IR lines vary very slowly on a time scale of 10 h. Particularly, the intensities of the lines 3,472 and 3,488 cm^{-1}

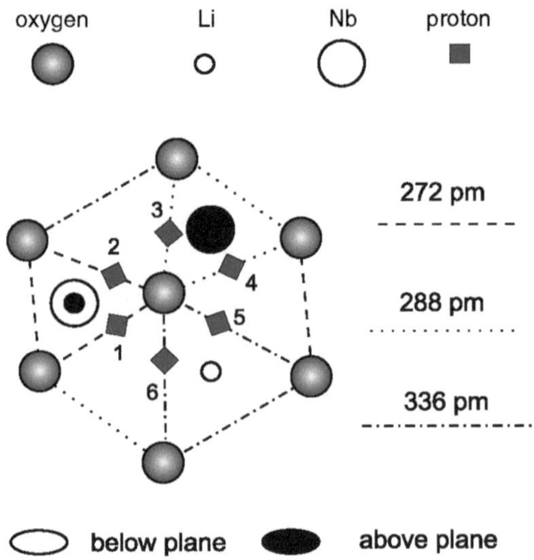

Fig. 2.10. A schematic drawing of the oxygen plane perpendicular to the *c*-axis, redrawn from [107]

increase at the expense of 3,465 and 3,479 cm^{-1}, the total area under the band (i.e., the total amount of hydrogen) being unchanged. Moreover, after growth of the crystal or annealing at about 1,000°C, the OH-vibrational spectrum reaches an equilibrium state during an extremely long time of the order of years [110]. The intensity ratio of two bands $R = I_{3,480}/I_{3,465}$ (where the subscripts denote the wavenumbers) slowly grows after annealing tending to a saturation during 10–15 months.

Because of an easy observation of the OH stretch mode by absorption spectroscopy, it serves as a tool for a quantitative determination of the hydrogen content required particularly for discussing the transport process. For today, in congruent undoped $LiNbO_3$ the most reliable averaged value of the cross-section is $\sigma_{OH} \approx 4 \times 10^{-19}\,cm^{-2}$ obtained by different methods in [109, 111, 112]. Therefore, the concentration range of hydroxyl ions may be estimated approximately as $c_{OH} = c_H \approx 2 \times 10^{18} cm^{-2} \alpha_{2.87\,\mu m}$, where $\alpha_{2.87\,\mu m}$ is the peak absorption value in cm^{-1} [100].

In as-grown congruent $LiNbO_3$ usually $\alpha_{2.87\,\mu m} \approx (1–2)\,cm^{-1}$, which corresponds to a hydrogen concentration in the range of $10^{18}\,cm^{-3}$. Other estimates give a value $(3–6) \times 10^{19}\,cm^{-3}$ [113, 114]. A change of the hydrogen content in the crystal is performed by different methods. The most commonly used method to enrich the hydrogen content is the annealing of a crystal in a water-vapour-rich atmosphere at temperatures between 400–700°C, sometimes at enhanced pressures. After a few hours of annealing, OH is distributed rather uniformly over the crystal bulk ([100] and references therein). A dehydration is achieved by vacuum annealing the crystal at temperatures of about 400°C. As this reduction procedure is accompanied by the formation of bipolarons, as explained earlier, a subsequent annealing of the crystal in dry oxidizing atmosphere is required, to restore the initial optical transparency. Hydrogen (or hydroxyl) ions play a significant role in all processes related to the charge transport (photorefraction, thermal fixation of holograms, etc.), because, as shown in Sect. 3.2.1, the dark conductivity in undoped and low-doped $LiNbO_3$ crystals is controlled by the protonic conductivity in the range from room temperature to 1,000°C

2.4 Photorefractive and Relative Impurity Ions in the $LiNbO_3$ Lattice

As known, the dominant mechanism of the photorefraction in $LiNbO_3$ is the bulk photovoltaic effect. Therefore, the photorefractive properties of $LiNbO_3$ in the first instance are defined by the presence of the photovoltaic-active ions, belonging to multi-charge transition metal (TM) impurities, mainly Fe, and as well Cu, Mn, Ni, etc., [115]. In contrast, the TM impurity Cr is evidently not photovoltaic-active. Doping with any photorefractive ion provides in the visible range of the absorption spectrum a strong broad band responsible for the photoexcitation of the free electrons, which is a necessary condition for the realization of the photorefraction. Recently when developing the two-color holographic recording (so-called gating regime, see below) a definite, although an obscure role of some lanthanide ions (Ln^{3+} = Pr, Tb) in the photorefractive process was established, too. Of course, the main interest in $LiNbO_3$:Ln^{3+} is due to the possible laser action of Ln^{3+} ions and self-doubling of the generated frequency provided by the nonlinear optical properties of the $LiNbO_3$ host. In this section, we discuss the incorporation of the photorefractive impurities, mainly of Fe ions in the $LiNbO_3$ lattice. At the end the present state of studies of Ln^{3+} in $LiNbO_3$ is mentioned briefly. Particular attention is given to the influence of the crystal composition, i.e., the concentration ratio [Li]/[Nb], on this ion incorporation, because, as shown later, this factor is crucial for the charge-transport scheme controlling the photorefractive process.

An incorporation of TM and Ln^{3+} ions into the $LiNbO_3$ lattice was a subject of long debates. As their concentration in the crystals is relatively low, X-ray diffraction methods cannot be applied for analyzing their lattice sites. The problem was solved using a combination of resonance and spectroscopic techniques, such as EPR, ENDOR (electron–nuclear double resonance), EXAFS (extended X-ray absorption fine structure), PIXE (particle induced X-ray emission), RBS (Rutherford backscattering spectroscopy), SSS (site-selective-spectroscopy), etc. (for references see the bibliographies in [115–123]). In principle, these ions might occupy either of three available octahedra sites (Li, Nb, and empty octahedron) or even interstitial sites. As the ionic radii of Nb^{5+} and Li^{+} are close, so from a qualitative reasoning of the charge compensation conditions, one could expect the incorporation of divalent impurity ions (Me^{2+} = Fe^{2+}, Cu^{2+}) onto the Li sites, of trivalent ones (Me^{3+} = Fe^{3+},

Cr^{3+}, Ln^{3+}, etc.) equiprobably onto the Li and Nb sites (with a possible self-charge compensation of $[Me^{3+}_{Li}]^{\bullet\bullet}$ by $[Me^{3+}_{Nb}]''$), and of Me^{4+} ions (e.g., Ti^{4+}) dominantly onto the Nb sites. However, according to the data obtained with methods listed above, the impurity ions independently of their charge state are localized dominantly on the Li sites. Some authors [117] and [119] qualitatively explain it by a larger dimension of the Li octahedron discussed earlier. In [117], a qualitative correlation was found between the localization of an impurity ion Me in the $LiNbO_3$ lattice and the bondlength Me–O in the oxide of the given element. If this bondlength exceeds an average distance Nb–O in $LiNbO_3$ (of about 200 pm), then the Me ion incorporates onto the Li site. According to this qualitative criterion, there are only a few impurity ions (for example, Ta^{5+}) substituting for Nb. Strictly speaking, this consideration is valid for the congruent, relatively low-doped $LiNbO_3$, because in what follows, other important factors governing the localization of impurity ions are the concentration ratio [Li]/[Nb] and the impurity concentration itself.

Among the TM photorefractive impurities the most strongest photovoltaic-active impurity Fe has been intensively studied in detail. The localization of Fe in $LiNbO_3$ was investigated by multiple methods; for the bibliography on spectral and resonance methods one may address the reviews [7,55,119,122]. The impurity Fe exists in $LiNbO_3$ and $LiTaO_3$ in two charge states Fe^{2+} and Fe^{3+} and forms localized states in the gap, leading to a dichroic optical absorption spectrum (Fig. 2.11). The concentration ratio $[Fe^{2+}]/[Fe^{3+}]$ in the congruent as-grown $LiNbO_3$ crystals is usually in the range of 0.1–0.2 and may be easily varied by an oxidation–reduction annealing [55,115,125].

The following absorption bands may be distinguished in the spectrum of $LiNbO_3$:Fe [124]. The high energy one at 3.1 eV overlapping the fundamental bandedge at 3.8 eV is assigned to the charge transfer from the π-oxygen orbitals (forming the valence band) to Fe^{3+} ions. A broad band with a maximum at about 2.6 eV corresponds to the intervalence transition Fe^{2+} - Nb^{5+} and is responsible for the photoexcitation of electrons to the conduction band formed by Nb^{5+} ions. The third pronounced band at 1.1 eV is attributed to an

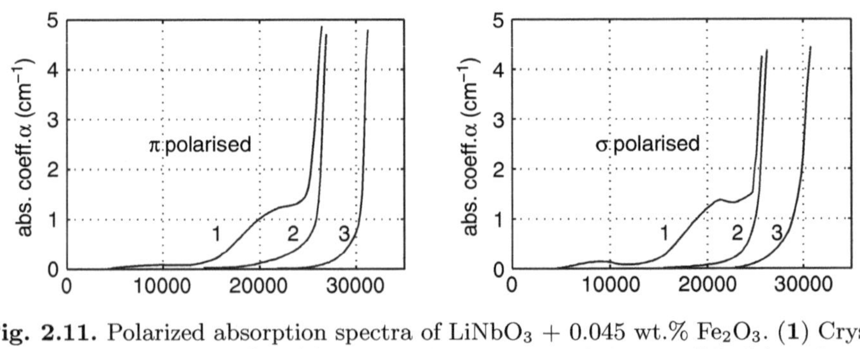

Fig. 2.11. Polarized absorption spectra of $LiNbO_3$ + 0.045 wt.% Fe_2O_3. (**1**) Crystal quenched from 1,100°C; (**2**) slow cooled and annealed at 600°C. (**3**) undoped $LiNbO_3$, data are taken from [124]

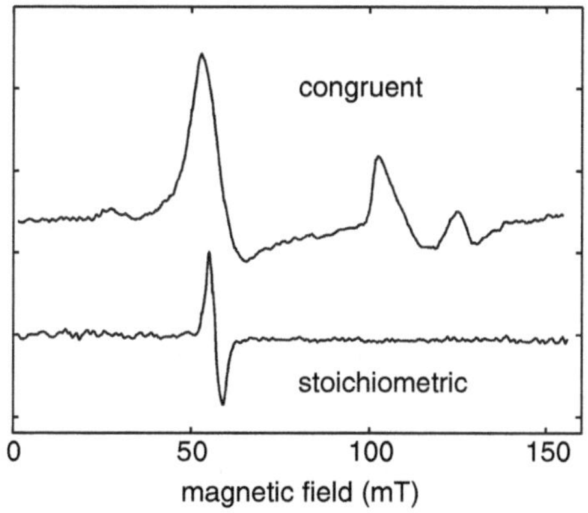

Fig. 2.12. Room temperature low field EPR spectra of Fe^{3+} in congruent and stoichiometric $LiNbO_3$, redrawn from [35]

internal d–d ($^5A \rightarrow {}^5E$) transition in Fe^{2+} ions. Additionally, one may detect two small lines at 2.55 and 2.95 eV owing to a spin-forbidden d–d transition of Fe^{3+} ions. The oscillator strengths of the bands 2.6 and 1.1 eV are 10^{-2} and 4×10^{-4}, respectively [115]. Therefore, the absorption bands 2.6 and 1.1 eV in $LiNbO_3$:Fe with intensities linearly scaling with the Fe^{2+} concentration, are tests for the presence of Fe^{2+}. The spectrum of $LiTaO_3$:Fe is very similar to that of $LiNbO_3$:Fe except for slightly different spectral positions of the band peaks. Particularly, the peak of the band related to the intervalence transition Fe^{2+}–Ta^{5+} is located at about 3 eV.

Fe^{3+} centers in $LiNbO_3$ and $LiTaO_3$ have axial symmetry, a characteristic EPR spectrum of Fe^{3+} in $LiNbO_3$ is shown in Fig. 2.12. As was unambiguously concluded, Fe^{2+} ions incorporate on Li sites [124, 126]. However, the incorporation of Fe^{3+} was a subject of a dispute. For example, some authors interpreted the EPR spectra of Fe^{3+} as evidences of its incorporation on the Nb sites [127–129]. Difficulties in the identification of EPR spectra of TM^{3+} (and Ln^{3+}) ions in $LiNbO_3$ are due to the fact that the local environment at the Li and Nb sites is very similar. Only a combination of the EPR data with the EXAFS and ENDOR measurements [130–132], permitted to conclude that Fe^{3+} incorporate onto the Li sites in the congruent $LiNbO_3$.

As seen from Fig. 2.12, the EPR spectrum of Fe^{3+} in congruent $LiNbO_3$ is strongly asymmetric. $LiNbO_3$ crystals have the point symmetry C_{3v}, thus an axial Fe^{3+} center should exhibit C_3 symmetry. So, an asymmetrical shape of EPR spectra of Fe^{3+} in congruent $LiNbO_3$ (Fig. 2.12) was attributed to the existence of a large concentration of intrinsic defects (Nb_{Li}, V_{Li}, V_{Nb}) perturbing the crystal field and reducing its symmetry to C_1 [35]. In agreement with this conclusion it was found [35], that in near-stoichiometric $LiNbO_3$ crystals (grown by means of adding 6 wt.% K_2O into the melt [34]) the Fe^{3+} EPR spectrum becomes more symmetrical (Fig. 2.12). Therefore in a stoichiometric $LiNbO_3$ crystal, free of defects, a more symmetrical shape of the Fe^{3+}-EPR spectrum fits the required C_3 symmetry of a nonperturbed crystal field. It should be emphasized that in the stoichiometric crystals Fe impurity keeps its traditional state Fe^{2+} and Fe^{3+}, the Fe^{2+} ion occupying the Li site again. However the concentration ratio $[Fe^{2+}]/[Fe^{3+}]$ in as-grown SLN:Fe crystals is lesser than in CLN:Fe, so the former crystals are weakly colored. For a detailed bibliography on the fundamental resonance studies in SLN doped with TM^{3+}, see for example [121, 123].

Analyzing the Fe^{3+} EPR spectrum in near-stoichiometric $LiNbO_3$:Fe crystals grown with K_2O in the melt, Malovichko et al. [133] found that additional to the traditional Fe^{3+}_{Li} centers, one may identify two new axial centers related to Fe^{3+} with the Spin-Hamiltonian parameters strongly differing from those in congruent crystals (e.g., $b_2^0 = 0.1768\,cm^{-1}$ for Fe^{3+}_{Li}, whereas for the two new centers $b_2^0 = 0.0495$ and $0.0688\,cm^{-1}$ [133]). They suggested that these new lines could be assigned to Fe^{3+}_{Nb} and Fe^{3+}_{Nb}–K^+ centers (where K ions, whose concentration in these crystals is in the range of 10^{-2}, are located either on the Li sites or within the empty O octahedra). Therefore, in the stoichiometric $LiNbO_3$ a partial incorporation of Fe^{3+} on the Nb sites was assumed. Following Donnerberg et al. [134] the self-charge compensation $[Fe^{3+}_{Li}]^{\bullet\bullet}$–$[Fe^{3+}_{Nb}]''$ may occur. Similarly, in the stoichiometric $LiTaO_3$:Fe the existence of Fe^{3+} centers on Ta sites was recently deduced from EPR/ENDOR studies [135].

Keeble et al. [136] investigated the EPR spectra of Fe^{3+} in the nearly stoichiometric $LiNbO_3$:Fe grown by the double-crucible method. A narrowing of the linewidth, a more symmetrical shape of the spectrum as well as Spin-Hamiltonian parameters very close to those in SLN:Fe, containing K admixture, were found [34, 133]. However, Keeble et al. [136] contend that the information obtained from a modified Fe^{3+} EPR spectrum is inadequate to deduce the alteration of the Fe^{3+} lattice sites. Anyhow, a possibility of the formation of Fe^{3+}_{Nb} in SLN crystals is of principal importance. As known, Fe^{3+}_{Li} serves in the congruent $LiNbO_3$ as the dominant electron trap. Thus, an alteration of the Fe^{3+} lattice site could fundamentally affect the charge transport scheme and possibly account for specials of the photorefractive process in SLN. This problem will be discussed in the appropriate Sects. 4.1 and 4.2.2.

Another photorefractive impurity studied in more or less detail in $LiNbO_3$ is Cu [115, 137]. Copper exists in $LiNbO_3$ in the charge states Cu^{2+} and Cu^+. The absorption spectrum of $LiNbO_3$:Cu (Fig. 2.13) reveals two bands. One of them overlapping the bandedge and stretching far into the visible is attributed to the intervalence transition Cu^+–Nb^{5+}. By means of an oxidation–reduction treatment its peak position was established to be at 3.3 eV [115]. Another band at about 1.2 eV is assigned to the transition $^2E \rightarrow {}^2T_2$ of Cu^{2+} ions. In photorefractive $LiNbO_3$:Mn crystals the absorption spectrum exhibits two broad bands at about 2.2 and 1 eV [138, 139]. According to ENDOR studies [140] Mn^{2+} ions occupy Li sites. The appropriate bibliography on other photorefractive impurities in $LiNbO_3$ is few in number [55].

Although Cr^{3+} is not a photovoltaic center, its spectral characteristics in $LiNbO_3$ are worthy of a short discussion, because they are strongly affected

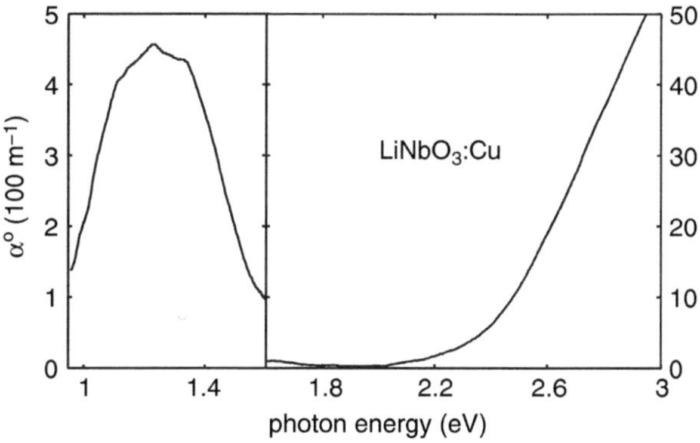

Fig. 2.13. Absorption coefficient α vs. photon energy for ordinarily polarized light in Copper-doped $LiNbO_3$, data taken from [137]

by the stoichiometry. A specific of the incorporation of Cr^{3+} ions into $LiNbO_3$ is the formation of several axial nonequivalent centers at the same lattice site, caused by a nonequivalent local environment. Difficulties in the interpretation of the EPR- and emission/absorption spectra are the cause of an extremely long discussion on the Cr^{3+} ion localization in $LiNbO_3$ [121,141–143]. According to combination of EPR, ENDOR, EXAFS and other data, in congruent crystals Cr^{3+} ions incorporate onto the Li site, at least in low-doped crystals. EPR studies in congruent $LiNbO_3$:Cr additional to the main axial Cr_{Li}^{3+} center detected several nonequivalent Cr^{3+} centers – dimer or even trimer complexes, satellite Cr centers and even nonaxial centers, which were accounted for by an interaction of Cr_{Li}^{3+} with the compensating intrinsic defects [121,144]. In near-stoichiometric crystals the linewidths of the Cr^{3+} EPR spectra are narrowed, the satellites disappeared and the spectrum is more symmetric [121,142]. The EPR and ENDOR studies in these crystals [142] revealed a family of new centers attributed to Cr^{3+} incorporation onto an Nb site. Regarding the compensation mechanism of the $[Cr_{Nb}]''$ center, a compensation by H^+ ions was assumed, whereas the self-compensation $[Cr_{Nb}]''-[Cr_{Li}]^{\bullet\bullet}$ was ruled out [142]. A similar conclusion was drawn from the spectroscopic measurements (absorption/emission spectra, site-selective-spectroscopy, etc.). As known, Cr doping in $LiNbO_3$ produces broad absorption bands related to vibronic transitions $^4A_2 \rightarrow {}^4T_1$ and $^4A_2 \rightarrow {}^4T_2$ with maxima at about 500 and 650 nm, respectively [145]. At the long wavelength shoulder of the latter band a group of weak lines exists in the range from 720 to 750 nm. They are related to $^4A_2 \rightarrow {}^2E$ transitions. This excited final state is the origin of radiative transitions, the so-called R-lines. A great number of data was recently presented on the absorption/emission spectra, particularly in the R-line range in near-stoichiometric $LiNbO_3$:Cr crystals, [146–152]. At increased Li content, additionally to the initial strong emission bands two more at longer wavelengths appear, whose intensities increase with increasing Li content, whereas the intensities of the initial bands decrease. A correlation to the EPR and ENDOR data led to the conclusion that these variations in emission spectra are associated with the appearance of Cr^{3+}_{Nb} and their coupling to other nonequivalent Cr^{3+} centers, e.g., the formation of a dimer Cr^{3+}_{Nb}–Cr^{3+}_{Li} [148].

Surprisingly in the isostructural $LiTaO_3$ crystal so far no evidences have been found of an impact of [Li]/[Ta] on the incorporation of TM ions. Some researchers performed detailed EPR studies of Fe^{3+} and Cr^{3+} in nearly stoichiometric $LiTaO_3$ crystals grown by the double-crucible method [153, 154]. The Spin–Hamiltonian parameters calculated for both centers at 300 K are very close to those in congruent material. Hence, it was concluded that both in the stoichiometric and congruent $LiTaO_3$ crystals Fe^{3+} and Cr^{3+} ions incorporate on Li sites.

We now briefly dwell on the incorporation of the lanthanide ions (Ln^{3+}) into the $LiNbO_3$ lattice. As mentioned earlier, according to recent measurements with various methods, all Ln^{3+} ions occupy the Li sites (for references see, e.g., [120]). Unlike TM ions, Ln^{3+} ions in Li octahedra are off-centered with respect to the regular Li site by 30–50 pm, the shift being dependent on the Ln^{3+} ionic radii, see Fig. 2.14. A speciality of the axial Ln^{3+} centers in $LiNbO_3$, like Cr^{3+} ions is the coexistence of several nonequivalent centers at the same lattice site. Practically, the most interesting material is $LiNbO_3$:Nd^{3+} in which a laser excitation, even with the self-frequency doubling has been reported since long ago [155–158]). The emission spectrum in this crystal reveals a band corresponding to a radiative $^4I_{\frac{9}{2}} \rightarrow {}^4F_{\frac{3}{2}}$ transition (860–960 nm), whose fine structure was formerly assigned to the existence of three nonequivalent Nd^{3+} centers [159,160] and then refined to six ones [161], for details see Fig. 2.15. At early stages, these nonequivalent Ln^{3+} centers were sometimes interpreted as Nd^{3+} both on the Li and Nb sites [159, 162]. After convincing arguments in favor of the Ln^{3+} localization on Li sites, all these six centers were related to Nd_{Li}. The fine structure of the excitation spectrum

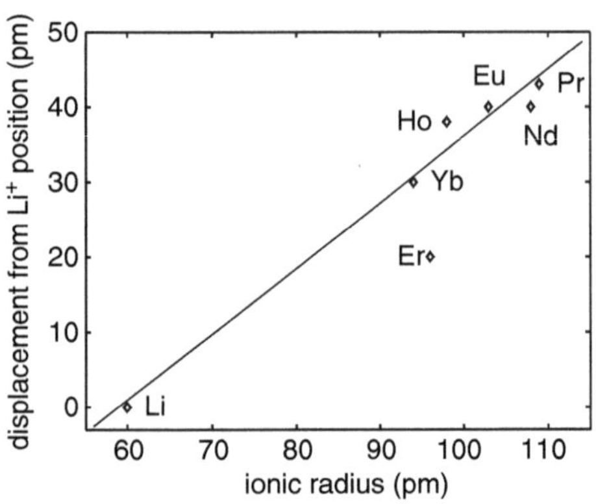

Fig. 2.14. Displacement from the Li^+ position in $LiNbO_3$ vs. the ionic radius of active RE^{3+} ions, data taken from [120]

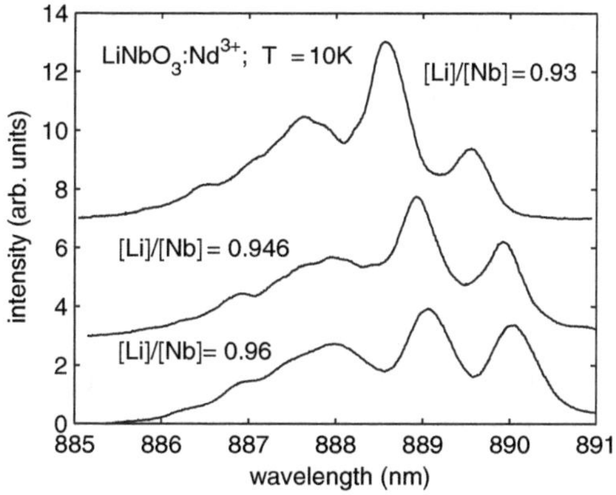

Fig. 2.15. Low temperature emission spectrum of the $^4F_{\frac{3}{2}} \rightarrow {}^4I_{\frac{9}{2}}$ transitions of Nd^{3+} ions in $LiNbO_3$, data taken from [120]

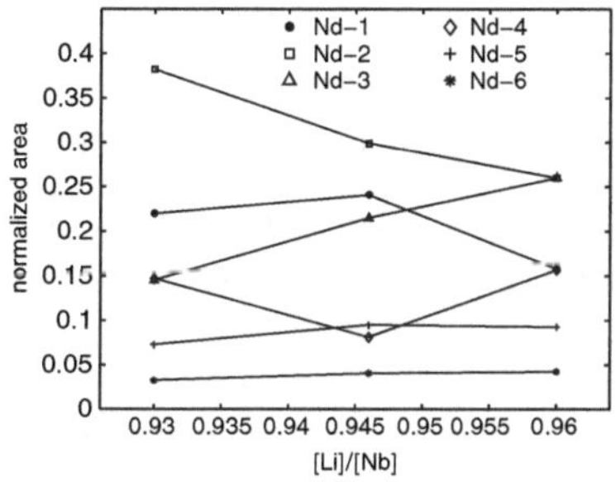

Fig. 2.16. Compositional dependence of the intensities of the six centers of the $^4F_{\frac{3}{2}} \rightarrow {}^4I_{\frac{9}{2}}$ transitions shown in Fig. 2.15, data taken from [120]

is sensitive to the crystal stoichiometry. As seen in Fig. 2.15 a slight variation in the [Li]/[Nb] ratio, achieved in [161] by controlling the [Li]/[Nb] ratio in the melt, affects the strengths of some lines, the spectral positions of the lines being almost unchanged. More quantitative information is given in Fig. 2.16, where the intensities in terms of the normalized area for each center as a function of the [Li]/[Nb] ratio is presented. The results of the spectroscopic studies in accordance with the EXAFS data [163, 164] indicate that Nd^{3+} ions keep the Li sites independently of the crystal stoichiometry. Comparative studies of Nd^{3+} in near-stoichiometric and congruent $LiNbO_3$ crystals were performed by the EPR method [165]. The Nd^{3+}-EPR band in congruent crystals is too broad to analyze its fine structure. An enormous narrowing of the EPR lines in stoichiometric $LiNbO_3$:Nd permitted to distinguish four nonequivalent centers, one of them being axial with a C_3 symmetry and others with the lower symmetry C_1.

An impact of the crystal stoichiometry on the excitation spectra was found also for Er^{3+} ions. As known, Er doping is of practical interest due to possible laser action at technically important wavelengths of about 1,500 nm (0.83 eV). Additionally, the Er^{3+} ion offers a variety of other transitions and excitation possibilities in the spectral ranges 2.26, 2.55, and 2.75 eV. According to detailed studies by EXAFS, RBS, ion-beam channeling methods, and others (for references see, e.g., [166]) the Er^{3+} ion occupies the Li site forming several non-equivalent centers. For example, several groups [167–169] when investigating the emission range at 2.26 eV, i.e., the transition ${}^4S_{\frac{3}{2}} \rightarrow {}^4I_{\frac{15}{2}}$, with the use of site-selective-spectroscopy found in SLN:Er^{3+} not less than 11 resolved peaks belonging to optically different Er^{3+} centers and in CLN:Er^{3+} even more (13 ones). The widths of all lines in SLN:Er^{3+} are strongly narrowed as compared to CLN:Er^{3+} and the line intensities are noticeably redistributed [166, 168], like in $LiNbO_3$:Nd^{3+}. Similar results were obtained for Eu^{3+} [170].

In summary, at present there are convincing data indicative of an alteration of TM ($TM^{3+} = Fe^{3+}$, Cr^{3+}) lattice site in the stoichiometric $LiNbO_3$, whereas Ln^{3+} ions seem to keep the Li sites independently of the crystal stoichiometry.

2.5 Optical-Damage-Resistant (Non-Photorefractive) Impurity Ions in the $LiNbO_3$ Lattice and Their Impact on the Incorporation of Other Impurity Ions

The effect of optical-damage-resistant impurities on the optical properties of $LiNbO_3$ is rather specific. The main role of these ions seems to vary the amount of the Nb antisites in the crystal. It is a reason of a pronounced dependence of many optical properties controlled by the intrinsic defects, on the doping with the optical-damage-resistant ions.

2.5.1 Incorporation of the Optical-Damage-Resistant Ions into the Lattice

As mentioned in Chap. 1, the most important practical task with respect to optical applications of $LiNbO_3$ is an improvement of its optical stability under intensive light, in particular a protection against the optical damage. The most efficient and technologically easy method is doping $LiNbO_3$ with relatively high concentrations of optical-damage-resistant impurities. The first optical-damage-resistant composition found was $LiNbO_3$:Mg, where a reduction of the optical damage was noticed when doping with 4.6% Mg [171]. In more details $LiNbO_3$:Mg was studied later [76, 172, 173]. These authors accounted for the optical damage resistance an increased photoconductivity and detected a critical concentration of about 5 mol.% MgO for the congruent melt, referred to as a *threshold*, above which the optical damage drastically falls off by more than two orders of magnitude. Several optical properties as well revealed anomalies at about this critical concentration. Additionally, it was found by

Sweeney et al. [76] and confirmed by Feng et al. [174] that this threshold is lowered to 3% in a crystal grown from a melt with a ratio of [Li]/[Nb] = 1.2. Just these results led to the assumption that Mg doping effects indirectly and is related to an action of Mg^{2+} ions on the intrinsic defect structure.

Starting from a chemical analogy of Zn^{2+} ions to Mg^{2+} ions a significant reduction of the optical damage was found in $LiNbO_3$:Zn [175, 176], with a particularly strong effect for ZnO concentrations exceeding 7 mol.% for the congruent melt. Zhang et al. [177] observed in congruent $LiNbO_3$:Zn a lower threshold of about 6 mol.% ZnO. The dependencies of many optical properties of $LiNbO_3$:Zn were very similar to those of $LiNbO_3$:Mg, as they revealed anomalies in the concentration range at about 7 mol.% ZnO. This similarity of properties of both $LiNbO_3$:Mg and $LiNbO_3$:Zn crystals permitted to conclude the existence of an optical-damage-resistant *family* of ions and to predict other members of this family on the base of the qualitative crystal-chemistry *principle of the diagonal rows* within the periodic table of the elements. These are in the first instance the trivalent ions In^{3+} and Sc^{3+}. Optical damage resistance was actually proved in $LiNbO_3$:In [178–180] and $LiNbO_3$:Sc [181,182]. In these cases, a drastic decrease of the photorefraction and anomalies of optical properties are observed at concentration below 2 mol.% oxides (i.e., 4 at.%) in the melt for congruent crystals. Recently, two new optical-damage-resistant impurities, Hafnium (Hf) and Zirconium (Zr), were found [183–186], whose influence on the photorefraction and optical properties qualitatively reminds of that of the former optical-damage-resistant ions. After subsequent refinement, the threshold concentration of Hf was established below 2% [187]. The most recently studied optical-damage-resistant impurity ion is Zr^{4+} [188], although discovered already 3 years earlier [183]. In $LiNbO_3$:Zr, the photorefraction drastically drops at a ZrO_2 concentration less than 2 mol.%. So, at present the family of the optical-damage-resistant impurities includes divalent (Mg, Zn), trivalent (In, Sc) and tetravalent (Hf, Zr) ions. Note that Hf is a full analog of Zr; it belongs to the group of scattered elements (Rb, Ga, Ge, etc.), forms no natural minerals and persists always as an admixture in zirconium-based natural minerals. The term family is appropriate, because an impact of these impurities on various properties of $LiNbO_3$ is qualitatively very similar in spite of different valency.

In $LiTaO_3$ doping with Mg lowers the photorefraction as well [189]. One more impurity suppressing the optical damage in $LiNbO_3$ is Na [190], but the results are still ambiguous. There are indirect indications on a similar role of Ga doping in waveguiding structures in $LiNbO_3$:Mg [191]. Other impurity ions lowering the optical damage in the $LiNbO_3$ waveguides are indiffused monovalent Ag [192] and protons [193].

In contrast to $LiNbO_3$ doped with the photorefractive impurities described and discussed in many monographs and reviews, material on optical damage resistance in $LiNbO_3$ has been summarized so far only in two reviews [194,195]. So, we discuss the optical-damage-resistant impurities in $LiNbO_3$ in more detail than the well-known photorefractive ones.

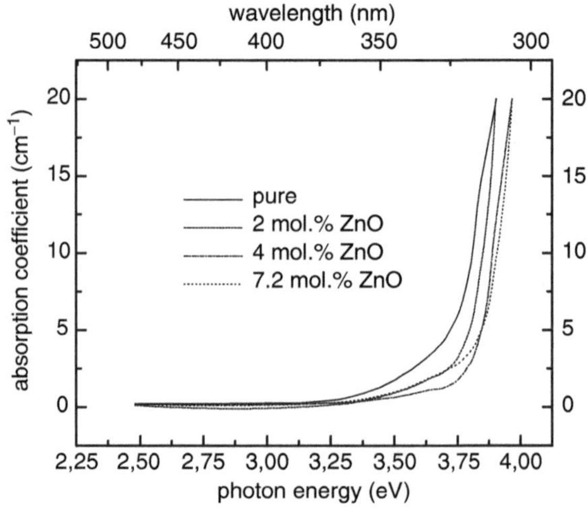

Fig. 2.17. Shift of the absorption edge of Zn-doped $LiNbO_3$, redrawn from [196]

At first a great similarity of the properties of optical-damage-resistant $LiNbO_3$ crystals manifests in their optical absorption spectra exemplified by $LiNbO_3$:Zn (Fig. 2.17). Doping with any member of this family (Mg [197,198], Zn [196], In [199], Sc [181], Hf [183], and Zr [183, 188]) induces no changes within the whole transparency range 0.35–4 µm except for a slight shift of the UV bandedge towards shorter wavelengths. The optical inactivity of these impurities is the basis for the great interest on them from the point of view of optical applications.

As mentioned earlier, a drastic decrease of photorefraction at impurity concentrations exceeding certain thresholds, constant for a given impurity, is accompanied by anomalies (extremes, kinks, steplike changes, etc.) of optical properties. Non-monotonic dependencies of the refractive indices and phase-matching temperatures on the concentrations of the optical-damage-resistant impurities will be described in Chap. 6 devoted to the characterization of the optical-damage-resistant $LiNbO_3$ compositions. Now we exemplify the threshold behavior of optical parameters by the steplike shift of the OH absorption band in $LiNbO_3$:Zr, see Fig. 2.18. Such a steplike shift of the IR band is observed in all optical-damage-resistant crystals, the value of the shift decreasing from di- to tetravalent ions, see Table 2.4.

Before discussing the threshold concentrations of optical-damage-resistant impurities and the concentration dependencies of the parameters, the following remark is worthy of noting. There are two ways to specify the impurity content, which with $LiNbO_3$:Mg as an example looks as follows:

$$c_{\mathrm{Mg}}^{(1)} = [\mathrm{MgO}]/([\mathrm{Li_2O}] + [\mathrm{MgO}] + [\mathrm{Nb_2O_5}]),$$
$$c_{\mathrm{Mg}}^{(2)} = [\mathrm{MgO}]/([\mathrm{LiNbO_3}] + [\mathrm{MgO}]).$$

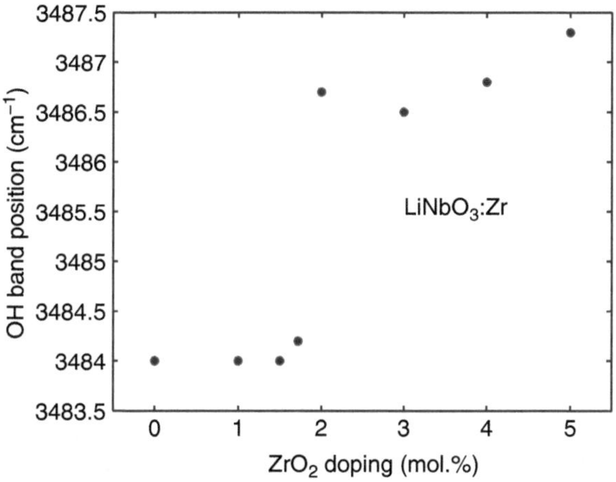

Fig. 2.18. Position of the OH stretch mode band as a function of the ZrO_2 doping in $LiNbO_3$, data taken from [188]

The first definition is adapted to the phase diagram Li_2O–MgO–Nb_2O_5 [200, 201]. These values are different, for example for usual MgO concentration $c^{(2)}_{Mg} \approx 2\,c^{(1)}_{Mg}$ [202]. Sometimes a scatter in the threshold concentration values and in concentration dependencies of various parameters is due to an ambiguity in the definition of the concentrations.

The effects of optical-damage-resistant impurities on numerous properties of $LiNbO_3$ are an increase of T_c (Fig. 6.3) and of the crystal density in $LiNbO_3$:Mg [200] and $LiNbO_3$:Zn [203], an increase of the temperature of the noncritical phase synchronism [173, 175, 176, 182], a UV shift of the optical absorption band (Fig. 2.17), qualitatively reminds us the results of increasing Li content, i.e., of an ordering of the structure. Particularly, it concerns the blueshift of the absorption edge (Fig. 2.17), because this shift in Li-enriched $LiNbO_3$ was attributed to the disappearance of intrinsic defects and subsequent hardening of the structure [204]. In the framework of the current model of the defect structure, an impact of the optical-damage-resistant ions resembles a decreasing Nb_{Li} content. Thus, it was a priori explained by a removal of Nb_{Li} from the lattice due to the entering the impurity ions onto Li sites.

We now trace the development of the proposed scenario of entering the optical-damage-resistant dopants into the $LiNbO_3$ lattice and their action on the intrinsic defect structure. As mentioned above, according to the current conclusion, the preferable sites of almost all impurity ions are the Li octahedra. In the case of $LiNbO_3$:Mg [201, 205, 206] the a priori expected incorporation of these ions onto Li sites was convincingly proved by different methods. A precise chemical analysis in $LiNbO_3$:Sc [207] and EXAFS, ion beam and hyperfine interaction methods in $LiNbO_3$:Hf [208, 209] led to the same conclusion for low Sc and Hf concentrations.

However, the incorporation mechanism and in particular its influence on the threshold concentrations was and still remains debatable. Because the threshold concentration of Mg (5.5%) in congruent crystals is close to the Nb_{Li} content [41], it was reasonable to relate it to a complete removal of Nb_{Li} from the lattice [134,205,210,211]. This assumption was supported by the fact that the threshold Mg concentration was lowered with increasing Li content in crystals [76, 174, 212]. From the same viewpoint, the most energetically favorable solution reactions underlying the different threshold concentrations for di- and trivalent impurities are according to [202]

$$5\,MgO + 2\,Nb_{Li}^{4\bullet} + 8\,V_{Li}' + LiNbO_3' \rightarrow 5\,Mg_{Li}^{\bullet} + 5\,V_{Li}' + Nb_2O_5, \qquad (2.13)$$

$$5\,Sc_2O_3 + 6\,Nb_{Li}^{4\bullet} + 24\,V_{Li}' \rightarrow 10\,Sc_{Li}^{\bullet\bullet} + 20\,V_{Li}' + 3\,Nb_2O_5. \qquad (2.14)$$

As seen from this simplified presentation, the number of trivalent cations required for a removal of Nb_{Li} is less than that of divalent ones, that means the threshold for the former is lower, because the ratio of the replacement is decreased from 5/2 to 5/3.

The relation of the threshold concentrations to the disappearance of Nb_{Li} seemed to be supported by spectroscopic tests described in Sect. 2.2. Recall that a reduction of Li-deficient (congruent) $LiNbO_3$ is accompanied by the appearance of a broad absorption band in the visible (Fig. 2.6) attributed to the formation of a bipolaron $(Nb_{Nb}–Nb_{Li})^{2-}$ [55]. The low temperature illumination of reduced crystals, leading to a dissociation the bipolaron by (2.10) and giving a characteristic 10-line EPR spectrum (Fig. 2.4) serves as a test for the presence of Nb_{Li}. Such experiments were performed in $LiNbO_3$:Mg [76], $LiNbO_3$:Zn [56,78] and $LiNbO_3$:In [179] crystals with impurity concentrations below and above the thresholds. The results are exemplified by $LiNbO_3$:In. For impurity concentrations below the thresholds in all these crystals the optical absorption spectra (Fig. 2.19, left) were qualitatively similar to those in congruent crystals (Figs. 2.4 and 2.6), thus evidencing the existence of Nb_{Li}. On the other hand, at impurity concentrations exceeding the thresholds, the

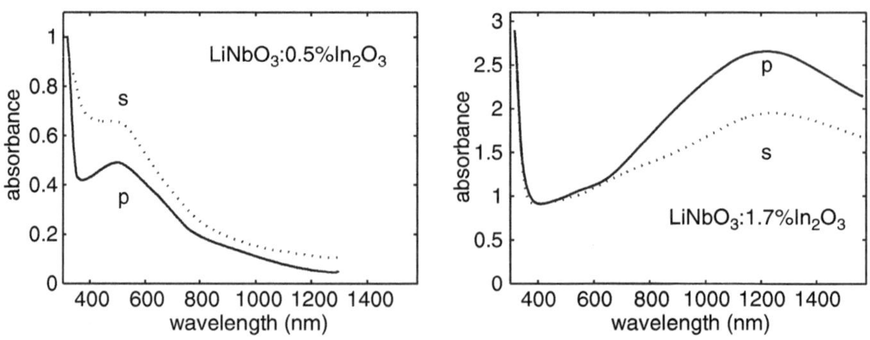

Fig. 2.19. Polarized absorbance (s, p) of reduced $LiNbO_3$ doped with indium below the threshold (*left*) and above (*right*)

spectrum of reduced material looks quite different (Fig. 2.19, right) and does not respond to illumination. The broad band at about 0.95 eV (1,200 nm) was assigned to a free small polaron Nb^{+4}_{Nb} [77, 78]. So, these results were indicative of missing Nb_{Li} for optical-damage-resistant impurity concentrations above the thresholds.

However, further studies in $LiNbO_3$:Mg [201, 206] revealed that the origin of the threshold required a refinement. A precise chemical analysis have shown that the ratio [Li]/[Nb] = 0.94 in $LiNbO_3$:Mg remains constant up to 2–3% MgO, whereas a further increase of the impurity concentration diminishes it [201, 206]. Both research groups started from the assumption of an occupation of 100% of oxygen sites, they accounted for a decrease of [Li]/[Nb] by an increase of the Li vacancy concentration. In the framework of the Li site vacancy model, the authors of [206] proposed a two-step scheme of the Mg incorporation, accordingly to which Nb antisites are completely substituted by Mg ions at [Mg] ~ 2%, whereupon Mg ions substitute for the regular Li ions [206]. The authors assumed that the appearing surplus of Mg_{Li} is compensated by V_{Li}, so at this stage their concentration increases again. Grabmaier et al. [201] discussed their data in terms of the Nb site vacancy model, so their interpretation was qualitatively different. Similar precise chemical analysis in $LiNbO_3$:Sc [207] discussed in the same way evidenced that Nb antisites are removed at 2% Sc, afterwards Sc substitutes for the regular Li ions. Table 2.2 presents a variation scheme of the defect structure of $LiNbO_3$:Mg following the two-step scheme proposed by Iyi et al. [206].

However, the calculations of [202] show that such a two-step Mg incorporation is energetically unfavored, therefore a more probable process seems to be a simultaneous substitution of Nb antisites and regular Li_{Li} ions by Mg. The same conclusion was drawn in [211]. So, up to now it is unclear whether Nb antisites disappear from the lattice at the threshold concentration in its traditional meaning, or at a lower concentration referred sometimes to as *the first threshold* [213]. According to the conclusions of [202] and [211] Nb antisites do exist in the crystal up to the usual thresholds of the optical-damage-resistant ions, although the process of substitution and the compensation mechanism vary in different concentration ranges. As described above, the existence of Nb antisites up to the threshold is supported by spectroscopic tests.

In addition to a removal of Nb antisites at threshold concentrations, the model calculations [55, 134] and indirect indications from NMR-spectra of the

Table 2.2. Incorporation of Mg ions into congruent $LiNbO_3$ (deduced from [206]). The symbol □ denotes the Li vacancy

% MgO	Formula (Li site vacancy compensation)
0.0	$[Li_{0.95}Nb_{0.01}\square_{0.04}][Nb]O_3$
3.0	$[Li_{0.94}Mg_{0.03}\square_{0.03}][Nb]O_3$
8.0	$[Li_{0.84}Mg_{0.08}\square_{0.08}][Nb]O_3$
28.0	$[Li_{0.79}Mg_{0.21}][Nb_{0.931}Mg_{0.069}]O_3$

Table 2.3. Incorporation of Zn ions into congruent $LiNbO_3$ [215–217]. The symbol □ denotes the Li vacancy

Zn (at.%)	Formula
0.0	$[Li_{0.940}Nb_{0.012}\square_{0.048}][Nb]O_3$
2.87	$[Li_{0.908}Nb_{0.007}Zn_{0.029}\square_{0.056}][Nb]O_3$
5.2	$[Li_{0.898}Zn_{0.052}\square_{0.05}][Nb]O_3$
7.6	$[Li_{0.939}Zn_{0.06}]\square_{0.001}][Nb_{0.98}Zn_{0.015}\square_{0.005}]O_3$
8.2	$[Li_{0.95}Zn_{0.06}][Nb_{0.98}Zn_{0.022}]O_3$

Mg nuclei [214] suggested a possibility of a partial incorporation of optical-damage-resistant ions onto the Nb sites after a complete removal of Nb_{Li} defects. From this viewpoint, the microscopic origin of the threshold concentration was investigated by structure studies in $LiNbO_3$:Zn and $LiNbO_3$:In [215–218]. The Zn impurity is the most convenient one for such studies, because it has the largest threshold (7.5%) among all damage-resistant impurities, thus providing the highest faithfulness of the estimates. The precise X-ray diffraction measurements were performed in single crystals and powders [215, 216]. Data on the occupancy coefficients of atoms in $LiNbO_3$:Zn obtained from the refinement of the structures, led to the chemical formulas of $LiNbO_3$:Zn shown in Table 2.3; Zn_{Li} was assumed to be compensated by Li vacancies. The main conclusion drawn from Table 2.3 is that at Zn concentrations close to the thresholds and above them, Zn ions incorporate partially onto the Nb sites. This conclusion was supported for $LiNbO_3$:8%Zn single crystals by neutron diffraction studies [217]. Similar results were obtained in $LiNbO_3$:In [218]. At a concentration of about 2.7 at.% (above the threshold) In ions were found on both the Li and Nb sites.

Therefore, these structure measurements established unambiguously the microscopic origin of the threshold concentrations, namely, the association of the threshold to a partial incorporation of the optical-damage-resistant ions onto the Nb sites. This fundamental unambiguous conclusion is obviously valid for all optical-damage-resistant dopants. Additionally to direct structure investigations in $LiNbO_3$:Zn and $LiNbO_3$:In [215–218], in Hf-doped $LiNbO_3$ the indirect data on the ion-beam and hyperfine interaction methods are evidences of the incorporation of Hf ions onto both Li and Nb sites in highly-doped crystals [209, 219]. The incorporation mechanism for the Hf impurity depends on the crystal stoichiometry: in congruent Hf_{Nb} were found at $[HfO_2] = 6$ mol.% (in the melt) and in near-stoichiometric $LiNbO_3$ at $[HfO_2]$ as low as 1%. This is in accordance with the well-known result on decreasing Mg threshold concentration in Li-enriched crystals [76, 174].

An incorporation of impurity ions simultaneously onto both cation sites requires a modification of the charge compensation conditions (2.13), (2.14). The model calculations [134] showed that for this case a self-charge compensation mechanism may occur. For a two-valence ion it looks as follows:

$$20\,ZnO + 8\,[Nb_{Li}^{4\bullet}–V_{Li}'] = 15\,Zn_{Li}^{\bullet} + 5\,Zn_{Nb}^{3\prime} + 4\,Nb_2O_5 \qquad (2.15)$$

and requires a concentration ratio of the self-compensating ions $[Zn_{Nb}^{3\prime}]/[Zn_{Li}^{\bullet}] = 1:3$. As seen from Table 2.3, this ratio is fulfilled in a rough approximation for $LiNbO_3$:Zn with the above-threshold concentrations 7.6% and 8.2% Zn, hence the mechanism (2.15) may be assumed. As it does not require more Li vacancies for the charge compensation, the threshold impurity concentration is followed by the disappearance of V_{Li} from the lattice. Chemical microanalysis measurements in $LiNbO_3$:Mg [201,206] also indirectly indicate an anomaly in the Li vacancy content at the threshold Mg concentration. Figure 2.20 presents a scheme of the defect structure variation produced by Zn doping based on the data of Table 2.3 [215–217]. Additionally, within certain assumptions for $LiNbO_3$:Zn and following the two-step mechanism of Mg incorporation [206] it was concluded that Nb antisites disappear from the lattice in the concentration range $3\% < Zn < 5\%$ (Table 2.3).

As seen from Tables 2.1 and 2.2 optical-damage-resistant ions Mg and Zn at relatively low concentrations may be regarded as controllers of the Nb_{Li} content, so their effect is qualitatively analogous to a Li enrichment. This is in a qualitative consistence, with a similarity between the blueshift of the band-edge in $LiNbO_3$:Zn (Fig. 2.17) [220] or $LiNbO_3$:Mg [197,198] and in Li-enriched crystals [204]. This shift saturates at $[Zn] \sim 4\%$, which according to (Fig. 2.20) corresponds to a disappearance of $[Nb_{Li}]$. These qualitative considerations and the results in Fig. 2.20 are consistent with the data on the lattice parameters in $LiNbO_3$:Zn vs. the Zn concentration Fig. 2.21 [221]. This dependence deviates from Vegards rule, expecting a smooth variation of the lattice parameters when increasing the concentration of a substituting impurity. Plateaus of the lattice parameters a and c as well as the volume V vs. [Zn] at low Zn concentrations transform to a steep slope at above-threshold Zn concentrations. Very similar plots were obtained for $LiNbO_3$:Zn [177,203], where these plateaus in the low concentration range (2–3% Zn) are more pronounced, so a and c even

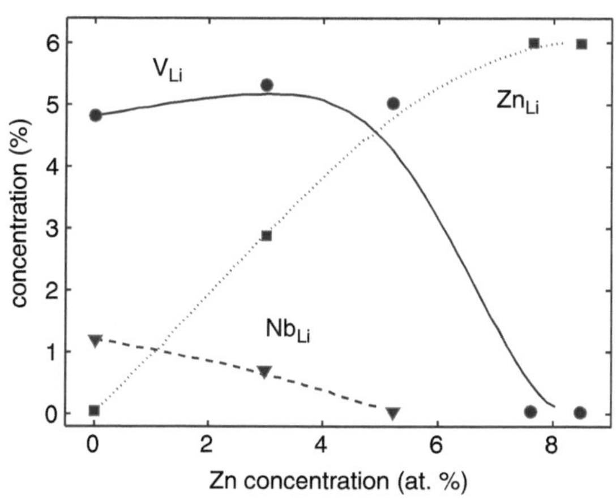

Fig. 2.20. Intrinsic defects in Zn-doped $LiNbO_3$ vs. Zn concentration

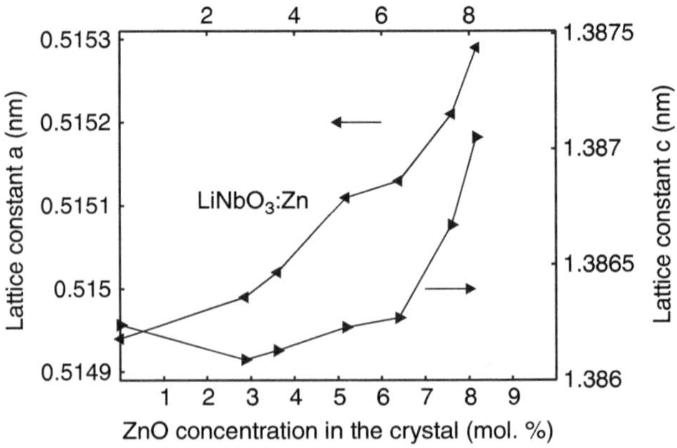

Fig. 2.21. Lattice constants of Zn-doped $LiNbO_3$ vs. ZnO concentration in the crystal. Data were taken from Abdi et al. [221]

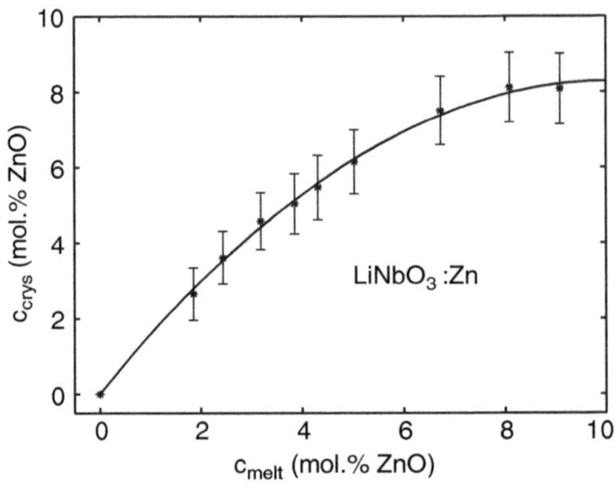

Fig. 2.22. Variation of the Zn concentration in the crystal as a function of the ZnO in the melt. The curve is a fit according to the equation in the text. Data were taken from Schlarb et al. [220]

tend to decrease. These plateaus argue of a hardening of the lattice due to a decreasing V_{Li} content and lowering Coulomb repulsion. Indeed, according to [203] the crystal density of $LiNbO_3$:2%Zn noticeably grows with respect to the undoped crystal (4.72–4.62 g cm^{-3}, respectively), whereas at further [Zn] increasing it varies slightly.

As repeatedly mentioned earlier, the preferable site of all impurity ions in $LiNbO_3$, is the Li octahedron. At low concentrations, the optical-damage-resistant ions incorporate in $LiNbO_3$ comparatively easy with distribution coefficients in the range $K_{eff} = 1.2$–1.3; at higher concentrations K_{eff} decreases to 1 for MgO [200] and ZnO [220] in the range 5–6 mol.% and to about 0.9 for In_2O_3 in the range 1.5–2 mol.% [222]. However, at above-threshold concentrations K_{eff} gradually saturates. This is illustrated in Fig. 2.22, which presents the dependence of the Zn concentration in the crystal c_{crys} on the concentration in the melt c_{melt}. The dependence is described with good accuracy by $c_{crys} = 1.649\, c_{melt} - 0.082\, c_{melt}^2$ [220]. One may see, that above the threshold the Zn concentration is close to a tolerable limit, because increasing c_{melt} does not increase c_{crys} more, as if an impurity incorporation onto both cation sites brings a given crystal lattice to a critical state. In contrast to di- and trivalent ions, the distribution coefficient for Zr seems not to depend on the concentration being of about 1.0 for the doping range from 1 to 5 mol.% ZrO_2 [188].

He et al. [223] discussed the incorporation and the thresholds of the optical-damage-resistant ions in terms of the influence of these ions on the chemical bonds within LiO_6 and NbO_6 octahedra. The consideration is based on the semi-empirical bond-valence approach, for references see, e.g., [224], according to which the stability of a structure may be established from calculations of the bond-valence sum for a given cation in a given environment. Within this concept, the stability of the $LiNbO_3$ lattice decreases with increasing optical-damage-resistant ion concentrations, so the location of such an ion is changed from an Li to Nb site provided that certain threshold of a global lattice instability is exceeded. Based on the traditional Nb_{Li}–V_{rmLi} concept and some experimental data, He et al. [223] deduce reasonable values of thresholds for di-, tri-, and tetravalent optical-damage-resistant impurities. Summarizing the data on the threshold concentrations of optical-damage-resistant impurities in congruent $LiNbO_3$ crystals, one may conclude that a more or less substantial microscopic model is elaborated for Mg and Zn ions, for which the thresholds are first related to a partial change of the occupancy from Li to Nb sites and second to a removal of Nb antisites due to the substitution by Mg or Zn ions. No speculation can be proposed concerning the incorporation of the newly found optical-damage-resistant ions Hf and Zr, except for the fact that by indirect evidences Hf at low concentrations incorporate onto Li sites [208,209], whereas at high concentrations it probably occupies both cation sites [219].

At present Li-enriched (near-stoichiometric) $LiNbO_3$:Mg is the most intensively studied optical-damage-resistant composition [225, 226]. In these crystals the threshold Mg concentration is lower than in congruent ones in accordance with [76,174]. In near-stoichiometric $LiNbO_3$:Mg the optical damage resistance occurs at [Mg] > 1–2%, that is close to the first threshold for congruent $LiNbO_3$:Mg. In terms of the above scenario of a two-step Mg incorporation, one may describe the first step in near-stoichiometric $LiNbO_3$:Mg as a complete disappearance of Nb antisites in the range ~1–2%. At a further increase of the Mg concentration the optical-damage-resistant Mg immediately starts to incorporate partially onto Nb sites. The value of the threshold Mg concentration in near-stoichiometric $LiNbO_3$ crystals has not been established yet. Formerly it was 2% [206], later a value of about 1% Mg was reported [225,227], for the crystals NSLN:Mg obtained with the use of the high-temperature top-seeded solution growth method a threshold value as low as 0.67 mol.% [228] and even 0.2 mol.% Mg was observed [229]. Iyi et al. [42] assume that the threshold concentration of Mg in $LiNbO_3$ in the framework of the Li-vacancy model may be taken as $2x$, where x is the Li-deficiency in the crystal. This may account for the observed scatter of the Mg threshold value in Mg-doped near-stoichiometric $LiNbO_3$ crystals because the degree of the near-stoichiometry strongly depends on the growth conditions and has not been specified.

Unluckily, to our knowledge no structure investigations in SLN:Mg have been performed so far, although it could clarify the origin of an extremely high optical damage resistance in these crystals.

At the end of this section, it seems appropriate to show the dependence of Li_2O and Nb_2O_5 on the MgO concentration in the congruent $LiNbO_3$ proposed in [211] in line with the MgO–Nb_2O_5–Li_2O phase diagram [200, 201] (Fig. 2.23). This dependence was constructed assuming an incorporation of Mg ions onto the Li sites with a simultaneous substitution for Nb_{Li} and regular Li_{Li} below the threshold and a partial incorporation of Mg onto the Nb sites above the thresholds with fulfilling the compensation conditions $[Mg_{Nb}^{3\prime\prime}] - 3\,[Mg_{Li}^{\bullet}]$. This scheme is in a satisfactory agreement with the concentration dependencies of many properties.

In conclusion, it is worthy of repeating again that in the framework of the currently accepted defect model of $LiNbO_3$, the optical-damage-resistant ions, at least Mg and Zn, may be regarded as controllers of the concentration of the intrinsic defects Nb_{Li} and V_{Li}. As shown in subsequent chapters, optical, photorefractive and even ferroelectric (coercive fields) properties of optical-damage-resistant $LiNbO_3$ compositions are discussed in terms of a variation in the intrinsic defect structure.

2.5.2 Effect of Optical-Damage-Resistant Impurities on the Incorporation of Other Impurity Ions

The effects of Mg and Zn doping on the optical properties of $LiNbO_3$ crystals activated by TM or Ln^{3+} dopants are numerously reported in literature. This

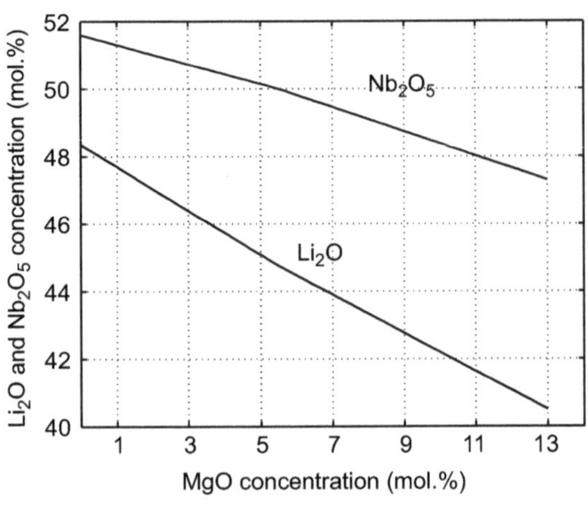

Fig. 2.23. Dependence of the Nb_2O_5 and Li_2O content in congruent Mg-doped $LiNbO_3$ crystals, redrawn from [211]

interest is provoked by the necessity of an insight into a microscopic origin of the optical damage resistance, and of obtaining optical-damage-resistant lasing structures on the basis of $LiNbO_3:Ln^{3+}$. As shown in Sect. 2.4, the incorporation of TM (Fe^{3+}, Cr^{3+}) ions into $LiNbO_3$ depends on the crystal composition, namely in Li-deficient (congruent) crystals they incorporate onto the Li sites only, whereas in the near-stoichiometric crystals evidently enter partially the Nb sites. In contrast, Ln^{3+} ions seem to occupy Li sites independently of the [Li]/[Nb] ratio. Below one will see that the doping with the optical-damage-resistant ions with the above-threshold concentrations affects the lattice site of TM ions and of some Ln^{3+} ions, too. We summarize later the available results of resonance and optical studies in $LiNbO_3$:Mg(Zn):TM and $LiNbO_3:Mg(Zn):Ln^{3+}$ not coming into details of the interpretation and dwelling on a qualitative description of the observed effects.

According to [174, 230, 231] Fe in Mg-doped $LiNbO_3$ exists in the traditional states Fe^{2+} and Fe^{3+}. Figure 2.24 shows the variations in the Fe^{3+} EPR spectra in $LiNbO_3$:Mg:0.01%Fe and $LiNbO_3$:Zn:0.01%Fe crystals when increasing the Mg or Zn concentration [175] (note the identity of the curves for Mg and Zn). One may see the abrupt changes in the shapes of the spectra, particularly the appearance of an isotropic EPR-line at Mg and Zn concentrations exceeding 5.5% and 7%, respectively. Similar change of the Fe^{3+} EPR spectrum in highly doped $LiNbO_3$:Mg was observed in [76,174,231,232]. From the parameters of the Spin–Hamiltonian in $LiNbO_3$:Fe:5%Mg the existence of a new axial center Fe^{3+}_{Nb} was deduced [174, 231]. Feng et al. [174] suggested that Fe^{3+}_{Nb} centers appear in the crystal after the complete substitution of Nb_{Li} by Mg. On the other hand, a detailed analysis of the dependence of the EPR spectra of Fe^{3+} on the Mg concentration in $LiNbO_3$:Mg:0.05%Fe [232] led to the conclusion that the spectrum of Fe^{3+}_{Li} transforms gradually within

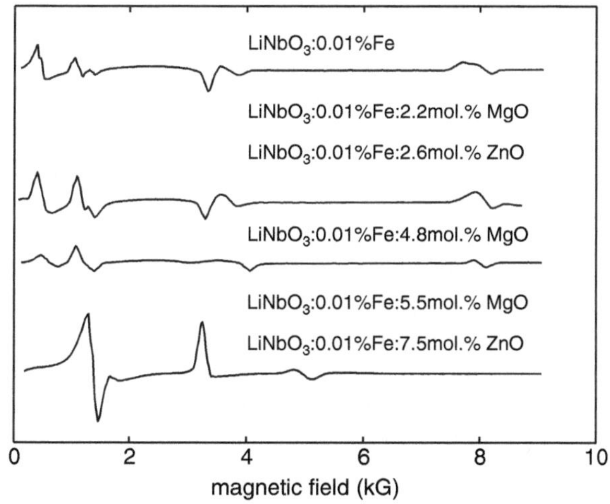

Fig. 2.24. EPR spectra of Fe-doped $LiNbO_3$ codoped with MgO or ZnO. The doping correspond to melt compositions, redrawn from Volk et al. [175]

a rather broad concentration range starting from [Mg] ≈ 4.6%. Therefore, a change of the Fe^{3+} lattice site occurs not in a step-like manner, and Fe^{3+}_{Li} and Fe^{3+}_{Nb} centers coexist in the concentration range 4.6–6% Mg. So, the EPR spectrum observed in a pre-threshold concentration range may be regarded as a superposition of two spectra. The change of the Fe^{3+} lattice site is of an important significance for the subsequent discussion of the microscopic origin of the optical damage resistance.

Similar variations in spectroscopic and resonance characteristics of Cr^{3+} ions at high Mg or Zn concentrations were observed, too. A sharp reconstruction of the Cr^{3+} EPR lines and the appearance of an isotropic line was detected in $LiNbO_3$:Cr highly codoped with Mg [233–237] and Zn [237]. At low Mg or Zn concentrations the spectrum was assigned to axial Cr^{3+}_{Li} centers [233,237], whereas a fundamental transformation of the spectrum at about 6% Mg was interpreted as appearance of axial Cr^{3+}_{Nb} centers. Similarly to Fe^{3+}_{Nb}, these centers arise at a pre-threshold concentration [Mg] > 4.6% [235,236] (for [Zn] > 4.7% [237]), the ratio of $[Cr^{3+}_{Nb}]/[Cr^{3+}_{Li}]$ smoothly increasing in the range from 4.6% to 6% Mg. On the basis of EPR and ENDOR measurements Corradi et al. [238], regard this new center as a complex of Cr^{3+}_{Nb} with an Mg^{2+} ion in proximity. The absorption/emission spectra in $LiNbO_3$:Cr are also affected by high Mg or Zn doping [236,239]. The broad absorption band $^4A_2 \rightarrow {}^4T_1$ is shifted towards longer wavelengths, while the band $^4A_2 \rightarrow {}^4T_2$ is sufficiently lowered and broadened. The emission spectra in the R-line range (720–750 nm) are modified as well. The intensities of the two original R-lines are redistributed, and two new emission lines shifted towards longer wavelengths appear and are attributed to the transition $^4A_2 \rightarrow {}^2E$. Qualitatively, the situation is very similar to the behavior of the R-line emission in the stoichiometric $LiNbO_3$:Cr described earlier. A new narrow p-polarized line at 781 nm is characteristic for the Cr^{3+} emission both in SLN:Cr and CLN:Cr:Mg [240]. From the correlation to the EPR data [233, 238] Camarillo et al. [239] attribute the reconstruction of R-lines to the formation of Cr_{Nb}-related centers. The authors of subsequent studies in $LiNbO_3$:Cr:Mg share this conclusion [241].

Variations of the optical absorption spectra in $LiNbO_3$:TM under an impact of the optical-damage-resistant ions affect the appearance of the above-threshold crystals. As a result of this reconstruction of the absorption spectrum in $LiNbO_3$:Cr:Mg, the highly doped crystals transform from green to pink. In $LiNbO_3$:Fe:Mg(Zn) with Mg(Zn) exceeding the thresholds the optical absorption at $\lambda \approx 500$ nm is sufficiently lower than in $LiNbO_3$:Fe or low-doped $LiNbO_3$:Fe:Mg(Zn), when compared for the same Fe concentration in the crystal. As the band at about 500 nm is related to Fe^{2+}_{Li}, so this observation means that at high Mg(Zn) concentrations the ratio $[Fe^{2+}]/[Fe^{3+}]$ in as-grown crystals is lesser than in $LiNbO_3$:Fe, as if the incorporation of Fe ions onto the Li sites is hampered. Consequently, the visual appearance of the above-threshold crystals is more light.

In [242] an impact of a trivalent optical-damage-resistant impurity Sc on the incorporation of Cr^{3+} ion was found. Studies by EPR and spectroscopic methods in $LiNbO_3$:Cr:Sc detected the two centers Cr_{Li} and Cr_{Nb} with optical and magnetic characteristics, very similar to those of $LiNbO_3$:Cr:Mg(Zn).

The effects of Mg doping on the incorporation of Ti ions are of importance with regard to the formation of Ti-induced optical waveguides. In congruent crystals Ti^{4+} substitutes for Li [119]. To investigate its incorporation by the EPR method, the crystals have to be reduced to transform Ti^{4+} to the paramagnetic Ti^{3+} state [243]. EPR and optical absorption spectra in reduced $LiNbO_3$:Ti were attributed to Ti^{3+} centers at Li sites [244, 245], independently of the [Li]/[Nb] ratio [246]. However, high Mg doping strongly affects the Ti incorporation. Studies with EPR and optical absorption spectra in reduced $LiNbO_3$:Ti:6%Mg found dominant new centers interpreted as Ti^{3+} at Nb sites [246, 247], whose concentration noticeably exceeds that of Ti^{3+}_{Li}.

Evidently, Mg doping affects the incorporation of the other optical-damage-resistant ion Hf. On the base of computer simulations of ion-beam channeling data it was concluded that at high Mg concentrations Hf^{4+} ions incorporate onto Nb sites [117, 119]. This action resembles the above mentioned evidences of the alteration of Hf lattice site in near-stoichiometric crystals [209, 219]. Here it should be mentioned that in the literature one may meet experiments on studies in $LiNbO_3$ doped with two types of optical-damage-resistant ions, e.g., (Zn + In) [248] or (Mg + In) [249], etc. Such combinations lead to interesting variations in physical properties, which are evidences of a nonadditivity of the effects on the defect structure. No a priori conclusion can be drawn concerning the intrinsic defect structure of these doubly doped crystals, because their incorporation onto the Li sites at low concentrations and onto both cation sites at high concentrations is concurrent. Thus one may expect a variation of the distribution coefficient of either partner depending on their concentration ratio in the melt.

As already mentioned, in the congruent $LiNbO_3$ several nonequivalent Ln^{3+} centers coexist on Li sites, particularly in Nd-doped crystals six ones were detected (Nd-1, Nd-2, etc.) [120]. Even at an early stage of studies in $LiNbO_3$:Nd:Mg the effect of a strong (about 5%) Mg codoping on the fine structure of the emission spectrum was observed (e.g., [162, 250, 251]). The authors of these works in agreement with the subsequent studies found in the emission spectrum of $LiNbO_3$:Nd:5%Mg two additional well separated lines in the range 875–890 nm corresponding to $I(^4I_{9/2}) \rightarrow R1, R2\,(^4F_{3/2})$ transitions. At that time they were attributed to a perturbation of Nd^{3+}_{Nb} by the nearest Mg_{Li} [250]. In more recent works the new lines appearing in the spectra of $LiNbO_3$:Nd highly codoped with Mg or Zn are assigned to new centers referred to as Nd–Mg or Nd–Zn, with a very prudent assumption of their relation to Nd_{Nb} [120]. At the same time, the authors of [164] on the base of the EXAFS measurements concluded that at a high Mg concentration Nd^{3+} ions partially change their lattice sites and occupy the empty oxygen octahedra.

The influence of a high codoping with Mg or Zn was studied in $LiNbO_3$:Er^{3+} [252] and $LiNbO_3$:Yb^{3+} [253] using EPR. In crystals without codopants the EPR spectra were attributed to Er^{3+} and Yb^{3+} ions, respectively, on the Li sites. At high Mg or Zn concentrations new EPR lines appeared, and were associated with new centers Er^{3+} (Yb^{3+}). The authors correlated these spectra with a variation of the Cr^{3+} EPR spectra in $LiNbO_3$:Cr:Mg, mentioned above and interpreted their results in terms of the location of Er^{3+} or Yb^{3+} ions at Nb sites in analogy to Cr^{3+}_{Nb}. Summarizing this section, one can see, that as the influence of optical-damage-resistant cations on the location of TM ions seems to be more or less understood, the situation with Ln^{3+} ions is still uncertain and requires further investigations.

2.5.3 OH-Spectra in Optical-Damage-Resistant $LiNbO_3$ Crystals

For an extended bibliography on this subject one may address the review [99]. The impact of optical-damage-resistant ions on the OH stretch mode characteristics present one more striking manifestation of the threshold phenomena, which are exemplified by the dependence of the IR band on the Zn concentration in $LiNbO_3$:Zn, see Fig. 2.25. Below the threshold the spectral position and the shape of the IR band is practically unchanged, above the threshold it is shifted to higher frequencies, and in a pre-threshold concentration range those bands coexist, which are characteristic for low- and heavily doped crystals. Particularly, in the pre-threshold concentration range in

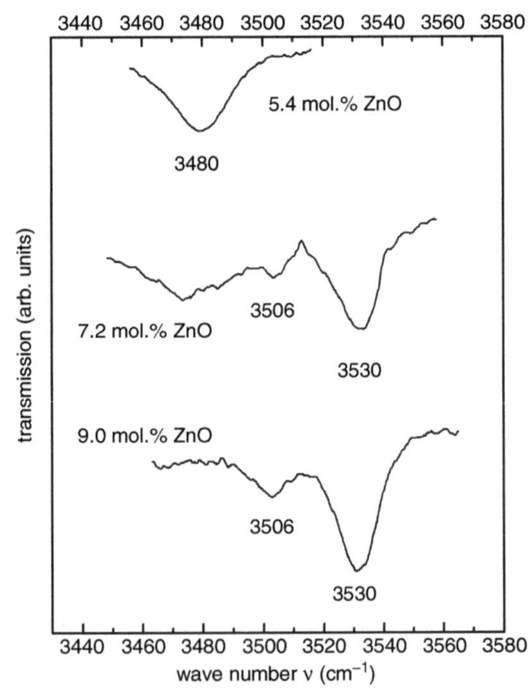

Fig. 2.25. IR transmission spectra of Zn-doped $LiNbO_3$ below the threshold (*top*), in the threshold region (*middle position*), and above threshold (*bottom*)

Table 2.4. Energies of the OH stretch mode in optical-damage-resistant $LiNbO_3$. TSSG means top seeded solution growth.

Composition/doping	Energies (cm^{-1})	Reference
Lithiumniobate undoped		
Congruent	3, 478, 3491	[255]
Stoichiometric	3, 466	[254, 256]
Lithiumniobate doped with damage-resistant elements		
$LiNbO_3$:Mg [Li]/[Nb] = 0.945		
<4.5% MgO	3,484	[172]
>4.5% MgO	3,534	[172]
	3,507, 3,537	[257]
$LiNbO_3$:Mg; [Li]/[Nb] = 1.1		
<3.5% MgO	3,485, 3,530, 3,540	[258], [259]
>3.5% MgO	3,540	[258], [259]
$LiNbO_3$:Mg; [Li]/[Nb] = 1.2		
<2% MgO	3,485, 3,530, 3,540	[258], [259]
>2.5% MgO	3,540	[258], [259]
$LiNbO_3$:Mg; [Li]/[Nb] = 1.38		
≃1% MgO	3,532	[226]
$LiNbO_3$:Mg; TSSG		
<1% MgO	3,466	[226]
>2% MgO	3,532	[226]
$LiNbO_3$:Zn		
<7% Zn	3,484	[260], [177]
>7% Zn	3,506, 3,535	
$LiNbO_3$:In		
>1.5%In	2,506, 3,508	[178], [249]
$LiNbO_3$:Sc	3,497	[261]
$LiNbO_3$:Hf [Li]/[Nb] = 0.945		
<4% HfO_2	3,484	[262]
>4% HfO_2	3,487	[262]
$LiNbO_3$:Zr [Li]/[Nb] = 0.945		
<2% ZrO_2	3,484	[188]
>2% ZrO_2	3,487	[188]

$LiNbO_3$:Mg and $LiNbO_3$:Zn the band consists of two or three peaks, whereas above the thresholds the low-frequency peak, characteristic for an undoped crystal disappears. Table 2.4 summarizes the literature data on the peak positions of the IR band in $LiNbO_3$ with different stoichiometry doped with the optical-damage-resistant impurities. As seen, the band-shift regularly decreases with increasing ion valence state, being a maximum for divalent Mg or Zn and negligible for tetravalent Hf and Zr. Additionally, the presented data illustrate the repeatedly mentioned fact that the threshold concentration decreases in Li-enriched crystals. Recently, the shift of the IR band in Mg-doped near-stoichiometric crystals was observed at Mg concentrations as low as 1.5 mol.% MgO; in crystals grown by the top-seeded solution method with K_2O and even less than 1 mol.% MgO [226, 229]. No consistent microscopic model has been proposed to account for the effects of the optical-damage-resistant ions on the OH-band in $LiNbO_3$, although some authors try to analyze the defect structure on the base of these spectra, e.g., [207, 254].

A more complicated variation of the IR band is observed in doubly-doped crystals such as $LiNbO_3$:Mg, Me (Me = Nd, Cr, Ti, Mn, In, Sc) [249, 263, 264] and $LiNbO_3$:Zn,In [248]. In $LiNbO_3$:Mg, Me crystals new long-wave broad-bands arise, whose intensities grow with the Me concentrations. The authors of [99] conditionally sort these crystals for two groups accordingly to the wavelengths of the new IR band. A band at about 3,507 cm^{-1} is characteristic for Sc and In, and for TM impurities Cr and Fe. The second group includes Ln^{3+}, particularly Nd in which the band appears at about 3,523 cm^{-1}. A qualitative conclusion may be drawn that the effects in doubly doped crystals are not additive, perhaps due to their concurrent incorporation onto the same lattice sites. The new bands in $LiNbO_3$:Mg and $LiNbO_3$:Mg,Me at high Mg concentrations are attributed to formation of complexes Mg–OH^- and Me–OH^-–Mg.

Note that the OH-band has no practical importance, but, since its shift correlates with the occurrence of optical damage resistance, it provides a convenient tool for crystal characterization. Namely, on the basis of the position of the IR band one may conclude beforehand, whether the given impurity concentration is below or above the threshold, or, in other words, either the given composition is optical-damage-resistant or not.

3

General Introduction to Photorefraction in $LiNbO_3$

Summary. The photorefractive properties of $LiNbO_3$ with an emphasis on the contributions from intrinsic defects are examined. After a short background of the photorefraction with the presentation of basic relations, the photoinduced charge transport in photorefractive compositions, mainly in $LiNbO_3$:Fe are discussed. The current concept of the dark conductivity is formulated assuming a concurrence between protons and the electronic conductivity dictated by the Fe concentration. The traditional one-center charge transport scheme based on the electron transfer between the donor Fe^{2+} and the deep trap Fe^{3+} is refreshed and the experimental data not fitting this model are listed. A refined two-center scheme based on the assumption of the Nb antisite involvement in the charge transport as a shallow electron trap is discussed and the experimental arguments supporting this model are summarized.

The photoinduced change of the refractive indices (originally called *optical damage*) is one of the three photoinduced effects limiting the traditional use of $LiNbO_3$ in optics as electro-optic modulators, optical frequency converters, etc. The first effect, the so-called laser-induced damage or optical breakdown, is a nonreversible mechanical destruction of the material. It is characteristic for a wide group of insulators and occurs at rather high light intensities, e.g., for $LiNbO_3$ in the range of $\mathrm{GW\,cm^{-2}}$ for $\lambda = 1{,}064\,\mathrm{nm}$ depending on the laser pulse duration, for example $14.6\,\mathrm{J\,cm^{-2}}$, $t_p = 1\,\mathrm{ns}$, $\lambda = 1.053\,\mu\mathrm{m}$ for $LiNbO_3$:1%Mg [265]. Note that the laser damage is formed of two parts – the surface and the volume, so the threshold of the optical breakdown depends on the surface treatment and defect state, respectively. This is a reason, why experimentally observed threshold values are scattered and often far below the literature ones. The second negative factor, the photoinduced coloration occurring both in the visible (denoted as *dark*- or *gray*-trace effect) and in the infrared (denoted as *GRIIRA* – green induced IR absorption) spectral ranges, is observed at light intensities in the range of $I \approx 100\,\mathrm{MW\,cm^{-2}}$ after multiple shooting the crystal by light pulses. The former shooting results in a noticeable reduction of the optical transmission in the visible, sometimes up to 10–15%.

This effect as a rule is reversible and may be optically erased or thermally annealed. In contrast to these two effects, the photoinduced change of the refractive indices (photorefraction or optical damage) appears at low light intensities (e.g., as low as $10\,\mathrm{mW\,cm^{-2}}$ for blue–green light for $LiNbO_3$:Fe) and shows no intensity threshold. The negative consequences of the photorefraction is the light-wave degradation (the wave-front damage) and an instability or loss of lasing when using, for example, $LiNbO_3$ doped with Ln^{3+} as a solid-state laser medium. Additionally, optical damage results in a strong scattering of a transmitting light-wave.

3.1 Background with Basic Relations

We start from a short formulation of the fundamentals of the photorefractive effect and of the holographic recording based on it as applied to $LiNbO_3$. The basics may be found in numerous monographs and reviews (e.g., [4, 9, 12–16, 266–269]). A photoinduced change of the refractive indices is caused by the formation of a photoinduced space-charge field E_{sc} in the illuminated part of the crystal, which in turn leads to a variation of the refractive indices via the linear electro-optic effect in noncentrosymmetric media

$$\frac{1}{\epsilon_{ij}} = \frac{1}{\epsilon_0} + r_{ijk}E_k, \tag{3.1}$$

where ϵ_{ij} and r_{ijk} are components of the tensors of the dielectric permittivity and linear electro-optic effect, respectively, ϵ_0 is the initial value, E_k is a component of the space-charge field. The formation of E_{sc} consists of the following steps: a photoexcitation of charge carriers to the conduction band, their spatial transport and a re-trapping by capturing centers. Usually a band-transport mechanism is assumed. Formerly, photorefraction was accounted to the photoexcitation from donor centers (extrinsic or intrinsic defects); however, recently a relation of the photorefraction to a band-to-band electron excitation was found, for example in near-stoichiometric $LiTaO_3$ [270, 271] and $LiNbO_3$:Mg [272] where an interband photorefraction induced by a deep UV light was obtained. The recent review [273] summarizes the available data on the interband photorefraction and proposes a model description. Additionally, the photorefraction may be induced by X- or γ-irradiation [274–276]. The photoinduced charge transport involves the following components:

$$j = j_{\mathrm{drift}} + j_{\mathrm{pv}} + j_{\mathrm{dif}}. \tag{3.2}$$

We now discuss one after the other the contribution of each component of expression (3.2) to E_{sc}. The first component is the Ohmic current

$$j_{\mathrm{drift}} = \sigma(E_{\mathrm{ext}} + E_{\mathrm{int}}) \tag{3.3}$$

arising under external E_{ext} or internal E_{int} fields. An example of E_{int} is given by the field of the pyroelectric current caused by a local heating of the crystal under illumination; in this case the photorefraction is owing to the screening of the pyroelectric field by photocarriers [277]. The second term in (3.2) is the bulk photovoltaic current, which is characteristic for noncentrosymmetric (piezoelectric and pyroelectric) crystals [4, 9]. The bulk photovoltaic effect is fundamentally related to the specific of the photoexcitation and recombination processes in noncentrosymmetric media, so the bulk photovoltaic current is observed even under a uniform illumination in contrast to the usual photovoltaic current that requires either a heterogeneity of the medium or a nonuniformity of the light field. The photovoltaic current is expressed as

$$j_{pv,i} = \beta_{ijk} e_j e_k I \; , \tag{3.4}$$

where $\beta_{ijk} = \beta ikj^{\star}$ is the complex photovoltaic tensor, e_j and e_k are unit vectors of the light polarization, I is the light intensity. In general case, the real and imaginary parts of this tensor are, respectively, symmetric and antisymmetric in the indices k and j; for linearly polarized light only the real β_{ijk} is taken into account. If symmetric, the photovoltaic tensor is identical to the piezoelectric tensor d_{ijk}, so for $LiNbO_3$ or $LiTaO_3$ (point symmetry group C_{3v}) contains four nonzero components $\beta_{311} = \beta_{322}$, β_{333}, $\beta_{222} = -\beta_{112} = -\beta_{121} = -\beta_{211}$, $\beta_{113} = \beta_{223}$. The component β_{113} characterizes a spatially oscillating current controlled by the phases of the transmitting ordinary and extraordinary waves and thus having an oscillation period equal to the coherence length $l_{\mathrm{c}} = \lambda/(2 * (n_{\mathrm{e}} - n_{\mathrm{o}}))$, where λ is the wavelength. In $LiNbO_3$ we have $|n_{\mathrm{e}} - n_{\mathrm{o}}| = 0.095 \approx 0.1$ and thus $l_{\mathrm{c}} \approx 5 * \lambda$. Therefore the spatially-oscillating currents contribute when an illuminated region is comparable to this value, for example for hologram recording in photorefractive optical waveguides [278].

Additionally, usage of the component β_{113} formed the base for holographic recording by orthogonally polarized light beams in $LiNbO_3$:Fe, often called anisotropic hologram recording, which permitted particularly the determination of this nondiagonal component [279]. Most often the photovoltaic current is presented in a scalar form

$$j_{\mathrm{pv}} = k_{\mathrm{G}} \alpha I, \tag{3.5}$$

where α is the optical absorption coefficient assumed to be isotropic, and $k_{\mathrm{G}} = \beta_{ijk}/\alpha$ is the so-called Glass constant characterizing the photovoltaic *activity* of the given impurity in the particular lattice. The stationary photovoltaic field corresponding to the diffusion–drift equilibrium in a scalar presentation is expressed as

$$E_{\mathrm{pv}} = \frac{j_{\mathrm{pv}}}{(\sigma_{\mathrm{ph}} + \sigma_{\mathrm{d}})} = \frac{k_{\mathrm{G}} \alpha I}{(\sigma_{\mathrm{ph}} + \sigma_{\mathrm{d}})}, \tag{3.6}$$

where σ_{ph} and σ_{d} are photo- and dark conductivities, respectively. From (3.5) and (3.6) one may see that, for relatively low light intensities corresponding to $\sigma_{\mathrm{ph}} > \sigma_{\mathrm{d}}$ the photovoltaic field saturates independent of I, assuming the usual dependence $\sigma_{\mathrm{ph}} \propto I$.

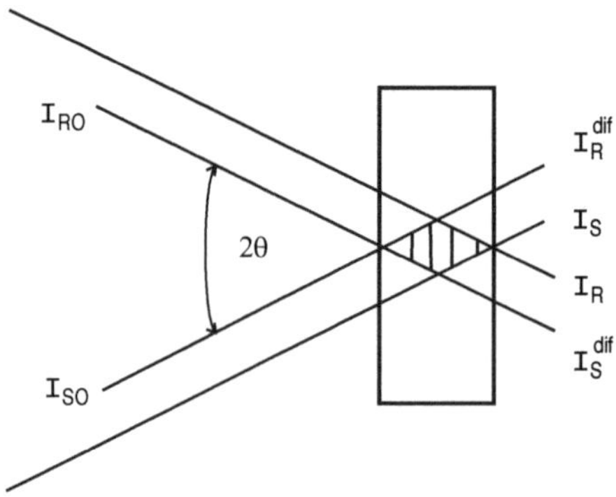

Fig. 3.1. The scheme of a two-beam hologram recording on transmission. I_{SO}, I_S^{dif}, I_S are initial, diffracted, and transmitted intensities of the signal beam, I_{RO}, I_R^{dif}, I_R are the same for the reference beam

The third component in (3.2) is the diffusion transport of photocarriers that occurs under a spatially nonuniform illumination and may be expressed as

$$j_{dif}(z) = e\left(D_n\frac{dn}{dz} + D_p\frac{dp}{dz}\right) = k_BT\left(\mu_n\frac{dn}{dz} + \mu_p\frac{dp}{dz}\right), \quad (3.7)$$

where D_n, μ_n, D_p, μ_p are diffusion coefficients and mobilities of electrons and holes, respectively, k_B the Boltzmann constant, and e is the unit charge. For a two-beam hologram recording (Fig. 3.1), a spatially varied light intensity owing to a superposition of two mutually coherent plane waves with incident intensities I_1 and I_2 is characterized by the coordinate dependence

$$I(z) = I_0 \sin(1 + m\cos Kz) \quad (3.8)$$

where $m = 2\sqrt{I_1I_2}/(I_1 + I_2)$ is the modulation index, $K = 4\pi\sin\theta/\lambda$ is the spatial frequency of the interference pattern, θ is the half-angle between the recording beams, λ the wavelength of the recording light. For the case of a mono-polar (electron) photoconductivity, a scalar presentation of the diffusion field corresponding to the intensity distribution (3.8), is

$$E_{dif}(z) = \frac{k_BT}{e}\frac{Km\sin Kz}{1 + m\cos Kz} \quad (3.9)$$

Principal distinctions of the diffusion field (3.9) from the photovoltaic field (3.6) are independent of E_{dif} of the light intensity and a shift of the field grating (3.9) by $\pi/2$ with respect to the interferogram (3.8). The photoexcited carriers transferred by either of the listed mechanisms or their combination are captured by traps (if any) to form a spatially modulated space-charge field E_{sc}. If the photovoltaic mechanism is dominant, then $E(z)$ is in phase with the interferogram, which is most often realized in doped $LiNbO_3$. If the diffusion mechanism prevails and the photoconductivity is monopolar, then

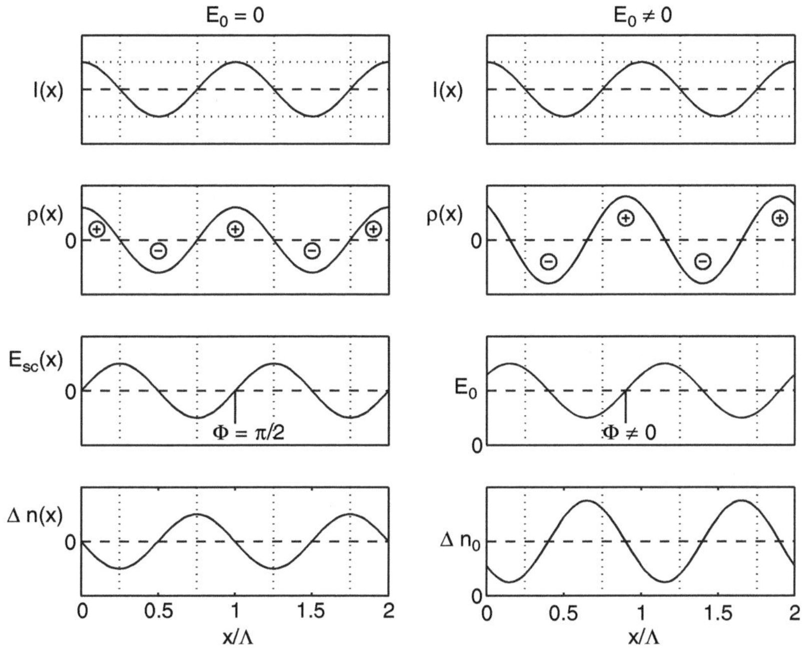

Fig. 3.2. Phase relations between the intensity distribution $I(x)$, the space-charge $\rho(x)$, the space-charge field $E_{sc}(x)$, and the refractive index variation $\Delta n(x)$ for the diffusion mechanism of the photorefraction without external field $E_0 = 0$ and in the presence of an external field $E_0 \neq 0$; Λ is the grating period, redrawn from [268]

the grating is shifted by $\varphi = \pi/2$, the so-called nonlocal response [280–282], see Fig. 3.2. If E_{pv} and E_{dif} are comparable, then the shift of the grating is less than $\pi/2$.

We now briefly mention the characteristic parameters of the recording used in the experiments. The phase grating $\Delta n(z)$ formed under the space-charge field is characterized by the diffraction efficiency η, the coupling gain factor Γ and the sensitivity S. For recording in the transparent regime (Fig. 3.1)

$$\eta = \frac{I_S^{dif}}{I_S}, \tag{3.10}$$

where I_S and I_S^{dif} are the intensities of the transmitted and diffracted beams, respectively. For equal intensities of the two recording beams the diffraction efficiency η for a hologram on transmission is predicted by the coupled-wave theory as [283]

$$\eta = \exp\left(\frac{-\alpha L}{\cos\theta}\right)\sin^2\left(\frac{\pi\Delta n L}{\lambda\cos\theta}\right), \tag{3.11}$$

where L is the hologram thickness, $\Delta n = r_{eff}E_{sc}$, the exponent takes into account the optical absorption of the crystal.

The two-beam coupling gain Γ describes an energy transfer between the recording beams. This parameter is given by the expression

$$\frac{I_S}{I_{S0}} = \frac{I_{R0}}{I_R} \exp\left(\Gamma - \alpha\right), \tag{3.12}$$

where I_{S0}, I_S, I_{R0}, I_R are initial and transmitted intensities of the signal and reference beams, respectively. The *signal* I_{S0} beam is a weak one amplified at the expense of an energy transfer from the intensive *reference* beam I_{R0}. If $I_{S0} \ll I_{R0}$, then one may neglect the depletion of the reference beam during energy pumping ($I_{R0} \approx I_R$). For these conditions and for $\alpha \ll \Gamma$, i.e., an optically transparent crystal we have

$$\Gamma = L^{-1} \ln\left(\frac{I_S}{I_{S0}}\right). \tag{3.13}$$

For a photorefractive grating recorded by the diffusion mechanism the expression for the stationary two-beam coupling gain is

$$\Gamma = \frac{2\pi}{\lambda} \frac{n^3 r_{\text{eff}} E_{\text{sc}}}{m \cos\theta}. \tag{3.14}$$

The sensitivity of recording is expressed by different ways. A sensitivity per incident radiation intensity (a *nonreduced* sensitivity) is

$$S_1 = \frac{1}{I}\frac{\mathrm{d}\Delta n}{\mathrm{d}t} = \frac{1}{I}\frac{\mathrm{d}\sqrt{\eta}}{\mathrm{d}t}, \tag{3.15}$$

$$S_1' = \frac{1}{IL}\frac{\mathrm{d}\Delta n}{\mathrm{d}t} = \frac{1}{IL}\frac{\mathrm{d}\sqrt{\eta}}{\mathrm{d}t}. \tag{3.16}$$

A sensitivity per absorbed radiation energy, which is more illustrative for crystals with a high optical absorption, is

$$S_2 = \frac{1}{I\alpha}\frac{\mathrm{d}\Delta n}{\mathrm{d}t} = \frac{1}{I\alpha}\frac{\mathrm{d}\sqrt{\eta}}{\mathrm{d}t}. \tag{3.17}$$

Additionally, a customary characteristic of sensitivity is the incident radiation energy required for recording 1% diffraction efficiency ($W_{\eta=1\%}$). For nonphotovoltaic crystals with recording by the diffusion mechanism the sensitivity is often defined as $\mathrm{d}\Gamma/\mathrm{d}t$. A kinetics of the space-charge field for the case of a narrow light stripe normal to the current direction $E(0) = 0$, the open-voltage conditions, is given by the expression

$$E(t) = -\frac{j}{\sigma_{\text{ph}} + \sigma_{\text{d}}}\left[1 - \exp\left(-\frac{t}{\tau}\right)\right] = -E_{\text{sat}}\left[1 - \exp\left(-\frac{t}{\tau}\right)\right]. \tag{3.18}$$

A kinetics of the space-charge field decay under a uniform illumination or in darkness, $E \neq 0$, the short-circuit conditions is

$$E(t) = E_{\text{sat}} \exp\left(\frac{-t}{\tau}\right). \tag{3.19}$$

E_{sat} is the steady-state value of the field and τ in the general case means the dielectric relaxation time $\tau = \tau_{\text{di}} = \epsilon\epsilon_0/(\sigma_{\text{ph}}+\sigma_{\text{d}})$, and $\epsilon\epsilon_0$ is the dielectric permittivity. Clearly, in darkness τ_{di} is dictated by σ_{d} and under illumination by σ_{ph} ($\sigma_{\text{ph}} \gg \sigma_{\text{d}}$) depending on the light intensity. If the photoexcitation and recombination times are comparable to τ_{di}, the expressions for τ are more complicated, e.g., [284]. The expression (3.19) immediately predicts the fundamental drawback of storage systems based on the photorefractive recording, namely, the erasure of the grating under the readout by a uniform Bragg beam. By definition S corresponds to the initial quasi-linear part of $E(t)$ for $t \ll \tau_{\text{di}}$. For a linear lux-ampere characteristic $\sigma_{\text{ph}} \propto I$ which that occurs in photorefractive crystals at moderate light intensities. Then, as seen from (3.16), the sensitivity when recording by the photovoltaic mechanism is presented as

$$S_1^{\text{pv}} = r_{\text{eff}} k_{\text{G}} \alpha / \epsilon\epsilon_0 \tag{3.20}$$

As a criterion for the sensitivity very often that of $LiNbO_3$:Fe is used which is $S_1' \sim 0.01$–$0.1\,\text{cm J}^{-1}$ for $\lambda = 633\,\text{nm}$ and 0.05–$0.3\,\text{cm J}^{-1}$ for $\lambda = 488\,\text{nm}$. A special approach was developed to characterize the process of successive recording (multiplexing) a large number of elementary holograms in the crystal volume. As illumination leads to a partial erasure of the hologram following (3.19), so when multiplexing, all pre-recorded holograms are partially optically erased on recording of each sequential one. A convenient parameter characterizing the diffraction efficiency of the multiplexed holograms with account for this erasure is the dynamical parameter $M\#$ (the Mok factor) [285] introduced by the following way. For a large number M of holograms the diffraction efficiency is $\eta \approx M^{-2}$ [286]. A recording of an elementary hologram and its decay are described by (3.18) and (3.19), respectively, denoted for simplicity as $A_0[1 - \exp(-t/\tau_{\text{rec}})] \approx A_0[1 - \exp(-t/\tau_{\text{er}})]$ (where A_0 is the saturation grating amplitude, τ_{rec} and τ_{er} are recording and erasure times). The final equalized diffraction efficiency of M holograms for short exposure times ($t \ll \tau_{\text{er}}$) may be expressed as where $M\# = (A_0/\tau_{\text{rec}})\tau_{\text{er}}$. It should be emphasized that this parameter may be applied only provided that the multiplexing was performed with the use of an appropriate schedule of recording (e.g., [287]) ensuring whenever possible equality of diffraction efficiencies of stored elementary holograms. A convenience of $M\#$ for characterizing the whole system of multiplexed holograms is due to the fact that the values of A_0 and τ_{rec} themselves are difficult to measure, whereas A_0/τ_{rec} and τ_{er} can be obtained from the slopes of $\eta(t)$ and $\ln(\eta(t))$ during recording and erasure, respectively, which immediately gives $M\#$. Therefore, factor $M\#$ is a measure of the average diffraction efficiency of multiplexed holograms. And vice versa, the dynamical parameter $M\#$ can be estimated from a single hologram recording and erasure [285] according to

$$\eta = \left[\frac{A_0}{\tau_{\rm rec}}\frac{\tau_{\rm er}}{M}\right]^2 = \left[\frac{M\#}{M}\right]^2. \tag{3.21}$$

$$M\# = \frac{d\sqrt{\eta}}{dt}\,\tau_{\rm er}. \tag{3.22}$$

for $t \ll \tau_{\rm er}$.

It is worthy of mentioning here that in addition to the hologram erasure when readout, there is another negative effect specific for the hologram recording in photorefractive media, especially in $LiNbO_3$:Fe. When recording or readout, a light scattering (fanning) arises caused (e.g., [288]) by a nonlinear interaction of recording waves with lightwaves scattered within the crystal bulk by nonuniformities of the refractive indices δn, defect clusters, etc. Note that for a two-wave coupling, a steady unidirectional energy transfer with a coupling gain Γ (3.13) takes plays only in the case of a nonlocal response, especially when recording by the diffusion mechanism (3.9) [289]. From this viewpoint a recording by the photovoltaic mechanism (a local response) is not accompanied by fanning. However, this is not valid for a multi-wave interaction. For example, according to Au and Solymar [290] in the case of a local response a three-wave interaction with a unidirectional energy transfer is possible, i.e., for recording by the photovoltaic mechanism. In other words, under two-beam recording, a multi-wave interaction within the medium bulk caused by a scattering of the recording beams is a reason of a strong fanning observed in the photovoltaic media. This effect results in photoinduced light scattering (fanning) from occurring multiple parasitic holograms and leads to a loss of the crystal transparency and to a reduction of the signal-to-noise ratio. A search for the photorefractive crystals with a suppressed fanning is desirable.

The often used geometry of recording is the grating vector parallel to the polar axis ($K \parallel z$), that is the grating grooves normal to it (Fig. 3.1). In this case, in $LiNbO_3$ and its symmetrical analogs, the expression for the effective electro-optic coefficient, $r_{\rm eff}$, used in (3.11), (3.14), and (3.20) is for the extraordinary polarization of the recording beams [291]

$$r_{\rm eff} = r_{33}\cos^2\theta - r_{13}\sin^2\theta + \frac{n_{\rm e}-n_{\rm o}}{n_{\rm e}}(r_{33}+r_{13})\sin^2(2\theta) \tag{3.23}$$

and for ordinary beams

$$r_{\rm eff} = r_{13}. \tag{3.24}$$

3.2 Photoinduced Charge Transport in $LiNbO_3$ Crystals

The main mechanism of the photorefraction in $LiNbO_3$ is the bulk photovoltaic effect. Actually, according to (3.9) the amplitude values of the diffusion fields for real grating periods $\lambda \approx 1\,\mu$m are approximately $1.5 \times 10^3\,{\rm V\,cm^{-1}}$.

In $LiNbO_3$:Fe and $LiNbO_3$:Cu the Glass constants are $k_G = 2.8 \times 10^{-9}$ and $5.5 \times 10^{-10}\,\mathrm{cm\,V^{-1}}$, respectively, and photoconductivities in as-grown crystals at moderate light intensities are in the range 10^{-11}–$10^{-12}\,\Omega^{-1}\,\mathrm{cm}^{-1}$ (for details see below), so the photovoltaic fields in saturation are of about 10^4 or even $10^5\,\mathrm{V\,cm^{-1}}$ [292]. In stoichiometric $LiNbO_3$ and in $LiNbO_3$ with optical-damage-resistant dopants Mg, Zn, etc., under certain conditions equal fields $E_{\mathrm{pv}} \approx E_{\mathrm{dif}}$ may occur [293, 294], so a combined mechanism of recording grating is observed. The photoinduced charge transport schemes have been extensively investigated in $LiNbO_3$ with transition metal (TM) multicharged impurities Fe, Cu, Mn, Ni, etc., the available enormous bibliography one may find, for example, in the reviews [115, 269, 295]. A search for photorefractive and photovoltaic impurities so far remains purely empirical. Note, that the Cr impurity in $LiNbO_3$ and $LiTaO_3$ is not photovoltaic active, although in literature is often mistakenly attached to those. We should emphasize that discussions of the charge transport schemes in $LiNbO_3$ are always based on an a priori assumption of the n-type photoconductivity, which in fact was experimentally proved only for congruent $LiNbO_3$ doped with Fe and Cu. In the literature we could find only a publication on studies of the photoconductivity in $LiNbO_3$ by the photo-Hall method [296] which were performed in strongly reduced $LiNbO_3$ and settled the n-type σ_{ph} with a Hall mobility of about $0.8\,\mathrm{cm^2\,V^{-1}\,s^{-1}}$ at room temperature. No studies of the σ_{ph} type in Li-enriched $LiNbO_3$ have been performed so far, so the consideration of the charge transport scheme in these crystals on the base of the electron photoconductivity seems to be rather speculative.

3.2.1 Dark Conductivity in $LiNbO_3$

Preparatory to discussing the photoinduced-charge transport in $LiNbO_3$, we briefly dwell on the current concept of the dark conductivity. The origin of the dark conductivity in $LiNbO_3$ was under a long discussion. First of all, the dominant role of the ionic conductivity was proved at $T > 1{,}000°\mathrm{C}$ [297–299]. Here it was shown that in the temperature range from 200°C to 1,000°C the conductivity obeys the single activation energy $E_{\mathrm{a}} \approx 1.1\,\mathrm{eV}$. A similar $E_{\mathrm{a}} \approx 1\,\mathrm{eV}$ was found from room temperature to 150°C in slightly doped $LiNbO_3$:Fe [300]. Therefore, in the whole temperature range the ionic conductivity is due to a unique type of ions. Originally, the ionic conductivity at high temperatures was ascribed either to oxygen vacancies V_O, or to Li-interstitials [297, 298]. As discussed in detail in Chap. 2, in as-grown congruent $LiNbO_3$ the formation of oxygen vacancies was ruled out, so V_O as possible charge carriers may be disregarded. At the same time, E_{a} of the Li-transport in $LiNbO_3$ is of about 0.4 eV [301] and is inconsistent with $E_{\mathrm{a}} \approx 1\,\mathrm{eV}$ estimated from direct conductivity measurements. The ultimate decision was made in [302], where a proportional relationship between the hydrogen concentration and the conductivity was demonstrated in $LiNbO_3$ at temperatures from 400°C to 1,000°C. This conclusion was supported in [109, 303] and extended to deuterium in

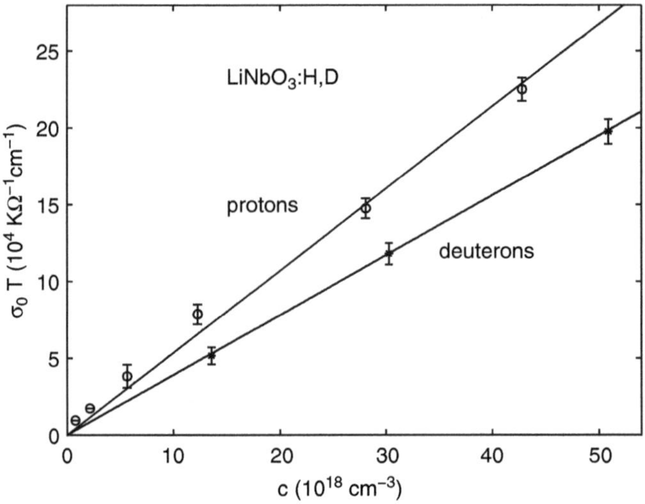

Fig. 3.3. Pre-exponential product $\sigma_0 T$ for protons and deuterons in $LiNbO_3$ as a function of their concentration in the crystals, redrawn from [109]

D^+-enriched $LiNbO_3$ [109], see Fig. 3.3. On the base of these data, the dominant protonic conductivity in $LiNbO_3$ may be taken as settled both at elevated and room temperature. However, this consideration of the conductivity is applicable to $LiNbO_3$, undoped or doped with low impurity concentrations. Of course, it has no relation to the reduced crystals, in which, as shown above, the dark electronic conductivity is greatly increased due to formation of bipolarons. An increase of the impurity concentration affects the transport mechanism of the equilibrium carriers. This was investigated especially systematically in $LiNbO_3$:Fe in view of the relation of the dark-storage times τ_s of recorded gratings to σ_d. We briefly dwell on the current state of this problem discussed in [300,304,305]. As σ_d in $LiNbO_3$ is extremely low; an extrapolation from elevated temperatures to room temperature gives values of 10^{-16}–$10^{-18}\,\Omega^{-1}\,cm^{-1}$ for undoped or slightly doped crystals, so the most reliable data may be obtained just from τ_s. The value of τ_s in low-doped $LiNbO_3$:Fe is up to year [306–309] and falls down to hours, minutes and even seconds with increasing iron concentration [308,309]. In [300,304,305], it was shown that in $LiNbO_3$:Fe with iron concentrations of 0.2–0.25% the activation energy of the dark conductivity decreases to 0.3–0.4 eV, thus the type of the majority charge carriers is changed. Moreover, according to [304,308,309] the dark decay of holograms in highly doped $LiNbO_3$:Fe is described by a stretched exponential rather than by a single value. On the basis of all these data, the authors of these works conclude that the dark conductivity in $LiNbO_3$:Fe is governed by two concurrent processes, namely by the proton transport and electron tunneling via Fe-sites with no band transport. At low Fe concentrations the former mechanism is dominant, whereas at iron concentrations of

0.2–0.25% the determining factor is the electron tunneling. The later conclusion on the dark conductivity relation to the electron tunneling via Fe-sites was recently expanded to $LiNbO_3$:Fe up to 3 wt.% Fe [310]. At elevated temperatures when the proton mobility increases, the decisive factor of σ_d is always the proton transport independent of the impurity state.

So, as applied for the optical storage, the requirements for a simultaneous high-sensitivity and stability in the dark are in certain contradictions, because an improvement of S by increasing the doping level is accompanied by a faster decay. An enhancement of the decay time in the dark in low-doped crystals may be achieved by dehydration of the crystal. According to [292, 300, 304] in $LiNbO_3$:Fe the highest acceptable doping is 0.1% Fe, otherwise a dark decay would be too fast. As the probability of the electron tunneling through the potential barrier decreases with increasing height of the barrier, in other words with the deepness of the dopant level, so a search for a photovoltaic impurity with a level deeper than Fe and comparable values of the photovoltaic coefficients is required. An appropriate candidate seems to be the Mn center [138, 139], which is deeper than Fe. The sensitivity of $LiNbO_3$:Mn is comparable to $LiNbO_3$:Fe.

Certain characteristic dependencies may serve in the recognition between the discussed mechanisms of the dark conductivity. If a purely ionic σ_d occurs, then a dark decay of the photorefraction is described by a single exponent analogous to (3.19). At a combined (proton–electron) conductivity, the dark decay is characterized by a stretched exponential, like $\delta\Delta n \sim \exp{(-t/\tau)^{\beta}}$. For the tunneling mechanism σ_d depends on the probability of tunneling between two neighboring traps. This in its turn has an exponential dependence on the mean distance between two traps, which is $C^{-1/3}$ (where C is the concentration of the traps). So, for the tunneling process the normalized conductivity is related to the concentration in the following way $\sigma_d/N_{\text{eff}} \sim \exp(-\gamma C^{-1/3})$ which is equivalent to $\ln(N_{\text{eff}}/\sigma_d) \sim C^{-1/3}$ ($\gamma > 0$ is a constant). For example, this criterion was applied for analyzing the type of the dark conductivity in $LiNbO_3$:Cu [311]. The σ_d was shown to be protonic up to $C_{\text{Cu}} > 7 \times 10^{19}\text{cm}^{-3}$ whereupon it sharply grows with $C^{-1/3}$ in accordance with the relations presented earlier (Fig. 3.4). In $LiNbO_3$:Mn the protonic conductivity dominates up to Mn concentrations of 0.2 at.% [139].

The values of the transport characteristics (activation energy E_a, diffusion coefficient D, mobility μ) of the protonic conductivity are noticeably scattered. The estimates of D_0 are varied in the range from 4 [312] to 0.01 [313] and $0.0014\,\text{cm}^2\,\text{s}^{-1}$ [314]. Note that the data for D were obtained both in bulk crystals and surface structures, such as proton exchanged optical waveguides, this may somewhat account for this scatter. The table exemplifies the dependence of E_a on $LiNbO_3$ crystal compositions, deduced either from direct conductivity measurements at elevated temperatures $\sigma_d(T)$ or from the dark decay times τ_{decay} of recorded holograms.

As seen from Table 3.1, the most obvious reason of a spread of E_a is its dependence on the [Li]/[Nb] ratio. Additionally, the scatter in D and E_a may

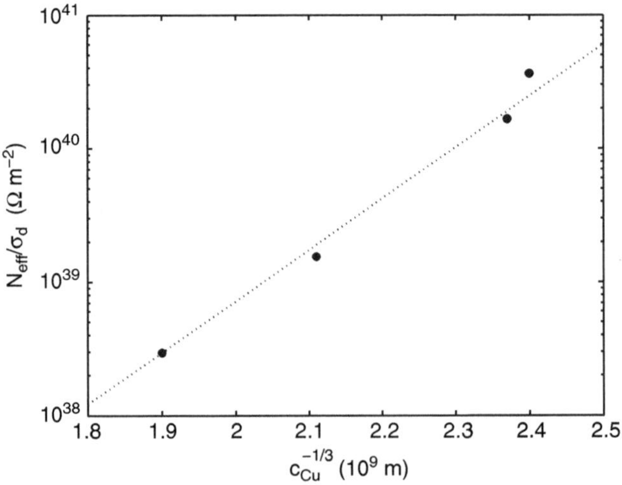

Fig. 3.4. The ration $N_{\text{eff}}/\sigma_{\text{d}}$ as a function of the Cu concentration $c_{\text{Cu}}^{-1/3}$ for crystals with strong copper doping $c_{\text{Cu}} > 7 \times 10^{25}\,\text{m}^{-3}$, redrawn from [311]

Table 3.1. Activation energy E_{a} as a function of the crystal composition and doping

[Li]/[Nb]	Impurity	E_{a} (eV) from $\sigma_{\text{d}}(T)$	E_{a} (eV) from τ_{decay}	Reference
Congruent		1.1		[299]
Congruent		1.23		[109]
Congruent	Fe	1.17		[109]
Congruent	Fe	1.13		[315]
0.99	Fe	0.95		[315]
Congruent	Mn	1.13	1.05	[315]
0.99	Mn	0.96	1.05	[315]

be related to a nonequilibrium distribution of hydrogen over several nonequivalent O-sites [108,316].

In summary, according to the presently accepted model, the dark conductivity in as-grown, low-doped $LiNbO_3$ is ascribed to the proton transport, whereas at moderate doping levels of about 0.1–0.2 at.% (at least transition metals like Fe) it transforms to the electron transfer, occurring by a tunneling mechanism.

3.2.2 One-Center Charge Transport Scheme in $LiNbO_3$

The microscopic mechanism of the charge transport in TM-doped $LiNbO_3$ was developed on the basis of the studies in $LiNbO_3$:Fe, $LiNbO_3$:Cu, and

$LiNbO_3$:Mn (e.g., [115, 116, 137]), in which the photoinduced charge carriers were proved to be electrons. Particularly, this conclusion has been drawn with the use of the original holographic method [317, 318], which is applicable to the case of the nonlocal response when the recording involves the diffusion mechanism. The method permits to determine the photo-carrier sign from the sign of the grating shift φ with respect to the recording light pattern, i.e., from the direction of the energy transfer between the recording beams; φ is positive for photo-electrons and negative, if the charge carriers are holes. Moreover, the method permits to detect a contribution from either type of photocarriers, provided that the photoconductivity is bipolar.

The one-center charge transport mechanism may be summarized in the following way. TM impurities as a rule are present in the $LiNbO_3$ lattice in two charge states (for example, Fe^{2+} and Fe^{3+}, Cu^{+} and Cu^{2+}, Mn^{2+} and Mn^{3+}, etc.) [4, 115, 137–139], which govern the transport of free photoexcited electrons e^{-} according to the scheme

$$\mathrm{Fe}^{2+} + \mathrm{h}\nu \Leftrightarrow \mathrm{Fe}^{3+} + \mathrm{e}^{-} , \tag{3.25}$$

where Fe^{2+} and Fe^{3+} (or Cu^{+} and Cu^{2+}) are electron donors and traps, respectively. The photoconductivity for such a one-center scheme in a simplified form may be presented as

$$\sigma_{\mathrm{ph}} = ne\mu = g\tau e\mu = \frac{qe\mu}{h\nu}\frac{s_{\mathrm{ph}}}{\gamma}\frac{N_{\mathrm{D}}}{N_{\mathrm{C}}}I = A\frac{N_{\mathrm{D}}}{N_{\mathrm{C}}}I, \tag{3.26}$$

where $g = q\alpha I/h\nu$ is the photoexcitation rate, $\tau = 1/\gamma_{\mathrm{r}}N_{\mathrm{C}}$ is the life-time of the photocarriers in the conduction band, μ the mobility, q the quantum efficiency, $\alpha = s_{\mathrm{ph}}N_{\mathrm{D}}$ the optical absorption coefficient, γ is the recombination coefficient, s_{ph} is the absorption cross-section, N_{C} and N_{D} are concentrations of the donors and traps, respectively. So, the photoconductivity for such a scheme is a linear function of the light intensity and of the concentration ratio of donors and traps. These relations were experimentally proved for $LiNbO_3$:Fe and $LiNbO_3$:Cu [115, 137]:

$$\begin{aligned} \frac{\sigma_{\mathrm{ph}}}{I} &= \frac{\mathrm{Fe}^{2+}}{\mathrm{Fe}^{3+}} \quad 10^{-12}\,\mathrm{cm}\,\Omega^{-1}\,\mathrm{W}^{-1}, \\ \frac{\sigma_{\mathrm{ph}}}{I} &= \frac{\mathrm{Cu}^{+}}{\mathrm{Cu}^{2+}} \quad 6.5\times10^{-12}\,\mathrm{cm}\,\Omega^{-1}\,\mathrm{W}^{-1}. \end{aligned} \tag{3.27}$$

They are fulfilled in a wide range of the impurity concentrations, however, as shown in [292], are no more valid in very highly doped $LiNbO_3$:Fe obviously because of a fundamental change of the transport mechanism. Thorough investigations of the spectral distribution of the photovoltaic current in $LiNbO_3$:Fe [72, 317] and $LiNbO_3$ [319] have found its broad maximum in the visible range and a smooth decrease when moving to the UV range from 364 to 334 nm. According to [115] j_{pv} in $LiNbO_3$:Fe (and $LiNbO_3$:Cu) depends on the donor concentration only

$$j_{pv} = k_G s_{ph} [Fe^{2+}] I. \quad (3.28)$$

From (3.27) and (3.28) one may see that the saturation value of the photorefraction $\delta \Delta n \approx E_{pv} \approx [Fe^{3+}]$ which was proved experimentally [320]. In the literature one may find the attempts to estimate the microscopic parameters of the charge transport in $LiNbO_3$:Fe. For example, in the review [116] the oscillator strength for the iron impurity is calculated. The data on the optical absorption in oxidized and reduced $LiNbO_3$:Fe [321] give the value of the Fe^{2+} absorption cross-section $S_{Fe^{2+}}$ of about $10^{-18}\,cm^2$ for $\lambda = 500$ nm. Similar experiments in [322] lead to the absorption cross-sections $S_{Fe^{2+}} = 10^{-17}\,cm^2$ and $S_{Fe^{3+}} = 8.5 \times 10^{-20}\,cm^2$ (for the blue–green spectral range). This permitted to estimate the number of transition metal ions (TM) $[TM^{2+}]_{\alpha(\lambda)}$ from the peak value of the absorption $\alpha(\lambda)$ at λ given in nm:

- $[Fe^{2+}]_{\alpha(477)} = (2.16 \pm 0.70)10^{17}\,cm^{-2}$, according to [116]
- $[Cu^{+}]_{\alpha(477)} = 2 \times 10^{17}\,cm^{-2}$ and
  $[Cu^{2+}]_{\alpha(1,040)} = 10^{18}\,cm^{-2}$, for details see [323, 324]
- $[Mn^{2+}]_{\alpha(577)} = 1.11 \times 10^{16}\,cm^{-2}$, reported in [139]

The transport scheme (3.25) in $LiNbO_3$:Fe may be fundamentally modified with the aid of certain external actions. For example, in strongly oxidized crystals, where $[Fe^{2+}] \approx 0$ and iron is in the state $[Fe^{3+}]$, the photoconductivity measured in the near UV becomes dominantly of the p-type [317, 318]. Recent studies in very highly doped $LiNbO_3$:Fe [310] also have found that a strong oxidation also transforms σ_{ph} to the p-type, however the existing photovoltaic current is due to the electron transport. Other possibilities of modifying the scheme (3.25) are provided by increasing the Li content or doping by optical-damage-resistant ions Mg or Zn. As shown in Chap. 2, in both cases the localization of the Fe^{3+}-ion in the lattice is changed which is accompanied by a change of the electron trap [325, 326]. Actually, in [327, 328] it was shown that in highly-doped $LiNbO_3$:Mg and $LiNbO_3$:Zn, respectively, the photocarriers are holes.

The relations (3.26)–(3.28) are valid in a wide range of light intensities, so under relatively low I (when $\sigma_{ph} > \sigma_d$) the photorefraction comes to saturation, which occurs in $LiNbO_3$:Fe even at $I \approx 1\,W\,cm^{-2}$ for blue–green light. From the expression (3.27), (3.28) we obtain the following relations for the saturation photovoltaic fields [115, 116, 329]

$$E_{pv} \propto [Fe^{3+}]; \quad E_{pv} \propto [Cu^{2+}] \quad (3.29)$$

for $LiNbO_3$:Fe and $LiNbO_3$:Cu, respectively.

The one-center scheme (3.25) formed the basis for a general description of the hologram recording in photorefractive crystals on a whole. This approach was proposed in [280–282] and in detail developed and supplemented

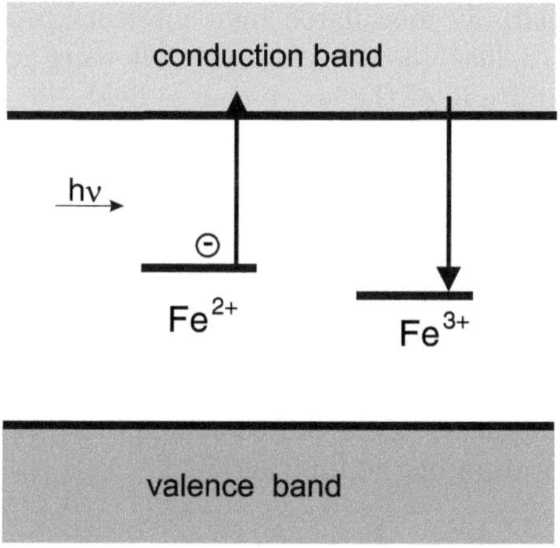

Fig. 3.5. Charge transport scheme in a one-center model. Fe^{2+} and Fe^{3+} are filled and empty electron traps

in numerical manuscripts and reviews [330–332]. The model assumes the band electron charge transport. The basic scheme consists of a single electron donor D which in an ionized state D^+ is the unique electron capture (recombination) center, i.e., for $LiNbO_3$:Fe we have $D \equiv Fe^{2+}$, and $D^+ \equiv Fe^{3+}$, see Fig. 3.5.

The calculations are performed assuming the case of a compensated semiconductor, where a hypothetical acceptor A providing the compensation condition is not involved in the charge transport. A coupled set of electrostatics and rate equations for a one-dimensional case looks as follows

$$\frac{\mathrm{d}n}{\mathrm{d}t} = \frac{\mathrm{d}N_\mathrm{D}^+}{\mathrm{d}t} - \frac{1}{e}\frac{\mathrm{d}j}{\mathrm{d}x}, \tag{3.30a}$$

$$\frac{\mathrm{d}N_\mathrm{D}^+}{\mathrm{d}t} = (s_\mathrm{ph}I + s_\mathrm{T})\left(N_\mathrm{D} - N_\mathrm{D}^+\right) - \gamma n N_\mathrm{D}^+, \tag{3.30b}$$

$$j = \sigma E - eD\frac{\mathrm{d}n}{\mathrm{d}x} + k_\mathrm{G} s_\mathrm{ph}\left(N_\mathrm{D} - N_\mathrm{D}^+\right), \tag{3.30c}$$

$$\epsilon\epsilon_0 \frac{\mathrm{d}E}{\mathrm{d}x} = e\left(n + N_\mathrm{A} - N_\mathrm{D}^+,\right) \tag{3.30d}$$

where n is the concentration of free electrons, N_D and N_D^+ are the donor and ionized donor densities, respectively, N_A is the density of a compensating passive acceptor, s_ph and s_T are the parameters characterizing the probability of the photoionization and thermal ionization of the donor, respectively, γ is the recombination coefficient, I is the light intensity, other notations were explained above. Accordingly to the Einstein relation $D_\mathrm{n} = \mu_\mathrm{e} k_\mathrm{B} T/e$ (where k_B is Boltzmann constant, D_n is the diffusion coefficient). As applied to $LiNbO_3$:Fe, $(N_\mathrm{D} - N_\mathrm{D}^+)$ and N_D^+ are the concentrations of Fe^{2+} and Fe^{3+}, respectively. Note that the electron current density (3.30c) takes into account both the diffusion and photovoltaic currents.

Assuming the spatially modulated light intensity presented by (3.8), the general solution to (3.30a)–(3.30d) gives the following general expression for the steady-state amplitude of the space-charge field

$$E_{\mathrm{K}} = \mathrm{i} m_{\mathrm{e}} \frac{E_{\mathrm{D}} - \mathrm{i} E_{\mathrm{pv}}}{1 + \frac{E_{\mathrm{D}}}{E_{\mathrm{q}}} - \mathrm{i} \frac{N_{\mathrm{D}}^{+}}{N_{\mathrm{D}}} \frac{E_{\mathrm{pv}}}{E_{\mathrm{q}}}}, \tag{3.31}$$

where

$$m_{\mathrm{e}} = m \frac{s_{\mathrm{ph}} I}{s_{\mathrm{ph}} I + s_{\mathrm{T}}}. \tag{3.32}$$

In (3.31) E_{D} is the amplitude of the diffusion field (3.9), E_{pv} is the photovoltaic field (3.6). The saturation or the limiting field

$$E_{\mathrm{q}} = e N_{\mathrm{t}} / K \epsilon \epsilon_0 \tag{3.33}$$

gives the maximum amplitude of the space-charge field provided by the electron transport for a given spatial frequency of the grating, that is it corresponds to a complete spatial separation of positive and negative charges captured by traps; $N_{\mathrm{t}} = N_{\mathrm{a}} N_{\mathrm{D}} / N$ is the so-called effective trap concentration.

We should mention a well-known analytical solution to the set (3.30) obtained neglecting the nonlinear effects and assuming the diffusion mechanism of the charge transport (thus the first and third terms in (3.30c) being zero). The space-charge field in this case is presented in the following way:

$$E_{\mathrm{sc}} = \frac{m E_{\mathrm{q}} E_{\mathrm{D}}}{E_{\mathrm{q}} + E_{\mathrm{D}}}. \tag{3.34}$$

This expression physically means that the E_{D} value given by (3.9) is limited by screening effects. The curve $E_{\mathrm{sc}}(K)$ passes over a maximum for $K = 2\pi/\lambda_0$, where $\lambda_0 \approx l_{\mathrm{d}}$, with $l_{\mathrm{d}} = \sqrt{(\epsilon \epsilon_0 K_B T)/(e^2_{\mathrm{N_{eff}}})}$ being the Debye screening length, see Fig. 3.6.

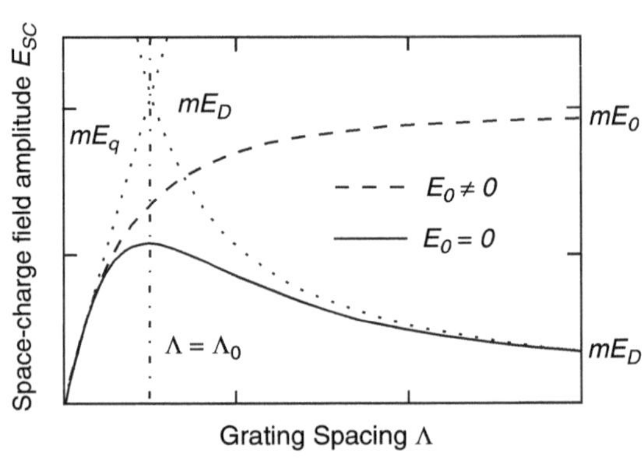

Fig. 3.6. Diffusion field vs. grating period, $j_{\mathrm{phv}} = 0$, redrawn from [268]

The angular dependencies of the diffraction efficiency $\eta(\theta)$ (3.11) and two-beam coupling gain $\Gamma(\theta)$ (3.14), following (3.34) pass over maxima at $\Lambda_0 \approx l_d$ and permit to estimate formally l_d and thus N_{eff} for a pure diffusion mechanism. For a combined recording mechanism ($j_{pv} \neq 0$), sometimes observed in $LiNbO_3$ [293, 294], the solution to the equation set (3.30) becomes more complicated, examples may be found in [330, 331]. Angular dependencies of E_{sc} in these cases depend on relations between E_D, E_{pv}, and E_q. General expressions for the recording kinetics and erasure by a uniform illumination have the forms

$$E(x,t) = E_K\left[1 - \exp(-\Phi_e t)\right]\exp(ikx) \tag{3.35}$$

and

$$E(x,t) = E_K \exp(-\Phi_e t)\exp(ikx), \tag{3.36}$$

respectively, where E_K is given by (3.31) and Φ_e is the characteristic rate, the same for recording and erasure. This rate may be presented by the following general expression:

$$\Phi_e = \Phi_{di}\frac{1 + \frac{K^2}{K_{Deb}^2} - i\frac{K}{k_{pv}}}{1 + \frac{K^2}{q_D^2}} \tag{3.37}$$

where $\Phi_{di} = \sigma/\epsilon\epsilon_0$ is the dielectric relaxation rate, K is the spatial frequency of the grating, K_{Deb} is the inverse Debye screening length, q_D is the inverse diffusion length, k_{pv} is a certain photovoltaic characteristic number. As reasonable parameters in the first approximation one may assume $\Phi_e \approx \Phi_{di}$. This means that the rates of the recording and optical erasure are linearly proportional to the photoconductivity σ_{ph}, and the decay rate in dark is proportional to the dark conductivity σ_d. The expressions (3.35), (3.36) are identical to (3.18), (3.19), respectively. An alternative characteristic time which manifests itself at very short τ_{di}, is the diffusion time $\tau_D = e/(K^2 k_B T\mu)$.

The one-center model shortly presented in this section was subsequently refined. For example, in [333] possible variations of the initial recording or erasure kinetics from (3.35), (3.36), particularly a nonequality of the recording/erasure times are discussed in view of a specific of the electron drift under the photovoltaic field (which is not identical to applying an external field). Another problem concerns the limits of the applicability of the whole concept. In principle, it is developed for so-called adiabatic or quasi-CW approximation, which means that the electron system is in the equilibrium state and the free electron density is by orders of magnitude lower than the trap density. However, this approximation is not valid for times shorter than those required for occurring this equilibrium, i.e., during an instantaneous rise of the free electron density. This should be taken into account for the fastest initial stage of recording or for recording by short pulses. According to the estimates performed in [334] the photoexcitation and recombination times of the photoelectrons are in the range of 10 ps, so when recording by light pulses shorter than 1 ps one could expect a violation of the given concept. There is very scant available material on hologram recording by ultrashort light pulses. Okamura [335] found in a comparison of recording gratings with moderate intensity at 532 nm by a CW-beam (72 mW cm^{-2}) and by pulses of 3.5 ps duration (68 mW cm^{-2}) no discernible distinctions in the temporal characteristics of the diffraction efficiency. In all other cases the recording by picosecond or femtosecond light pulses using extremely high light intensities in the range of 50–250 GW cm^{-2} (e.g., [336] and references therein) lead to an extremely involved scenario of the photoinduced change of the refractive indices which does not permit to analyze the applicability of the adiabatic approximation. We mention that Sturman et al. [336] detected no contribution from the photorefraction in the initial meaning of this term, i.e., related to the EO-effect under the internal field, when studying the femtosecond holography at 388 nm and $t_p = 0.22$ ps in $LiNbO_3$:Fe.

In summary, we again list the usual characteristic features of the photorefraction occurring via the one-center charge transport scheme (3.25): the lux–ampere characteristics are linear $\sigma_{ph} \propto I$; the characteristic times of the recording and erasure are equal for equal I; the sensitivity does not depend on I; the photorefraction comes to a saturation at relatively low intensities corresponding to $\sigma_{ph} \gg \sigma_d$.

3.2.3 Two-Center Charge Transport Scheme in $LiNbO_3$: The Intrinsic Defects as *Secondary* Electron Traps

At the first stage of studies of the photorefraction on a whole, the experimental data on the photoinduced charge transport fitted into the one-center scheme developed by an example of $LiNbO_3$:Fe and described in the preceding section. However, at high light intensities ($I > 10^2$ W cm^{-2}) peculiarities were found (e.g., [337,338] which departed from the requirements of the one-center model, for example, the recording and erasure kinetics could not be described by single exponents (32), (33) [339]). The most pronounced evidence of a nonvalidity of the one-center scheme was finding of sublinear lux–ampere characteristics $\sigma_{ph} \propto I^x$ ($x < 1$), first in $BaTiO_3$ [340, 341] and then in other photorefractive crystals even at rather low light intensities. These results led to necessity of a new approach to the charge transport scheme and its consideration on the base of a two- (or more) center model. The most simple and obvious is a transport scheme assuming the participation of an additional (secondary) shallow electron trap, that is empty in the equilibrium state (the appropriate balance equations may be found in any classical monograph on the photoconductivity in semiconductors). A relation of photorefraction anomalies to the existence of secondary traps was supported by the discovery of a photoinduced increase of the optical absorption $\Delta\alpha$ under high light intensities in $BaTiO_3$ [342] that was attributed to a nonequilibrium population of a shallow trap [343].

A system of material equations based on the set (3.30a–d) was proposed by Valley [344] for a photorefractive medium with two photoactive nonphotovoltaic centers. A set of coupled equations for two types of centers in a photovoltaic crystal was discussed in [345]. On the base of the model [344] the calculations of sublinear lux–ampere characteristics and kinetics of the photorefraction were performed for several nonphotovoltaic crystals ($BaTiO_3$, $KNbO_3$, SBN, BSO, BGO; e.g., [346–349]). As no information on the origin of involved centers was available, these calculations used mainly fitting parameters.

In the course of the development of two-center charge transport models in photorefractive crystals it became clear, that in $LiNbO_3$ the recording grating under high light intensities, especially in undoped and low-doped crystals also could not be confined to the one-center description. For example, the photorefractive sensitivity S_1 was found to depend on I [350,351]. Recall that accordingly (3.20), the dependence $S_1(I)$ may be related to a photoinduced change of the optical absorption only. Consequences of this dependence were anomalously high photorefraction values at $I > 10^3\,\mathrm{W\,cm^{-2}}$ when recording by pulsed light [352,353] and in optical waveguides [354]. In pure and low-doped $LiNbO_3$ the dependencies of the photorefraction on the duration of the recording light pulses were found [94,355–357] which also cannot be explained in terms of the one-center concept.

The scope of these data led to necessity of a refined charge transport scheme in $LiNbO_3$. This problem was discussed in a series of publications [98,337,357,358] in which two-center models of the charge transport are proposed for $LiNbO_3$, the resulting variations of the photorefractive kinetics are discussed and the origin of *secondary* centers is analyzed. This consideration may be exemplified by the scheme proposed by Jermann and Otten [98] for $LiNbO_3$:Fe containing additionally to the traditional centers Fe^{2+} and Fe^{3+} a hypothetical shallow trap X with the concentration N_X which is empty in the equilibrium state (that is completely ionized in darkness, $N_X = N_X^+$), see Fig. 3.7.

When populated, the center X is assumed to be photovoltaic active. On the base of this scheme the authors interpret the photorefractive characteristics of $LiNbO_3$:Fe at high light intensities. The set of equations looks as follows

$$\frac{\mathrm{d}n}{\mathrm{d}t} = \frac{\mathrm{d}N_D^+}{\mathrm{d}t} + \frac{\mathrm{d}N_X^+}{\mathrm{d}t} + \frac{1}{e}\frac{\mathrm{d}j}{\mathrm{d}x}, \tag{3.38a}$$

$$\frac{\mathrm{d}N_D^+}{\mathrm{d}t} = (s + s_{DX})\left(N_D - N_D^+\right) I$$
$$-\gamma n N_D^+ - \gamma_{XD}\left(N_X - N_X^+\right) N_D^+, \tag{3.38b}$$

$$\frac{\mathrm{d}N_X^+}{\mathrm{d}t} = \left(\beta_X + s_X I + \gamma_{XD} N_D^+\right)\left(N_X - N_X^+\right)$$
$$-\left[\gamma_X n + s_{DX}\left(N_D - N_D^+\right) I\right] N_X, \tag{3.38c}$$

$$j = \sigma E + k_G s\left(N_D - N_D^+\right) I + k_X s_X\left(N_X - N_X^+\right) I, \tag{3.38d}$$

$$\epsilon\epsilon_0 \frac{\mathrm{d}E}{\mathrm{d}x} = e\left(n + N_A - N_D^+ + N_X - N_X^+\right), \tag{3.38e}$$

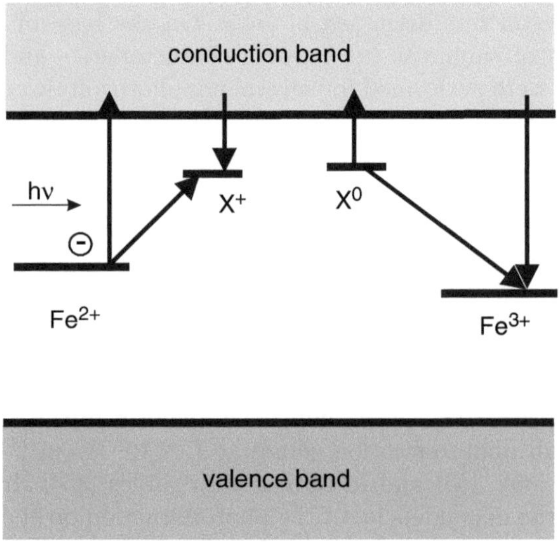

Fig. 3.7. Charge transport scheme in a two-center model. Fe^{2+} and Fe^{3+} are filled and empty deep electron traps. X^0 and X^+ are *filled* and *empty* shallow electron traps

where according to Fig. 3.7 the balance equations (3.38a)–(3.38c) take into account the electron photoexcitation from the donor D and from nonequilibrium populated center X^0 and the electron capture (recombination) by traps D^+ or X^+; β_X and s_X are probabilities of the thermal and photoexcitation of X, the notations for the center D are the same as in (3.30). Additionally, the authors suggest the direct electron transfer from Fe^{2+} to X^+ to form X^0 with a probability s_{DX} and from X^0 to Fe^{3+} with a recombination coefficient γ_{XD}. This assumption was taken in order to substantiate the observed linear dependence of the photoinduced optical absorption increase on the concentration of donors Fe^{2+}. The current equation (3.38d) takes into account the photovoltaic current only, the photovoltaic centers being both the donor Fe^{2+} and X^0 with the Glass constants k_G and k_X, respectively. Note that the necessary condition of the transfer Fe–X is a smallness of the inter-center distance, or, in other words a high concentration $N_X > 10^{20}\,\mathrm{cm}^{-3}$.

Not coming into details of this work, which is more or less repeated in other publications of this type, let recognize the most important qualitative conclusions. At intensities $I > 10^3$–$10^4\,\mathrm{W\,cm}^{-2}$ both under pulsed and CW pumping, a nonequilibrium population of X leads to a photoinduced increase $\Delta\alpha \propto N_X$ and a super-linear lux-ampere characteristic $\sigma(I)$. Both for a CW and pulsed radiation the dependencies of σ_{ph}, S_1 and E_{sc} were obtained. The main parameter determining the population of the trap X and the consequent anomalies of the photorefraction kinetics is the concentration of the *pumping* donor Fe^{2+}. For example, the expressions for the photoconductivity, and the space-charge field in saturation look as

$$\sigma_0 = e\mu \frac{s_{Fe} N_{Fe^{2+}} I + [\beta_X + (s_X - s_{Fe}) I] N_X^f}{\gamma_{Fe} \left(N_{Fe^{3+}} + N_X^f\right)}, \tag{3.39}$$

$$E_{sat} = \frac{k_{Fe}\, I\, N_{Fe^{2+}} + (K_X - k_{Fe})\, N_X^f\, I}{\sigma_0}, \tag{3.40}$$

where N_X^f is the saturation density of nonequilibrium filled secondary trap

$$N_X^f = N_X - N_X^+ = \frac{s_{FeX} N_X\, N_{Fe^{2+}}}{\beta_X + \gamma_{XFe} N_{Fe^{3+}}} I. \tag{3.41}$$

A contribution from the shallow trap X manifests itself only under peculiar conditions, such as very high I, when the concentration of nonequilibrium populated X^0 becomes comparable to $[Fe^{2+}]$, the concentration of Fe^{2+}, or at very low Fe, when $N_{Fe^{3+}} \ll N_X^+$. In accordance with this scheme, the authors of [358] observed in congruent $LiNbO_3$:Fe a significant photoinduced increase of the optical absorption at $\lambda \geqslant 633$ nm under illumination by an intensive light $\lambda = 532$ nm, the value of $\Delta\alpha$ being proportional to the concentration of Fe^{2+}; a nonlinear lux–ampere characteristic for $\lambda = 532$ nm is fitted by a second-order polynomial

$$\sigma_0 = \sigma_l I + \sigma_q I^2\ , \tag{3.42}$$

where the coefficients σ_l and σ_q are proportional to the concentration ratio $[Fe^{2+}]/[Fe^{3+}]$. As a result, the maximum values of $\delta\Delta n$ for each intensity at $I > 10^3$ W cm^{-2} are growing with I, so no complete (absolute) saturation of the photorefraction is achieved up to highest I in distinct from the usual linear case.

Similar calculations for nominally undoped $LiNbO_3$ with Fe of about 30 ppm and an accidental shallow trap with similar concentration [345] and at recording by a pulsed light lead to a dependence of the photorefraction both on the pulse duration and interval between pulses even at relatively low I. The latter dependence is owing to a partial thermal ionization of the nonequilibrium populated secondary trap during the dark interval. According to [357] the dependence of the photorefraction on the pulsed light at $I = 300$–800 W cm^{-2} is observed in congruent crystals up to the pulse frequencies of about 100 Hz, whereupon $\delta\Delta n$ comes to a constant value which is by a factor of two lower than for the CW light of the same intensity. Examples of these dependencies for several free of iron $LiNbO_3$ compositions are presented in Fig. 3.8.

A photoinduced increase of the optical absorption may lead to arising of a nonstationary (transient) component of the photovoltaic current during recording owing to an optical erasure of $\Delta\alpha$

$$j_{tr}(t) = k_G \Delta\alpha(t) I. \tag{3.43}$$

This current may manifest itself as a deviation of the photorefraction kinetics from the usual exponent (3.18), or even in appearance of extrema in $\delta\Delta n(t)$ at

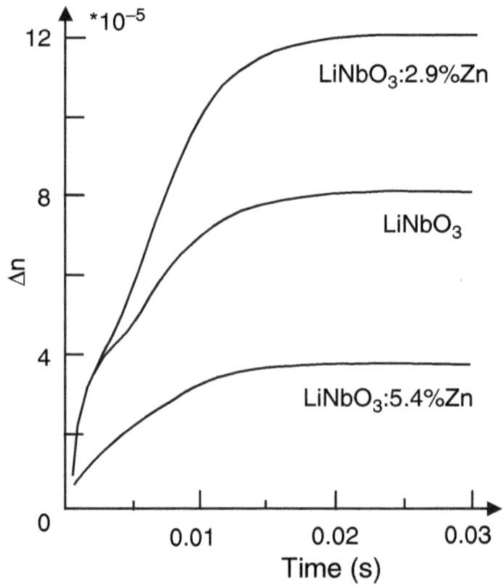

Fig. 3.8. The dependence of the saturation value of the photorefraction on the recording pulse duration in undoped $LiNbO_3$ and Zn-doped $LiNbO_3$ for $\lambda = 488\,nm$, redrawn from [345]

certain conditions [345]. In [359] a superlinear dependence of the photovoltaic current on I was predicted for a two-center scheme in $LiNbO_3$:Fe.

Experimentally observed anomalies of photorefraction in $LiNbO_3$ crystals of different compositions and calculations of relevant two-center schemes responsible for these anomalies always led to a conclusion of a very high density of secondary traps involved in the charge transport, $N_X \approx 10^{18}$–$10^{19}\,cm^{-3}$, corresponding to 0.001–0.01 at.%. Examination of compositions ruled out the presence of accidental impurities with such concentrations. Therefore, the most reasonable was to suggest that the secondary electron traps in $LiNbO_3$ are Nb_{Li} intrinsic defects, whose concentration in congruent crystals according to the structure investigations [41, 42] is actually in the range 10^{19}–$10^{20}\,cm^{-3}$. An alternative electron trap in oxides on a whole is the twice ionized O-vacancy. However, remind, that according to the current defect model of $LiNbO_3$ the probability of formation of oxygen vacancies is negligible. Additionally, the model calculations performed in [48] have shown that the energy of the electron capture by Nb_{Li} is $E_t = -1.4\,eV$, whereas the same value for an ionized O-vacancy gives $E_t = -0.16\,eV$, therefore centers $V_O^{\bullet}$ (if any) would be completely unstable. So, accordingly to the model calculations the most probable electron trap is $Nb_{Li}^{4\bullet}$. Therefore, in the two-center model developed in [98, 337, 357, 358] the X^0 center in the language of the polaron model is the small polaron Nb_{Li}^{4+}.

This consideration was recently supported by studies of short-living photoinduced centers in $LiNbO_3$ which was discussed in detail in Sect. 2.2 [58, 86, 96, 97]. Recall that an intensive visible light induces in $LiNbO_3$ crystals a short-living absorption band at about 740 nm (Fig. 2.8). Its characteristics (the spectral position and bandwidth) do not depend on the crystal composition, i.e., it is due to existence of certain centers common for all compositions. In the framework of the polaron model this band was ascribed to the formation of a small polaron Nb_{Li}^{4+}, or in other words, to capturing electrons by traps Nb_{Li}. The lifetime of this band is controlled by the crystal stoichiometry, namely, the lesser is the concentration of the traps Nb_{Li} (the more stoichiometric is the crystal), the more stable is the band. Qualitatively it may be explained as follows. In a rough approximation the lifetime of the band τ_X is inversely proportional to the probability of the direct recombination γ_{FeX}, which in its turn depends on a distance between Fe^{3+} and Nb_{Li} and decreases with decreasing Nb_{Li} content, i.e., τ_X for a given $[Fe^{3+}]$ is increased in stoichiometric crystals. The measurements of the lifetimes of photoinduced IR bands in $LiNbO_3$ are in agreement with this consideration. According to [86] the lifetime of the photoinduced polaron in congruent $LiNbO_3$:Fe is at room temperature in the range of microseconds, whereas in reduced near stoichiometric $LiNbO_3$ the lifetime of the photoinduced band at about 800 nm interpreted as well as the polaron band caused by the bipolaron dissociation [84,87], is as long as 10–100 ms at room temperature. Computer simulations of the dependence of τ_X on the distance between a donor and a nonequilibrium filled trap [86] concluded that the relaxation kinetics of a photoinduced band decaying by this mechanism should obey a stretched exponential law

$$\Delta\alpha(t) = \Delta\alpha_0 \exp\left[-\left(\frac{t}{\tau}\right)^{\beta}\right] \tag{3.44}$$

with $\beta < 1$. Actually, experimental results of [86] found that in darkness the relaxation of the photoinduced band occurs via (3.44).

We now mention other indirect experimental evidences of a contribution from Nb_{Li} to the charge transport. A degree of reduction of $LiNbO_3$ crystals strongly depends on the crystal stoichiometry. A criterion of the reduction degree in $LiNbO_3$ is the intensity of the absorption band at about 2.6 eV (Fig. 2.6) which appears after a reductive annealing and is assigned to electron capturing by Nb_{Li} to form a bipolaron [48, 70, 72]. As shown in [68], this absorption band in reduced stoichiometric crystals is significantly lower than in congruent ones. This difference may scarcely be interpreted in terms of the O-vacancy formation and seems to indicate a correlation between the content of Nb_{Li} and the degree of reduction: the lesser is Nb_{Li} concentration, the weaker is the band at 2.5 eV, i.e., the lesser is the concentration of centers responsible for its formation.

In the next chapter a growing photoconductivity in stoichiometric $LiNbO_3$ will be discussed in detail, which also corroborates the role of Nb_{Li} as an electron trap.

Nowadays, although the presented consideration is rather qualitative and speculative, studies of the transport scheme in $LiNbO_3$ by direct methods of the semiconductor physics are missing, nevertheless the two-center model of the photoinduced charge with Nb_{Li} as a *secondary* electron trap, is generally accepted and is always attracted for interpretation of the results on the photoconductivity and photorefraction in $LiNbO_3$.

4

Photorefraction in $LiNbO_3$ Crystals with Different Stoichiometry and/or Doped with Optical-Damage-Resistant Impurities

Summary. The photorefractive properties of $LiNbO_3$ with different [Li]/[Nb] ratios or/and doped with optical-damage-resistant impurities (ME = Mg, Zn, Sc, In, Hf, and Zr) are overviewed. The dependencies of the photorefraction on the composition caused mainly by variations in the photoconductivity are discussed in the framework of the two-center charge transport scheme. The origin of the photoconductivity decrease is attributed to a variation in the shallow electron trap (Nb antisite) concentration or/and an alteration of Fe^{3+}-like ion incorporation into the lattice, namely their partial localization on the Nb sites. An emphasis is given on an indubitable, although incomprehensible increase of the photovoltaic effect in Li-enriched and ME-doped $LiNbO_3$. The photoinduced coloration ('dark trace' and GRIIRA effects) in $LiNbO_3$ is considered in terms of the contribution from intrinsic defects. An improvement of the photorefractive properties in $LiNbO_3$ at low ME doping is demonstrated and an appropriateness of these compositions for practical applications is examined.

As usual in $LiNbO_3$, the optical damage is caused by the photovoltaic field, so according to (3.6), its value may be reduced by several ways. It may be achieved by an increase in σ_d as found in strongly reduced $LiNbO_3$ [360] or Ag-doped [192]. In H-exchanged optical waveguides, an observed optical damage resistance is due to an increasing σ_d [192]. Optical damage reduction may be provided by a suppression of the photovoltaic current. For example, because of an impurity origin of the photovoltaic effect in $LiNbO_3$, the spectral distribution of j_{pv} shows a maximum in the visible range and decrease in the UV, which results in a much lower value of the photorefraction recorded by the UV radiation compared to one induced by the blue–green light [361]. Moreover, as mentioned above, a strong oxidation of low-doped $LiNbO_3$:Fe strongly reduces j_{pv} in the UV-spectral range obviously because of altering the photoinduced charge carriers from electrons to holes [317, 318]. As discussed later, recent studies have found that the photovoltaic properties of $LiNbO_3$ are noticeably affected by a variation in the [Li]/[Nb] ratio. The microscopic reason of these effects has not been studied yet.

The most efficient tool to increase the optical damage resistance is, increasing the photoconductivity σ_{ph}, which provides a suppression of δn by two to four orders of magnitude. The reason of the optical damage resistance caused by an increased σ_{ph} and studied in more details in $LiNbO_3$ doped with optical damage resistant impurities, may be more or less adequately interpreted in terms of the current model of the defect structure.

4.1 Photorefractive Properties of Near-Stoichiometric $LiNbO_3$ Crystals

The photorefractive properties and photoinduced charge transport in relatively recently grown near stoichiometric $LiNbO_3$ and $LiTaO_3$ crystals (hereafter sometimes NSLN and NSLT, respectively) are investigated not in details. The available scant material is summarized in the recent reviews [29,195]. For the first time, a dependence of the optical damage on the stoichiometry of $LiNbO_3$ was reported by Anghert et al. [362], who observed an increase in the optical damage with increasing Li content. Later studies in $LiNbO_3$ with varied stoichiometry performed by several research groups [35, 357, 363, 364] confirmed a dependence of the optical damage on the crystal stoichiometry.

The summary of these studies is as follows. In the range of low and moderate light intensities the results confirm the data of Anghert et al. [362], who observed that δn increases with increasing [Li]/[Nb] ratio. This conclusion is exemplified by Figs. 4.1 and 4.2. Figure 4.1 presents temporal dependencies of photorefraction in $LiNbO_3$ with different [Li]/[Nb] ratio obtained under

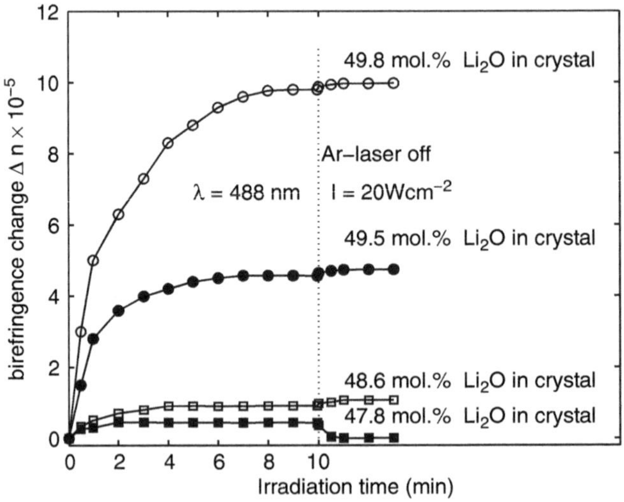

Fig. 4.1. Optical damage in $LiNbO_3$ of different Li content; $\lambda = 488\,\text{nm}$, $I = 20\,\text{W}\,\text{cm}^{-2}$ according to [365]

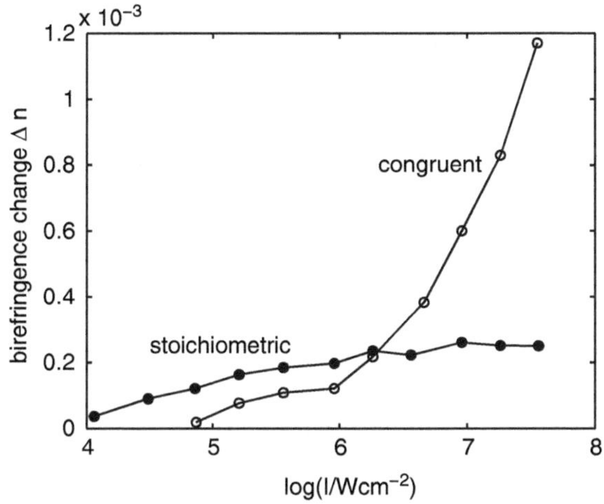

Fig. 4.2. Intensity dependencies of the optical damage in congruent and stoichiometric $LiNbO_3$ [35]. The stoichiometric crystal was grown by the Czochralski technique by adding 10.9 mol.% K_2O into the melt

$I = 20\,\mathrm{W\,cm^{-2}}$ for $\lambda = 488\,\mathrm{nm}$; the near-stoichiometric crystals were grown from the melt with $[Li]/[Nb] > 1$.

As seen, the optical damage resistance increases with decreasing Li concentration, since the value of δn is the highest in NSLN and negligible in the Li-deficient sample with $[Li]/[Nb] < 0.94$. The results shown in Fig. 4.2 were obtained in NSLN grown from the K_2O-modified melt [35,357] and confirmed in $LiNbO_3$ crystals grown by the VTE technique [357].

Again, the congruent $LiNbO_3$ is more optical damage resistant than the near-stoichiometric one up to a certain intensity threshold [35,357], whereupon the situation is reversed and CLN becomes less damage resistant than NSLN. This is due to the fact that the value of photorefraction in NSLN increases smoothly with the light intensity in the whole intensity range, while in CLN at $I > 100$–$200\,\mathrm{W\,cm^{-2}}$ the birefringence δn starts to grow very steeply. A steep slope of $\delta n(I)$ is a reason of a relatively low damage resistance of CLN at high intensities reported in [35, 357]. These results produced a widespread opinion of an enhanced optical damage resistance of NSLN at all, although this may be valid for high intensities only.

Later, the data obtained at low and moderate light intensities again and again confirmed the conclusion of a lower optical damage resistance in NSLN. When recording holograms at $I = 0.08\,\mathrm{W\,cm^{-2}}$, the diffraction efficiency is the highest in stoichiometric crystals and decreases with decreasing $[Li]/[Nb]$ ratio [293]. Measurements of the photorefractive two-beam coupling gain [366, 367] also established an increase of optical damage in Li-enriched $LiNbO_3$:Fe as compared to congruent $LiNbO_3$:Fe. In NSLN doped with 270 ppm Fe grown from the melt with $[Li]/[Nb] > 1$, the gain was about $25\,\mathrm{cm^{-1}}$ and exceeded the value in the congruent $LiNbO_3$:Fe with the same Fe concentration [366]. A similar result was obtained in NSLN:Fe crystals grown with K_2O in the melt [367].

We now discuss the reason for an increasing optical damage with increasing [Li]/[Nb] ratio observed in the range of moderate intensities. In the context of (3.6), the most obvious explanation would be a decreased photoconductivity in near-stoichiometric crystals. However, on the contrary, all experiments found an increase of σ_{ph} in NSLN. This follows from faster kinetics of the hologram recording and erasure in stoichiometric crystals [35,225,357,366–368]. A rough estimate gives $\sigma_{ph}/I \approx 10^{-15}$; $(3\text{–}4) \times 10^{-15}\,(\Omega\,\text{cm}\,\text{W})^{-1}$ for congruent and stoichiometric $LiNbO_3$, respectively [357]. The corresponding values observed by Furukawa et al. [225] are $\sigma_{ph} = 10^{-15}$ and $2 \times 10^{-14}\,(\Omega\,\text{cm})^{-1}$ for $\lambda = 488\,\text{nm}$, $I = 10\,\text{W}\,\text{cm}^{-2}$. A similar increase of σ_{ph} at increased Li content was found in crystals doped with Fe [366].

So, one observes in NSLN an inconsistency between an increasing $\delta\Delta n$ and enhanced σ_{ph}. An alternative reason eliminating this apparent mismatch, could be an increase of the photovoltaic effect in stoichiometric crystals. The first attempt to analyze the dependence of j_{phv} on the [Li]/[Nb] ratio was performed in [369]. The authors operated with the reduced $LiNbO_3$ crystals of different stoichiometry and found a difference in the spectral dependencies of j_{phv} in Li-enriched and near-stoichiometric $LiNbO_3$ accounted for, by the observed difference in the optical absorption spectra, but no data on j_{phv} were reported. Furukawa et al. [225, 227] performed combined measurements of the photovoltaic currents and photoconductivity in $LiNbO_3$ with different [Li]/[Nb] ratios in undoped and Mg-doped $LiNbO_3$. A surprising increase of the Glass constant was found in a pure stoichiometric crystal, more than by the order of magnitude as compared to k_G in congruent crystals. The results of this group presented in [29] are shown in Table 4.1. One may pay attention to extremely low values of k_G in undoped crystals almost by three orders of magnitude lower than in $LiNbO_3$:Fe and $LiNbO_3$:Cu.

It is noteworthy that, accordingly to [29], the Glass constants in Fe-doped CLN and Fe-doped NSLN are practically the same. This means, that the observed change of the photovoltaic properties of stoichiometric crystals (Table 4.1) is caused not by an influence of the lattice defect state on the photovoltaic ions, but may be related to a direct contribution of the intrinsic defects to the photovoltaic effect. So, at present, an increased optical damage in

Table 4.1. Comparison of the photovoltaic properties and photoconductivity in CLN and NSLN

	Congruent $LiNbO_3$		Stoichiometric $LiNbO_3$	
Light polarization	Ordinary	Extraordinary	Ordinary	Extraordinary
Glass constant 10^{-12} $(\text{cm}\,\text{V}^{-1})$	1.5	2.5	41	49
σ_{ph} 10^{-15} $(\Omega^{-1}\,\text{cm}^{-1})$	2	2.2	26	27

near-stoichiometric $LiNbO_3$ crystals may be accounted for, by a significantly increased photovoltaic current, which eliminates the effects of an increased photoconductivity.

Interesting investigations of the photorefractive process in NSLN crystals grown with K_2O in the melt are reported in [368]. When recording grating, a large stationary energy transfer was observed with $\Gamma = 25\,\mathrm{cm}^{-1}$ and an asymmetric photoinduced light scattering (*fanning*), which is characteristic for the recording by the diffusion mechanism. Nevertheless, the dominant mechanism of recording was the photovoltaic one, because the diffusion fields E_{D} corresponding to the used range of grating periods Λ could not provide the observed values of δn. The authors explained this paradox by a low concentration of filled traps N_{D} (see (3.33)). Accordingly to (3.31), if neglect E_{D}, then the angle of the shift φ between the interferogram and the recorded grating may be expressed as $\tan\varphi = E_{\mathrm{pv}}/E_{\mathrm{q}} \approx E_{\mathrm{pv}}/(\Lambda N_{\mathrm{D}})$. In other words, certain concurrence between E_{pv} and the saturation field E_{q} resulted in a nonzero stationary Γ.

We now try to discuss the range of high light intensities (Fig. 4.2), corresponding to an enhanced optical damage in congruent $LiNbO_3$. The situation seems to be ambiguous. In principle, the observed steep growth of δn in CLN may be due either to a photoinduced increase of α or to a strongly sublinear lux–ampere characteristic, or to a combination of the both factors, in other words invokes an explanation in the framework of a two-center charge transport scheme. Measurements of $\sigma_{\mathrm{ph}}(I)$ in CLN and NSLN performed in [357] found almost linear plots, namely $\sigma_{\mathrm{ph}} \propto I^{0.95}$ and $\sigma_{\mathrm{ph}} \propto I^{0.9}$, respectively, thus ruling out in CLN, a special relationship of $\delta n(I)$ to $\sigma_{\mathrm{ph}}(I)$. Therefore, the only possible reason for an increasing δn is a photoinduced optical absorption caused by a nonequilibrium population of certain shallow traps. Note that the observed steep increase of δn more than, by order of magnitude at about $1\,\mathrm{kW\,cm}^{-2}$ (Fig. 4.2) would require the corresponding drastic increase of α.

So, the question is raised, which centers could provide a photoinduced short-living absorption in the visible, responsible for an increasing $\delta\Delta n$ at moderate intensities of a blue–green CW-light? A photoinduced short-living optical absorption in the visible in undoped congruent $LiNbO_3$ was repeatedly reported, starting from the basic publications [63, 64] concluding a recent one [96]. Of course, the most attractive would be to relate it to the intrinsic defects and to debate a striking difference in $\delta\Delta n(I)$ graphs in CLN and NSLN (Fig. 4.2) by the presence or lack of Nb_{Li} and V_{Li}. Such an explanation was proposed, e.g., for the X-ray- or two-photon-induced absorption in $LiNbO_3$ [55, 63, 66] according to which the photoexcited interband electrons and holes are captured by Nb_{Li} and O^{2-}, respectively, to form small bound polarons and the corresponding short-living band extending from the near IR to near UV. However, this consideration may be applied to photoexciting with intensive light pulses, but not for the range of moderate CW-intensities shown in Fig. 4.2. Therefore, we believe that the observed steep increase of the photorefraction in CLN depicted in Fig. 4.2 is not an inherent property of the congruent $LiNbO_3$. It seems to depend on nonintentional admixtures

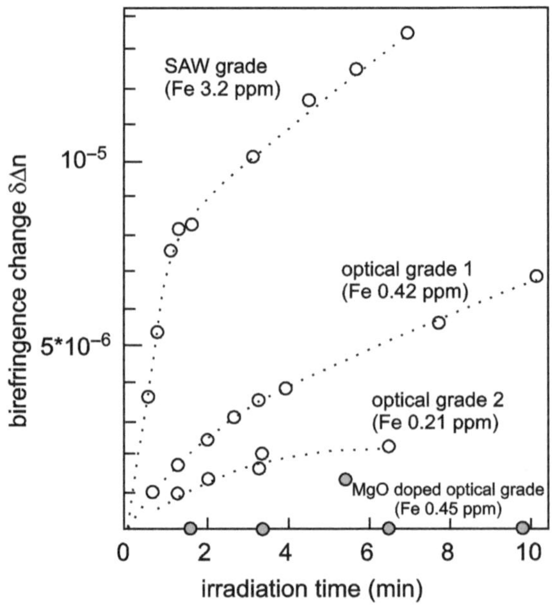

Fig. 4.3. Exposure dependencies of the photorefraction in $LiNbO_3$ with accidental admixtures of Fe, redrawn from [370]

and a photoinduced nonequilibrium repopulation of these accidental shallow levels which, as known from the physics of semiconductors, may occur even under low light intensities.

Our assumption is supported by the following experimental facts. The amount of photorefraction in congruent $LiNbO_3$ is very sensitive to the crystal purity (e.g., Fig. 4.3) and it increases with the concentration of Fe-like admixtures [370]. As seen from Fig. 4.2, in the crystals under investigations in [35, 357], the value of δn at relatively low intensities is in the range of 10^{-4}, which evidences a high concentration of accidental impurities. Under high light-intensities all accidental traps may be nonequilibrium populated, leading to an intensity dependence of the photovoltaic current, for example, to a superlinear plot $j_{\mathrm{pv}} \propto I^x (x > 1)$ [345,359]. So, it is reasonable to ascribe an anomalous growth of δn in the curve of Fig. 4.2 to the presence of accidental shallow traps. Such effects, controlled by a real spectrum of capture centers and particularly specific for high light intensities, are obviously responsible for an unpredictable behavior of the optical damage. That means to make any comparative estimates, a knowledge of the crystal purity is required.

A question is raised, why in near-stoichiometric crystals, grown from a not specially purified charge and thus also containing a large amount of admixtures, no nonmonotonous growth of δn is observed with increasing I as occurs in CLN (Fig. 4.2). This will be discussed in more detail in the section devoted to the photoinduced charge transport in optical-damage-resistant $LiNbO_3$

crystals. For the moment, we only mention that as mentioned in Sect. 2.4 (Fig. 2.12) in stoichiometric $LiNbO_3$ crystals, the Fe^{3+} – (and other TM^{3+}) ions change their lattice positions and are partially localized on the Nb sites. As a result, the capture cross-section of electrons by these centers is decreased and they do not serve as electron traps. So, we suggest that in NSLN crystals due to a change of the lattice sites of TM^{3+}-ions, the density of potential accidental electron traps is lesser than in CLN, which is consistent with the conclusions of Chen et al. [368]. Therefore, the photoinduced $\Delta\alpha$ in NSLN would be lower than in CLN with the same concentration of admixtures, and a dependence $\delta n(I)$ would be less pronounced than in CLN.

It should be mentioned that in the relative crystal $LiTaO_3$, the dependence of the photorefraction on the crystal stoichiometry is different from $LiNbO_3$ and more regular [371]. In $LiTaO_3$ crystals with different [Li]/[Ta] ratios grown by the VTE technique, a regular lowering of δn with increasing Li content was found in the whole intensity range up to $2.5\,\mathrm{kW\,cm^{-2}}$ (see Fig. 4.4). This dependence correlated quantitatively with the observed significant growth of σ_{ph} when approaching the stoichiometric composition [371]. These effects were discussed in terms of decreasing content of the hypothetical electron traps Ta_{Li}. In closing, we repeat again that there is a lack of detailed studies of photorefractive process in stoichiometric $LiNbO_3$ crystals. Unambiguously, the optical damage in the range of relatively low light intensities in stoichiometric crystals is higher than in congruent ones in spite of a somewhat enhanced photoconductivity. This seems to be related to an increased photovoltaic constant, which was proved experimentally but not discussed microscopically. For high light intensities the situation is yet unclear.

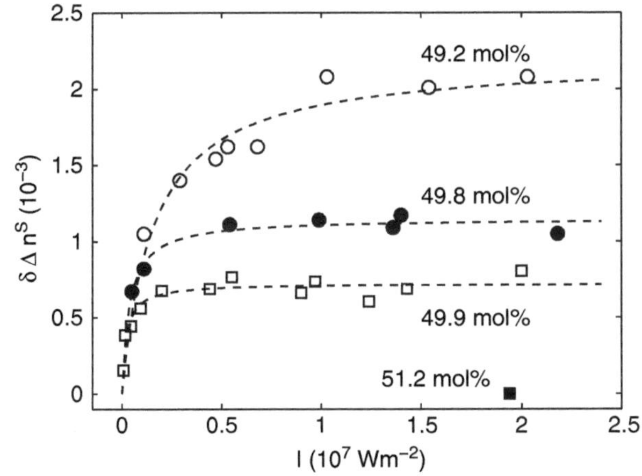

Fig. 4.4. Saturation values of light-induced birefringence changes $\delta\Delta n^S$ vs. pump-light intensity I in $LiTaO_3$. The Li content of the powder after the VTE treatment varied from 49.2 to 51.2 mol.%. The *dashed lines* are empirical fits, redrawn from [371]

4.2 Photorefraction in $LiNbO_3$ Crystals Doped with Optical-Damage-Resistant Ions

4.2.1 Photorefractive Properties of Optical-Damage-Resistant Compositions

Properties of $LiNbO_3$ crystals doped with optical-damage-resistant impurities Mg, Zn, In, Sc are described in reviews [194,195]; information on the photorefraction in Li-enriched $LiNbO_3$:Mg, one may find in [29,195]. Recently the family of the optical-damage-resistant ions was extended by Hf^{4+} [183,184,186] and Zr^{4+} [183,188].

As already briefly mentioned in Sect. 2.5.1, for the first time a possibility of suppression of the optical damage together with the preservation of the optical transparency was discovered in $LiNbO_3$:Mg [171] and studied in more detail later [76,172,173]. Bryan et al. [173] found a stability of $LiNbO_3$:Mg against pulse-laser intensities up to 100 MW cm^{-2}. The optical damage resistance was accounted for by an increased photoconductivity and a critical threshold concentration of 5 mol.% MgO was found for the congruent melt, above which a steep decrease of the optical damage occurred almost by two orders of magnitude. Interestingly, that in $LiNbO_3$:Mg with a very high purity, a comparable reduction of optical damage was observed even at 3 mol.% MgO [370]. In chemically analogous $LiNbO_3$:Zn crystals, an increased resistance against optical damage was found by Volk et al. [175,176], with a particularly strong effect at ZnO concentrations exceeding 7 mol.% for the congruent melt. Later, in congruent $LiNbO_3$:Zn a lower threshold of about 6 mol.% ZnO was found [177]. Figure 4.5 shows the dependence of the photorefraction on the Zn concentration and the dependency of $\delta\Delta n$ on the intensity in several $LiNbO_3$ crystals

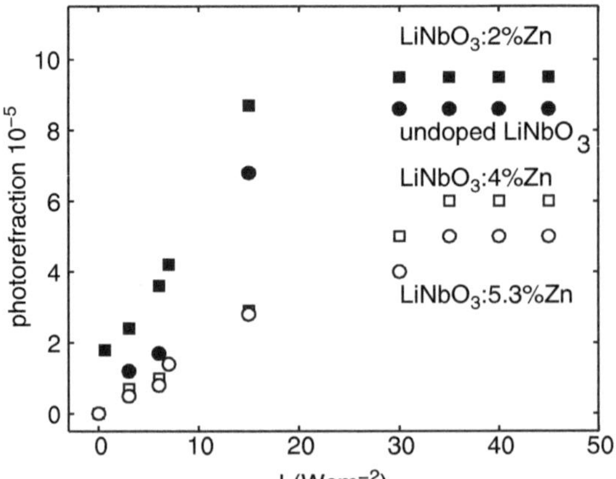

Fig. 4.5. Photorefraction as a function of the light intensity for $LiNbO_3$ with various Zn concentrations, data were taken from [175,176,180]

with different Zn concentrations, data were taken from [175, 176, 180]. Highly-doped $LiNbO_3$:Zn crystals revealed no optical damage under irradiation with 532 nm pulses of 10 ns duration up to intensities of 100 mW cm^{-2} [175, 372]. The next members of the optical-damage-resistant family were the trivalent ions In^{3+} [178–180] and Sc^{3+} [181, 182, 373]. In these cases, a drastic decrease of the photorefraction occurs already at concentration below 2 mol.% oxides (that is 4 at.% of the elements) in the melt for the congruent compositions. The effect of 1.7 mol.% In_2O_3 [178] and 1.5 mol.% Sc_2O_3 [182] on the value of $\delta\Delta n$ is comparable to the effect of 4.5–5 mol.% MgO. Data for stoichiometric LN doped with Sc have been reported in [374]. In the recently found optical-damage-resistant $LiNbO_3$:Hf composition [183–186] (see Fig. 2.18) the photorefraction falls down abruptly above a threshold concentration of $\approx$3% HfO_2 in the melt. Recently, the threshold concentration for Hf was refined to about 2 at.% in the crystal [187]. In the most recently investigated optical-damage-resistant composition $LiNbO_3$:Zr [188], the amount of photorefraction drastically drops at a ZrO_2 concentration less than 2 mol.%. So, one may conclude the threshold concentrations for both tetravalent optical-damage-resistant ions Hf and Zr to be about 2 at.% or even lower. As shown in Sect. 2.5.1, a difference in the threshold values of di-, tri- and tetravalent ions may be qualitatively explained in terms of the charge compensation effects. A reduction of the photorefraction in $LiNbO_3$:Hf and $LiNbO_3$:Zr with concentrations of Hf [186] or Zr [188] above the thresholds is even more pronounced than in $LiNbO_3$:6.5%Mg for equal light intensities. According to Kong et al. [188], $LiNbO_3$:2%Zr crystals show no optical damage up to intensities of 20 MW cm^{-2}.

At the first steps of studies of optical-damage resistance, it was shown that codoping $LiNbO_3$:Fe or $LiNbO_3$:Ln^{3+} with above-threshold concentrations of optical-damage-resistant ions, also strongly reduces the photorefraction. This was ascertained in $LiNbO_3$:Nd:Mg [375], $LiNbO_3$:Fe:Mg [230], $LiNbO_3$:Fe:Zn [56, 376] (see Fig. 4.6). Later on, these studies in doubly- or multiply-doped compositions were expanded owing to practical tasks. First, a stability of Ln^{3+}-doped $LiNbO_3$ should be analyzed in view of lasing applications. Second, as will be discussed in Sect. 4.2.4, codoping of photorefractive crystals, like $LiNbO_3$:Fe by an optical-damage-resistant ion promotes an improvement of the photorefractive properties, e.g., an increase of the sensitivity or the recording speed. Additionally, as discussed in Chap. 2, a slight doping with Fe^{3+}, Cr^{3+}, etc., is a very convenient probe for investigating the intrinsic defect structure. A considerable quantity of studies was recently been performed in such multiply-doped crystals, for example, in $LiNbO_3$:Cr:Zn [377], $LiNbO_3$:Nd:In [378], $LiNbO_3$:Fe:In [379], $LiNbO_3$:Mn,Fe:Mg [380], $LiNbO_3$:Fe:Hf [262], $LiNbO_3$:Cr:Mg [381], $LiNbO_3$:Ce:Cu:Sc [382]. In all cases, codoping with an optical-damage-resistant ion led to a significant decrease of the optical damage as is exemplified by Fig. 4.6.

As repeatedly mentioned, some authors [76, 174] observed in Li-enriched $LiNbO_3$:Mg crystals a decrease of the threshold Mg concentration, which finds

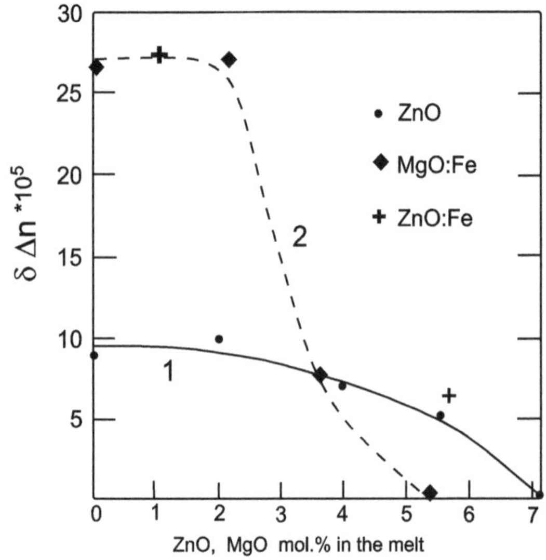

Fig. 4.6. Optical damage $\delta\Delta n$ in $LiNbO_3$ and $LiNbO_3$:Fe vs. Zn and Mg concentrations. The graphs 1 and 2, guides for the eye, correspond to $LiNbO_3$ and $LiNbO_3$:0.02%Fe:Mg, respectively. The two *plus markers* represent data for $LiNbO_3$:0.02%Fe:Zn. Optical damage was induced by an Ar-ion laser, $\lambda = 488\,\mathrm{nm}$, $I = 20\,\mathrm{W\,cm^{-2}}$

an adequate explanation in terms of the scheme of Mg incorporation into the $LiNbO_3$ lattice as discussed in Sect. 2.5. Actually, the higher is the Li content, the lower is the content of Nb antisites, thus the lower is the substitutional impurity concentration required to replace them. These observations stimulated a series of works [225–227,383] aimed to investigations of photorefraction in Li-enriched near-stoichiometric $LiNbO_3$ crystals doped with Mg. It is of a special importance for practice, because congruent $LiNbO_3$ crystals heavily doped with Mg exhibit a poor optical quality, whereas, according to the assertion of this research group [227], the near-stoichiometric $LiNbO_3$ doped with 1–2 mol.% Mg are optically more uniform. Later on, a great deal of results in Li-enriched $LiNbO_3$ doped with various optical-damage-resistant ions was published, but a detailed investigation of the photorefractive properties was reported for Mg-doped near-stoichiometric $LiNbO_3$ (NSLN) only.

Figure 4.7 from [225], compares the dependencies of the space-charge field on MgO concentration in NSLN and congruent $LiNbO_3$ (CLN). The former were grown by two techniques – either by the top-seeded solution with addition of 10.6% K_2O into the melt following [36] (the sample denoted as NSLN1), or by the double-crucible method from an Li-enriched melt [384] (denoted as NSLN2). Both Mg-free NSLN compositions corresponded to [Li]/[Nb] ≈ 0.99; while introducing Mg this ratio decreased to 0.96 and 0.98 in NSLN1:Mg and NSLN2:Mg, respectively. In the congruent crystal, the ratio [Li]/[Nb]

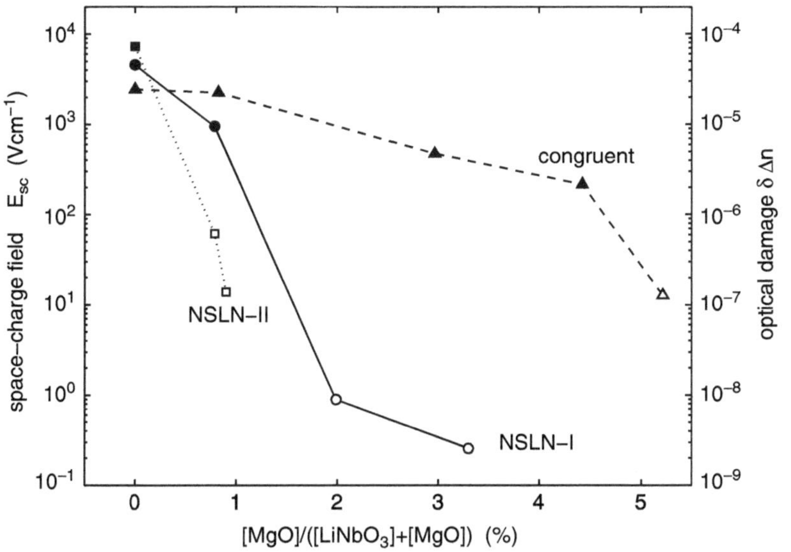

Fig. 4.7. Saturated space-charge field and optical damage vs. MgO concentration in congruent and stoichiometric (NSLN-I and NSLN-II) $LiNbO_3$ crystals, data taken from [225]

decreased from 0.943 in the pure Mg-free composition to 0.873 for $LiNbO_3$:5% Mg. One may see a drastic fall-down of the photorefraction in NSLN:Mg even at 1% Mg, where $\delta\Delta n$ drops to 10^{-8}. Therefore, it is more than by two orders of magnitude lower than in CLN doped with 5% Mg. One may conclude that a combination of Li-enriching and Mg doping is not simply additive. The question of the threshold Mg concentration in NSLN crystals seems to be still open. For example, in NSLN:Mg crystals grown by the high-temperature top-seeded solution growth method, a threshold value as low as 0.2 mol.% Mg was reported [229], deduced from measurements of the IR spectra. Furukawa et al. [225, 227] found an optical damage resistance for NSLN:1%Mg crystals up to intensities of 2–8 MW cm^{-2}. Moreover, according to [385, 386], the VTE-grown NLSN:Mg crystals are optically stable up to 26 MW cm^{-2}.

4.2.2 Fundamental Changes of the Charge Transport Process Produced by ODRI in $LiNbO_3$: Charge Transport Process in $LiNbO_3$ Crystals Doped with ODRI

The main reason for a decreased optical damage is an enhanced photoconductivity. The dark photoconductivity in $LiNbO_3$ doped with 5.5% Mg was shown to increase as well [387], but this effect is not of importance.

Figure 4.8 illustrates the concentration dependencies of the photoconductivity reduced to the light intensity σ_{ph}/I for Mg-, Zn- and In-doped congruent crystals. The photoconductivities were calculated from the characteristic

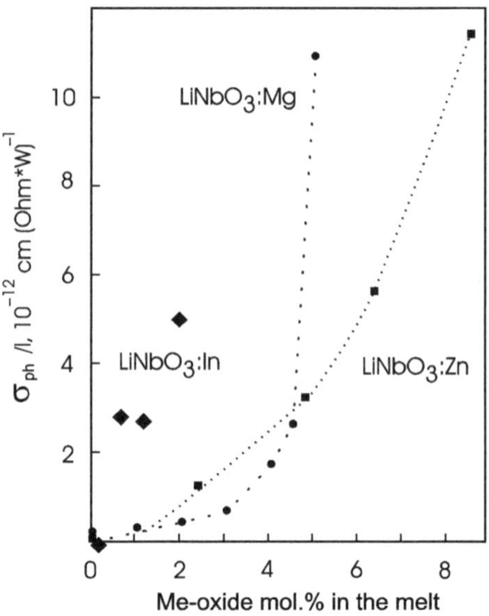

Fig. 4.8. The effect of Mg, Zn, In on the photoconductivity in $LiNbO_3$

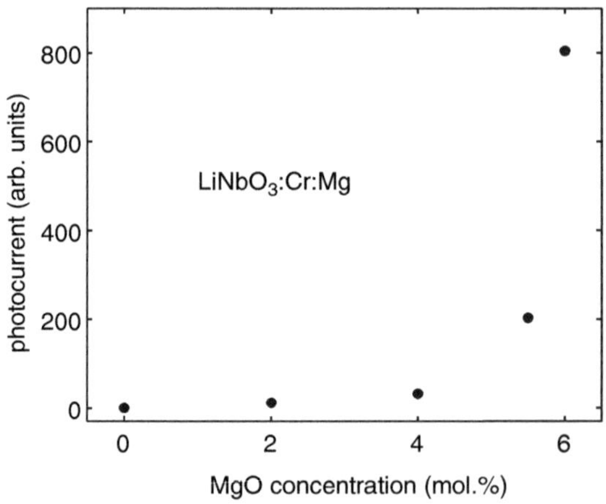

Fig. 4.9. Photocurrent under 410 nm illumination at room temperature in doubly-doped $LiNbO_3$:Cr:Mg, redrawn from [381]

times of grating erasure according to $\delta\Delta n(t) = \delta\Delta n_{\rm sat}\exp(-t/\tau_{\rm er})$, where $\sigma_{\rm ph} = \tau_{\rm er}/\epsilon\epsilon_0$. The intensities of the erasing Ar-ion laser beam ($\lambda = 488$ nm) were in the range of 30 W cm^{-2}. As seen, doping with 1.7 mol.% In_2O_3 produces the same effect, as a doping with MgO or ZnO of about 4.5 mol.%. Figure 4.9 demonstrates a steep increase of the photocurrent at an Mg concentration of about 5 mol.% in $LiNbO_3$:Cr:Mg.

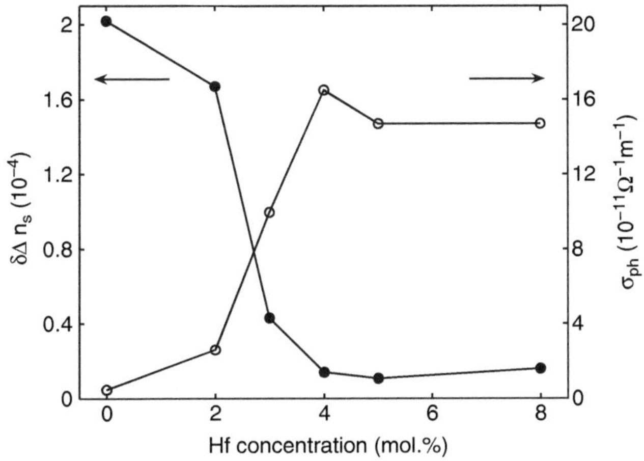

Fig. 4.10. Dependence of photorefraction and photoconductivity on the HfO_2 concentration in the congruent melt, measured at 532 nm and $310\,\mathrm{W\,cm^{-2}}$, redrawn from [184]

Figure 4.10 shows the dependence of the photoconductivity on the Hf concentration calculated from the kinetics of the holographic recording $\delta\Delta n(t) = \delta\Delta n_{\mathrm{sat}} \exp(-t/\tau_{\mathrm{rec}})$; the recording was performed for $\lambda = 532\,\mathrm{nm}$ and $I = 310\,\mathrm{W\,cm^{-2}}$. All these curves manifest a steep growth of σ_{ph} at impurity concentrations approximately above the thresholds.

In NSLN crystals doped with Mg, a correlation between the photorefraction and photoconductivity is observed as well [225]. Figure 4.11 shows the dependence of the photoconductivity on the Mg concentration in the set of crystals presented in Fig. 4.7 and reported [225]. A steep growth of σ_{ph} in NSLN crystals occurs already at [MgO] $\approx$ 1–2 mol.% in accordance with the dramatic drop of the photorefraction in Fig. 4.7.

The dependencies $\sigma(I)$ in Mg-, Zn- and In-doped congruent $LiNbO_3$ crystals are slightly sublinear $\sigma_{\mathrm{ph}} \approx I^x$ ($x = 0.8$–0.9) [294], which evidences of a two-center charge transport scheme. In contrast, even a slight codoping with Fe, transforms the lux–ampere characteristics to a strictly linear function [294]. It is analogous to a straightening of an initially slightly sublinear $\sigma_{\mathrm{ph}}(I)$ in congruent undoped $LiNbO_3$ after a weak doping with Fe. In terms of the current concept, this effect may be regarded as the transformation from a two-center to an one-center charge transport scheme owing to the dominant role of Fe^{3+} as an electron trap. Note that in the presentation given in Fig. 4.8, slight sublinearities of $\sigma_{\mathrm{ph}}(I)$ were neglected assuming the plots of $\sigma_{\mathrm{ph}}(I)$ to be linear. The lux–ampere characteristics for the newly found optical-damage-resistant ions Hf and Zr have not been published yet. According to [186], the saturation value of $\delta\Delta n$ in $LiNbO_3$:Hf very slightly grows with the light intensities up to $600\,\mathrm{W\,cm^{-2}}$, so one may assume a slightly sublinear lux-ampere characteristic similarly to $LiNbO_3$:Mg or $LiNbO_3$:Zn [294]. However, no data

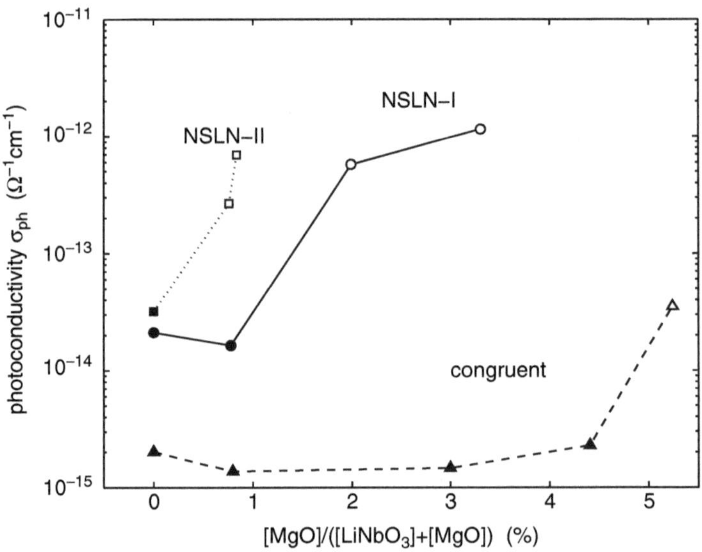

Fig. 4.11. Photoconductivity vs. MgO concentration in congruent and stoichiometric (NSLN-I and NSLN-II) $LiNbO_3$ crystals, redrawn with data from [225]

have been published yet for the $\sigma_{ph}(I)$ dependencies in near-stoichiometric $LiNbO_3$ both undoped and doped with Mg.

For the first sight, a correlation between decreasing $\delta\Delta n$ and a steep growth of $\sigma_{ph}(I)$ in the same concentration ranges gives an exhaustive explanation for the optical damage resistance. However, a detailed comparison of these curves detects one more reason, less pronounced but more fundamental, of the falling down of $\delta\Delta n$. The calculations for NSLN:Mg show that the observed increase of σ_{ph} would correspond to the $\delta\Delta n$ magnitude by one to two orders higher than those measured experimentally. This impelled Furukawa et al. [227] to assume and then to show experimentally [225, 383] that, Mg doping strongly affects the photovoltaic current as well (see Fig. 4.12). Note that this figure illustrates the fact mentioned above, that the j_{pv} value in the undoped stoichiometric crystals is significantly higher than in undoped congruent ones (see Table 4.1). When doping NSLN with Mg, the value of j_{pv} decreases drastically so that in NSLN:2%Mg it is more than two orders of magnitude lesser than in the undoped NSLN. This steep decrease of j_{pv} accounts for the enormous drop of the photorefraction to 10^{-8}. In this composition, the highest of the so far reported optical damage resistance is due to a combination of both effects, namely of a dramatic growth of σ_{ph} and of a simultaneous significant decrease of j_{pv}. Moreover, as seen from Fig. 4.12, in the congruent $LiNbO_3$:Mg crystals j_{pv} also decreases with increasing Mg concentration, although much less pronounced than in near-stoichiometric ones. Table 4.2 collects the data on the photorefractive parameters in CLN and

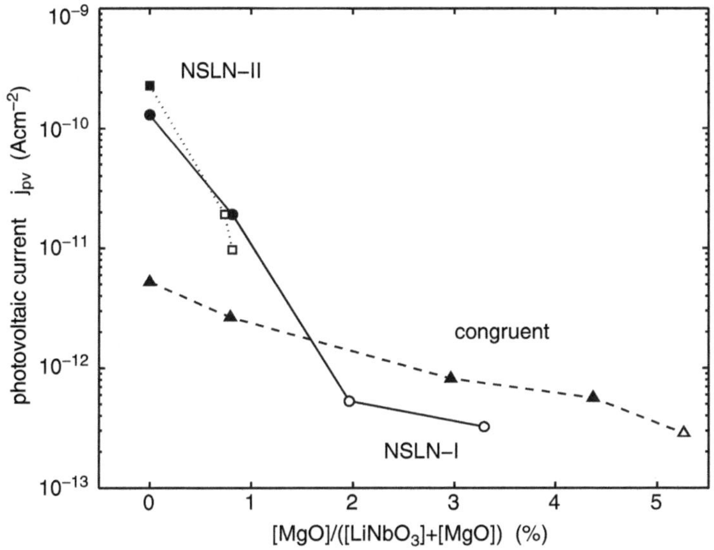

Fig. 4.12. Photovoltaic current vs. MgO concentration in congruent and stoichiometric (NSLN-I and NSLN-II) $LiNbO_3$ crystals, redrawn from [225]

Table 4.2. Comparison of the photoconductivity (σ_{ph}), photovoltaic current (j_{pv}) and space-charge fields (E_{sc}) in congruent (CLN), stoichiometric (NSLN) and optical damage resistant CLN:5.1%Mg and NSLN:2%Mg ($I = 10\,W\,cm^{-2}$, $\lambda = 488\,nm$). The data for CLN and CLN:5.1%Mg are taken from [56], the data for NSLN and NSLN:2%Mg are taken from [225]. The bottom line presents the Glass constant calculated according to $k = E_{sc}\sigma_{ph}/(\alpha I)$ assuming absorption coefficients $\alpha = 0.025$–$0.03\,cm^{-1}$ for $\lambda = 488\,nm$ (according to experimental estimates)

Property	CLN	NSLN	CLN:5.1%Mg	NSLN:2%Mg
σ_{ph} ($10^{-14}\,\Omega$)	0.2	2–3	3	60
j_{pv} ($10^{-12}\,A\,cm^{-2}$)	6	200	0.3	0.5–0.6
E_{sc} ($V\,cm^{-1}$)	2,500	7,000	10	1
k_G ($10^{-11}\,cm\,V^{-1}$)	2.4	80	0.1	2

NSLN with and without Mg. In the congruent $LiNbO_3$:Zn crystals [294] certain mismatch between experimentally measured dependencies σ_{ph} and $\delta\Delta n$ on the Zn concentration was also detected, which led to a conclusion that j_{pv} nonmonotonically varies with increasing Zn concentration.

The dependencies of the photovoltaic effect on this type of doping, qualitatively recalls the above-described influence of the Li content on the value of j_{pv}. In undoped congruent and near-stoichiometric crystals, the photovoltaic currents are significantly different (Fig. 4.12, Table 4.1), whereas according to [29] in Fe-doped CLN and NSLN they become equal. Similarly, measurements of the photovoltaic currents in Mg-doped $LiNbO_3$:Fe [230] and

Zn-doped $LiNbO_3$ [175] have found that the Glass constant is the same as in $LiNbO_3$:Fe (with the same $[Fe^{2+}]/[Fe^{3+}]$ ratio). In other words, even if a lowest Fe concentration is admixed, then the photovoltaic properties in $LiNbO_3$, $LiNbO_3$:Mg, $LiNbO_3$:Zn are controlled by the Fe-admixture, whereas in compositions free of Fe the photovoltaic current seems to be governed by other microscopic mechanism.

The microscopic origin of the increased photoconductivity in $LiNbO_3$ doped with optical-damage-resistant ions will be discussed in the Sect. 4.2.3. We now speculate about the effects of these ions on the photovoltaic properties. One of the most reasonable explanations could be an alteration of the photocarrier type resulting from an incorporation of optical-damage-resistant ions, or, in the language of the defect model, when decreasing Nb antisite and Li vacancy concentrations. Assume that the removal of Nb antisites and Li vacancies is followed by an appearance of a hole photoconductivity. It was actually supported by finding a photohole conductivity in highly doped $LiNbO_3$:Mg [327] and $LiNbO_3$:Zn [328] by means of the holographic method proposed in [317, 318]. Then, the value of (σ_{ph}) would grow with the Mg concentration at the expense of growing the hole component. At the same time, j_{pv} may decrease due to a decreasing electron component. Based on the data of Orlowski and Krätzig [317,318], one may suggest that the photovoltaic current in $LiNbO_3$ is related to the photoelectrons only. A relation of the photoconductivity and photovoltaic currents to different charge carriers is probable, for example, a coexistence of a hole photoconductivity and electron photovoltaic current was established in quartz [9].

We now discuss the microscopic reasons for an increased photoconductivity in $LiNbO_3$ crystals doped with optical-damage-resistant impurities with the above-threshold concentrations. In general, a photoconductivity growth may be caused by increasing either the mobility μ or the lifetime or free charge carriers. For an one-center charge transport scheme, a simplified expression (3.26) for the photoconductivity gives the following relation $\sigma_{ph} \approx \mu(s_{ph}/\gamma)(N_D/N_C)I$, where $s_{ph} = \alpha/N_D$ is the absorption cross-section of the electron donor, γ is the recombination coefficient, N_D and N_C are the concentrations of donors and traps, respectively. Let us begin the discussion with the case of an one-center scheme assuming that it is realized in $LiNbO_3$:Fe:Mg and $LiNbO_3$:Fe:Zn similarly to $LiNbO_3$:Fe. This assumption is supported particularly by the linearity of $\sigma_{ph}(I)$ plots in these crystals [230, 294]. The case of a two-center scheme involving the shallow trap Nb_{Li} will be discussed later. First, we should rule out the effects of $[N_D]/[N_C]$ variations. As mentioned in Sect. 2.5.2, iron in $LiNbO_3$:Mg:Fe exists in the traditional states Fe^{2+} and Fe^{3+} According to measurements of the optical absorption [387] and Mößbauer spectra [388] in $LiNbO_3$:Mg:Fe, the ratio $[N_D/N_C] = [Fe^{2+}]/[Fe^{3+}]$ at high Mg concentrations decreases by a factor of about 2, so one would more likely expect a lowering rather an increasing of σ_{ph}.

Additionally, the optical absorption coefficient α in $LiNbO_3$:Fe:Mg [387] and $LiNbO_3$:Fe:Zn [389] abruptly falls down above the thresholds that evidences a lowering of the Fe^{2+} concentration. Therefore an increase of σ_{ph} above the threshold cannot be accounted for, by concentration arguments. Other reasons of an increasing σ_{ph} as applied to expression (3.26), might be either a growth of μ, or a decrease of γ. In principle, the value of μ depending on the scattering of the carriers by defects might be changed in Mg- or Zn-doped crystals. This mechanism was proposed by Sommerfeldt et al. [230]. Arizmendi and Agulló-López [390] also suggest an enhancement of the electron mobility for Mg concentrations above the threshold. Adibi et al. [391] discussed a possible effect of increasing μ on the holographic recording in $LiNbO_3$. However, so far, there are no direct or indirect experimental proofs for this assumption. Moreover, the temperature dependencies of the conductivity performed in $LiNbO_3$, $LiNbO_3$:5%Mg and $LiNbO_3$:5%Mg:Fe [392] argue against it. According these measurements in all three compositions, the conductivity may be described by an identical expression

$$\sigma = 1,620 \exp\left(\frac{-E_a}{kT}\right) \tag{4.1}$$

with the same activation energy $E_a = 1.2\,\text{eV}$ independent of the composition. From this it was concluded that an increase of σ_{ph} in no case is related to a change of μ. We now discuss an alternative reason of increasing σ_{ph}, a decrease of γ, thus a reduction of a capture cross-section of electrons by Fe^{3+} traps at high Mg (or Zn) doping. This mechanism seems to be more substantiated due to the following reasons. As discussed in detail in Sect. 2.5.2, the Fe^{3+}-like ions in $LiNbO_3$:Fe:Mg (or Zn) with Mg (Zn)concentrations above the thresholds, change their lattice sites and incorporate, at least partially, onto the Nb sites; this was deduced on the base of EPR measurements. It is qualitatively clear that the center $[Fe^{3+}_{Nb}]^{2\prime}$ is a hardly probable electron trap. Actually, estimates of [392] have shown that the main reason of a drastic growth of the photoconductivity at high Mg concentrations is a decrease of the capture cross-section $S = \gamma_r/v$ of electrons by Fe^{3+} traps by more than two orders of magnitude. A rough estimation for $\left[Fe^{3+}_{Li}\right]^{2\bullet}$ gives $S = 1.3 \times 10^{-14}\,\text{cm}^2$ [393] and for $\left[Fe^{3+}_{Nb}\right]^{2-}$, the value $S < 10^{-15}\,\text{cm}^2$ [392].

The conclusion of a drastic decrease of γ in highly-doped $LiNbO_3$:Mg:Fe was supported by the measurements of the X-ray-induced absorption spectra in these crystals [325, 326].

Figure 4.13 presents the X-ray-induced spectra in $LiNbO_3$:Mg:0.01%Fe with Mg concentrations below and above the threshold. According to [230] and more recent data, the shape of the characteristic absorption band of Fe^{2+} is not affected by Mg doping; actually, the corresponding peak at about 2.6 eV (480 nm) exists in all $LiNbO_3$:Mg:Fe and $LiNbO_3$:Zn:Fe compositions independently of Mg or Zn concentration. X-ray irradiation of $LiNbO_3$:Mg:Fe with Mg concentrations below the threshold results in an increase of this

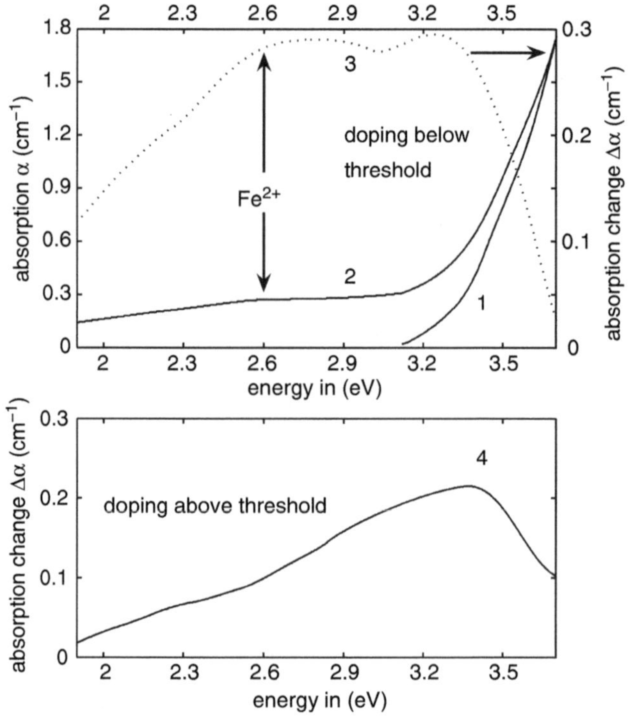

Fig. 4.13. X-ray-induced absorption of Mg- or Zn-doped below (*upper*) and above (*lower*) threshold concentrations in $LiNbO_3$ at room temperature. *Lines 1* and *2* represent the absorption before and after the X-ray irradiation, *line 3* shows their difference $\Delta\alpha$. This band is stable in the darkness, while that represented by *line 4* vanishes within several minutes [325]

band, see Fig. 4.13 (upper part). The same is observed in $LiNbO_3$:Fe [394]. This X-ray-induced band is accounted for, by formation of extra (nonequilibrium) Fe^{2+} centers owing to capture the X-ray-induced interband electrons by the trap Fe^{3+} following the scheme $Fe^{3+} + e^- \rightarrow Fe^{2+}$. This conclusion is supported by EPR measurements according to which the appearance of the X-ray-induced band is accompanied by a simultaneous decrease of the Fe^{3+} concentration [394]. The X-ray-induced band is stable and may be optically erased, with a simultaneous restoring of the initial Fe^{3+} EPR spectrum and the initial absorption peak at 2.6 eV. A quite different behavior is observed in $LiNbO_3$:Fe:Mg with Mg concentrations above the threshold, see Fig. 4.13 (lower part). In these crystals, X-ray irradiation induces no peak at 2.6 eV, thus the Fe^{2+} concentration stays unchanged. Instead of that, the spectrum contains a new unstable band with a broad peak at about 360 nm quickly decaying in the darkness. The same band is induced by UV irradiation [325, 395, 396]. Qualitatively similar, but a much more pronounced difference in X-ray-induced spectra in low- and highly-doped $LiNbO_3$:Fe:Mg and

$LiNbO_3$:Fe:Zn were observed at the low-temperature X-ray irradiation [326]. Missing X-ray-induced Fe^{2+} centers indicate that the Fe^{3+} centers are no more electron traps. Thus, we conclude that in $LiNbO_3$:Mg:Fe and other relative crystals with above-threshold concentrations of optical-damage-resistant ions, the charge transport scheme (3.25) is violated and the origin of electron traps is changed; we emphasize that Fe^{2+} remain electron donors. Note that this consideration may be extrapolated to a near-stoichiometric $LiNbO_3$:Fe, in which the measurements of the two-beam coupling gain [366] found an increased σ_{ph} compared to the congruent $LiNbO_3$:Fe. According to [126, 397], Fe^{3+} in stoichiometric $LiNbO_3$ incorporates partially onto the Nb site, so one may suggest that in Fe-doped NSLN, the center Fe^{3+} loses its acceptor properties and the charge transport scheme (3.25) does not give an adequate explanation of the photoelectric properties of Fe-doped stoichiometric crystals.

According to the structure data, the threshold concentrations of optical-damage-resistant ions correspond to their partial incorporation onto the Nb sites that assuming the self-compensation mechanism $3[Mg_{Li}^{\bullet}] - [Mg_{Nb}^{'''}]$, is followed by a minimization of the V_{Li} content, see Fig. 2.20. So, at above-threshold concentrations of these ions, the $LiNbO_3$ intrinsic defect structure is fundamentally modified, namely the deep Fe^{3+} and shallow Nb_{Li} electron traps disappeared and the Li vacancy content is decreased. These events would necessarily lead to a change of the charge transport scheme. Such a rather speculative consideration finds an experimental support. An analysis of the sign of the photocarriers using a holographic method [317,318] in $LiNbO_3$:Mg [327] and $LiNbO_3$:Zn [328] has shown that at low impurity concentrations, the photoconductivity is of n-type, whereas above the thresholds it changes its sign and becomes dominantly p-type, the contribution from the hole component increasing with Mg or Zn concentrations. The alteration of the photocarrier sign indicate that the traditional scheme (3.25) is actually violated. According to Donnerberg et al. [398], Li vacancies in $LiNbO_3$ are the most probably hole traps. As the Mg or Zn concentrations increase above the thresholds, the V_{Li} content decreases (Fig. 2.20), thus resulting in an increase of the hole photoconductivity.

We now turn to the case of a two-charge transport scheme that occurs in $LiNbO_3$:Mg and $LiNbO_3$:Zn crystals nominally free of Fe. Evidences of a two-center charge transport scheme in them are particularly given by nonlinearities of the $\sigma_{ph}(I)$ plots [294] and, for example, by a pronounced dependence of $\delta\Delta n$ on the duration of the recording light pulses, see Fig. 3.8. In terms of the conventional two-center model, it may be attributed to the participation of electron traps Nb_{Li}, which are concurrent with Fe^{3+} provided that the concentration of the latter is low. The expression (3.39) deduced for a particular case of a two-center scheme gives an example of a dependence of $\sigma_{ph}(I)$ on the concentrations N_C of deep traps (Fe^{3+}) and N_X of shallow traps (Nb_{Li}); as a priori clear, $\sigma_{ph}(I)$ increases with decreasing N_X and/or N_C. In the Figs. 4.5, 4.8, and 4.6 one may see a gradual growth of $\sigma_{ph}(I)$ and a decrease of $\delta\Delta n$ already at low Zn concentration. This may be accounted for by the conclusion drawn from many experiments, cited in Sect. 2.5.1, that Nb_{Li} traps are removed from the lattice starting already at low Mg or Zn concentrations, see Fig. 2.20. A more abrupt drop of $\delta\Delta n$ in Fe co-doped $LiNbO_3$:Mg or $LiNbO_3$:Zn (Fig. 4.6), is due to a dominant contribution from Fe^{3+} and a secondary role of removing Nb_{Li}. A low concentration of any optical-damage-resistant impurity affects the photoconductivity in $LiNbO_3$:Fe crystals provided that the Fe concentration is low. This is supported by the data reporting an enhancement of the photorefractive sensitivity of $LiNbO_3$:(0.01–0.03)% Fe achieved by a relatively low doping with an optical-damage-resistant impurity, e.g., Mg [399], Zn [380] or In [400].

4.2.3 Dark-Trace and GRIIRA (BLIIRA) Effects in Optical-Damage-Resistant $LiNbO_3$ Crystals

It is reasonable to discuss separately some aspects of the influence of optical-damage-resistant impurities on the photoinduced coloration in $LiNbO_3$, in order to introduce a bridge between the microscopic scenarios of the photocoloration and the photorefraction, and the negative phenomena known from practice. Additionally, the set of levels involved in the photorefraction and photocoloration seems to be similar, which is proved by similar influence of the optical-damage-resistant ions on both effects.

Formerly, when analyzing the appropriateness of $LiNbO_3$ crystals for nonlinear optical applications, it was found that a pulsed-IR irradiation with intensities of tens of $MW\,cm^{-2}$ or higher, induces in $LiNbO_3$ crystals broad absorption bands in the visible and near infra-red spectral ranges. The origin of this coloration in the visible was under discussion. Some authors [401, 402] interpreted it as a result of a two-photon absorption. Other authors [64] attributed its formation at room temperature to a nonequilibrium repopulation of traps under the visible radiation of the second harmonic. These interpretations are not mutually excluded and a contribution from any of them depends mostly on the used intensity range. As mentioned in Sect. 2.2, a photoinduced coloration in the visible actually occurs under intensive blue–green pulses, so the assumption of von der Linde et al. [64] was reasonable. The band induced in the visible was referred to as a "dark-" or "gray-trace" effect, and sometimes one may see by a naked eye a gray track within the crystal bulk after an intensive IR pulse shooting. A photoinduced band in the infrared was referred to as GRIIRA or BLIIRA (green- or blue- induced IR absorption) effect. The dark-trace and GRIIRA effects usually occur simultaneously and are typical for many ferroelectric crystals such as $LiNbO_3$, $LiTaO_3$, KTP, $KNbO_3$, etc., for references see, e.g., [403].

The dark-trace appears as a nonstructured absorption band with a weakly-pronounced peak at about 500 nm. In congruent $LiNbO_3$ it is stable and may be erased optically or thermally, similarly to the photorefraction. The negative consequence of the dark trace is a loss of the optical transparency, for example, in [402] after IR shooting $LiNbO_3$, a photoinduced increase of the optical absorption coefficient in the visible from 0.03 to 0.15–0.2 cm^{-1} was reported, which corresponds to a reduction of the transparency to 10–15%. Similar loss in the transparency of more than 20% were observed in $LiNbO_3$:Mg and $LiNbO_3$:Zn [404]. A loss in the transparency may, for example, break the required relation between the powers of the fundamental and second harmonic $P_{2\omega} \sim P_{\omega}^2$. The negative consequence of the GRIIRA (BLIIRA) effect is a local heating and a resulting nonuniform temperature profile in the active volume of the crystal, so to say a thermal lens, which may lead to a disturbance of the phase matching in the optical frequency conversion.

After the discovery of the optical-damage resistance in $LiNbO_3$:Mg, the studies of the dark trace effect started in highly-doped (optical-damage-resistant) $LiNbO_3$:Mg [404–409] and $LiNbO_3$:Zn [404, 409]. As mentioned above, a dark trace appears in these compositions as well, in $LiNbO_3$:Zn its value is lower than in $LiNbO_3$:Mg [176, 404, 410]. Particularly, Nightingale et al. [405] and Kozlovski et al. [406] accounted for a limitation in the conversion efficiency of $LiNbO_3$:Mg by the optical losses caused partially by the dark-trace effect. Yao et al. [407] also detected this photocoloration in $LiNbO_3$:5%Mg , but did not observe its influence on the parameters of the optical frequency conversion. The characteristics of the dark-trace in optical-damage-resistant compositions sharply differ from those in congruent ones. For Mg or Zn compositions above the thresholds, the photoinduced band in $LiNbO_3$:Mg and $LiNbO_3$:Zn is unstable and decays in darkness during tens of minutes [175,404,409]. Additionally, its broad peak seems to be slightly blue-shifted. Therefore, the dependence of the photoinduced band in visible on Mg or Zn doping reminds qualitatively to the behavior of the X-ray-induced band in the same crystals shown in Fig. 4.13; for Mg or Zn concentrations above the thresholds, the band becomes unstable and is shifted to the UV. Therefore, we may account for this behavior by similar reasons formulated in Sect. 4.2.2.

Following von der Linde et al. [64] it is most reasonable to assume that the main reason of the dark-trace is a photoinduced nonequilibrium repopulation of levels related to unintentional impurities like Fe, Mn, Ti, etc., with concentrations of tens of ppm. The coloration, for example, may be related to a formation of metastable Fe^{2+}, Mn^{2+}, etc. Note that in literature to our knowledge, no results have been published yet on the dependence of the dark-trace effect on the crystal purity. As follows from the detailed spectroscopic investigations discussed in Sect. 2.3, when doping $LiNbO_3$ with above threshold Mg or Zn concentrations; Fe^{3+}-like ions incorporate partially onto the Nb sites, thus loosing their acceptor properties. This change of the electron traps discussed in Sect. 4.2.2 as applied to the photorefraction in highly doped $LiNbO_3$:Mg(Zn), seems to be the most obvious qualitative reason accounting for an instability of the dark-trace in these crystals. At the end of Sect. 2.2, a new approach to the photoinduced coloration was mentioned, accordingly to which a contribution from bound hole polarons O^- also contributes to the photoinduced band in the visible. However, this fundamental explanation cannot be applied to the case of the dark-trace, because the band owing to

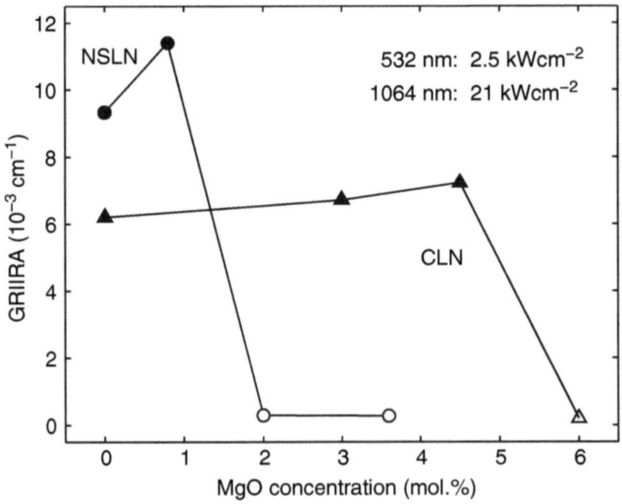

Fig. 4.14. Green-induced infrared absorption (GRIIRA) vs. MgO concentration in NSLN and CLN samples, redrawn from [411]

hypothetical O^- centers decays during times not exceeding seconds, whereas the dark-trace in congruent $LiNbO_3$ is stable and in $LiNbO_3$:Mg(Zn), it decays within 10–20 min.

We now remark on the photoinduced band in the near infrared (GRIIRA or BLIIRA effects). From recent results of studies on the spectral and kinetic characteristics of such a band discussed in Sect. 2.2 (e.g., [86], Fig. 2.8), an unambiguous conclusion may be drawn of its relation to the electron capturing by Nb_{Li} defects to form small bound polarons , $Nb_{Li}^{5+} + e^- \rightarrow Nb_{Li}^{4+}$. Therefore studies in GRIIRA are informative for the development of the polaron model. Particularly, Furukawa et al. [411] analyzed the dependence of GRIIRA in Mg-doped congruent and near-stoichiometric $LiNbO_3$ on the Mg concentration (Fig. 4.14). The photoinduced optical absorption measured at 1,064 nm and pumped by 532 nm with intensities in the range of kW cm^{-2}, decayed practically instantaneously on the turn-off of the pumping beam.

As seen in Fig. 4.14, the value of the photoinduced absorption falls drastically at Mg concentrations exceeding the thresholds, of about 5% and 1.5% Mg in CLN and NSLN, respectively. The simultaneous measurements of the photorefraction at 532 nm in accordance with the GRIIRA dependencies found a drastic fall down of $\delta\Delta n$ in the same concentration ranges.

These results are consistent both with the proposed explanation for the threshold concentrations of Mg-like impurities and with the interpretation of the photoinduced IR band (see Sects. 2.2, 2.4, and Table 4.3). Actually, the suggested removal of Nb antisites at the threshold concentrations eliminates the possibility of a formation of small bound polarons. These results agree with conclusions drawn from the spectroscopic and EPR measurements in

Table 4.3. Photorefractive damage threshold for Mg-doped congruent (CLN) and nearly-stoichiometric (NSLN) $LinbO_3$ crystals. The photorefractive damage threshold is defined as the cw green-light intensity, where the transmitted laser beam is distorted as a result of photorefraction after 10 min of irradiation, data taken from [411]

Crystal	Threshold intensity ($kW\,cm^{-2}$)	OH-mode (cm^{-1})
CLN	1	3,485
CLN:3%MgO	10	3,485
CLN:4.5%MgO	75	3,485
CLN:6%MgO	>8,000	3,532
NSLN	0.1	3,466
NSLN:0.6%MgO	10	3,466
NSLN:1.8%MgO	>8,000	3,532

optical-damage-resistant compositions, e.g., [55, 56]. At the same time, the dependence of GRIIRA on Mg concentration below the thresholds does not fit the proposed scenario of the Mg incorporation. Assuming a substitution for Nb antisites starting from low Mg concentrations, one could expect a decrease of GRIIRA with [Mg] rather than a plateau and even an increase observed in the graphs for CLN and NSLN, respectively (Fig. 4.14). From the practical viewpoint the scope of these results evidences that the optical damage resistance in Mg-doped CLN and NSLN crystals is accompanied by their improved stability against the photoinduced coloration.

At the end of this section, it is worthy of presenting recent comparative studies of the BLIIRA in several popular crystals [403]. One may see the improved stability of CLN:Mg and NSLN:Mg compared to KTP at intensities exceeding $1\,GW\,cm^{-2}$; the data were obtained both for single domain and periodically poled crystals (Fig. 4.15).

4.2.4 $LiNbO_3$ with Optical-Damage-Resistant Impurities as Media for the Photorefractive Recording

A search for optical-damage-resistant compositions in $LiNbO_3$ was stimulated by traditional applications of this material in optical schemes using intensive visible light. Later on, it was found that $LiNbO_3$ crystals doped with optical-damage-resistant impurities (hereafter denoted by ME), offer properties appropriate for alternative aims, namely for the photorefractive grating recording. First, at relatively low ME-doping these crystals combine a sufficiently increased photoconductivity with a large enough photorefraction (see Figs. 4.5, 4.6 and 4.8), altogether enhancing the photorefractive sensitivity and the recording speed. Second, a shift of the bandedge towards shorter wavelengths (Fig. 2.17) provides a possibility for UV-recording gratings. We

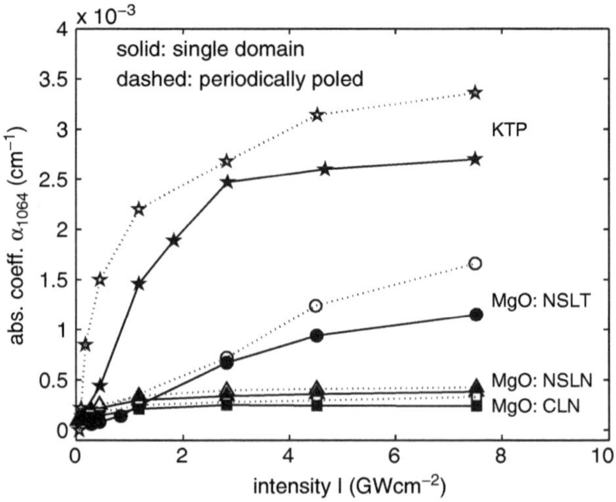

Fig. 4.15. Blue light induced absorption coefficients at 1,064 nm in single domain and periodically poled KTP, MgO:NSLT, MgO:NSLN, and MgO:CLN, redrawn from [403]

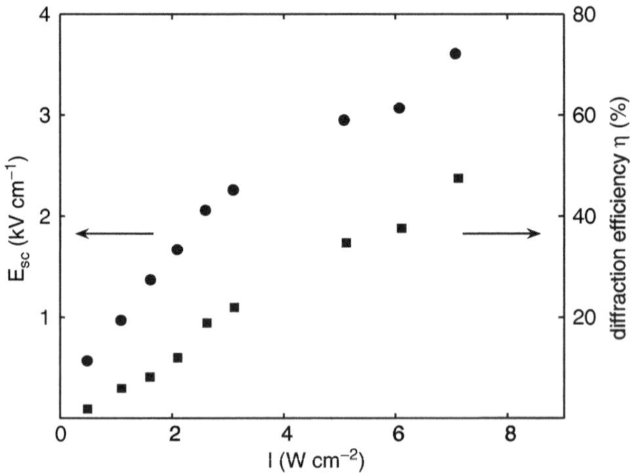

Fig. 4.16. Diffraction efficiency vs. light intensity in $LiNbO_3$:2%Zn at a grating period of about 5 μm, according to [199]

shortly dwell on the photorefractive recording in $LiNbO_3$:ME based on these potentials. Campbell and Yeh [412] recorded and thermally fixed holograms in $LiNbO_3$:4%Mg with a recording speed exceeding that in undoped crystals by an order of magnitude. Volk et al. [294] investigated the recording in $LiNbO_3$:Zn crystals with Zn concentrations below the threshold. In a 2 mm thick $LiNbO_3$:2%Zn a diffraction efficiency as high as 50% was achieved at 488 nm during several seconds (see Fig. 4.16), therefore the recording speed and sensitivity was comparable to those in low-doped $LiNbO_3$:Fe. The storage time of holograms in $LiNbO_3$:2%Zn is indefinitely long, in other words, not shorter than in $LiNbO_3$. The mechanism of recording in $LiNbO_3$:Zn unlike in $LiNbO_3$:Fe depends on the recording conditions, namely, at a given grating period Λ the recording at low intensities occurs via the diffusion mechanism, whereas at $I \approx 10\,\mathrm{W\,cm^{-2}}$ the dominant one is the photovoltaic mechanism. For medium I and Λ the gratings are recorded by a combined mechanism. This is mainly due to the fact that an increased photoconductivity in these crystals reduces E_{pv} to make it comparable with E_{dif}.

Neither $LiNbO_3$:4%Mg [412], nor $LiNbO_3$:2%Zn [294] revealed the photoinduced light scattering (fanning), which is typical for $LiNbO_3$:Fe. This specific of $LiNbO_3$:ME may be accounted for by a model of fanning in crystals with the photovoltaic mechanism of recording [413]. Assuming the fanning to originate from a forward three-wave interaction, the authors conclude that the fanning intensity is controlled by the photovoltaic field E_{pv} and increases at a certain threshold field, which by their calculations for $LiNbO_3$ is approximately $4\times10^4\,\mathrm{V\,cm^{-1}}$. Actually in $LiNbO_3$:Fe crystals, E_{pv} exceeds this value even at low light intensities, whereas in $LiNbO_3$:ME the field E_{pv} is within $10^4\,\mathrm{V\,cm^{-1}}$ [294].

The calculations performed by Zhang et al. [413] were supported by experiments in $LiNbO_3$:Fe, $LiNbO_3$:ME and $LiNbO_3$:Fe:ME with [Fe] = 0.05–0.1%, [Mg] = 2–4 mol.%, [Zn] = 2–6 mol.%, [In] = 0.06–1.2 mol.% [414, 415]. For each composition, a pronounced threshold intensity I_{thr}, i.e., a threshold field E_{pv} was found for impinging radiation. Recording with $I > I_{thr}$ is accompanied by a drastic (by orders of magnitude) increase of the fanning intensity. I_{thr} is the lowest in $LiNbO_3$:Fe (about $0.05\,\mathrm{W\,cm^{-2}}$) and increases when co-doping with ME: the higher the ME concentration, the larger is I_{thr} and the lower is the maximum fanning intensity. Most efficient was co-doping with In, because in $LiNbO_3$:Fe:1.8%In I_{thr} increased to $0.24\,\mathrm{W\,cm^{-2}}$. All compositions with $I < I_{thr}$ had a high recording speed, a diffraction efficiency close to 50% and a high signal-to-noise ratio. Additionally, these data suggest optimum recording intensities for obtaining the highest light amplification in each composition. Note that Kamber et al. [414, 415], according to the consideration of Zhang et al. [416], assign the observed specific of E_{pv} in $LiNbO_3$:Fe:ME to a strong increase of the dark conductivity σ_d with the ME concentration. However, as σ_d in $LiNbO_3$:Fe with [Fe] < 0.2% is dominantly protonic [300, 304] (see Sect. 3.2.1), so from this viewpoint the effect of ME ions on σ_d is unconvincing. The model presented in [413, 414] neglects the role of the diffusion mechanism and does not take into account the relation $\sigma_{ph} \approx I^x$ with ($x < 1$) in $LiNbO_3$:ME (see Sect. 4.2.2) leading to an intensity dependent E_{pv} in a large intensity range [186, 294]. This specific of $E_{pv}(I)$ in $LiNbO_3$:ME together with an apparent participation of the diffusion field [294] obviously contribute to the interesting threshold behavior of fanning found in [414, 415].

The potentials of $LiNbO_3$:ME and $LiNbO_3$:Fe:ME as recording media have been demonstrated in the references [294, 412, 414, 415] and gave rise to a large body of research in this direction; below we list some of them.

The photorefraction sensitivity at 488 nm was found to increase considerably when doping $LiNbO_3$ even with very low [In] = 0.1–0.2 at.% (in the melt) [417]. The achieved values of S exceeded those in low-doped $LiNbO_3$:Fe; such a low In doping did not affect the optical absorption and the optical quality of the crystals. A sufficient improvement of holographic properties was achieved by means of codoping with In, the most appropriate compositions for recording $LiNbO_3$:0.03%Fe and $LiNbO_3$:0.06%Fe [400]. For hologram multiplexing at 633 nm, the sensitivity S in $LiNbO_3$:0.03%Fe:In sufficiently increased with the In concentration by a factor of 3 for $[In_2O_3] = 3\%$ in the melt. The best results were demonstrated in $LiNbO_3$:0.06%Fe:0.5%In. In a 4 mm long sample the gained parameters are $\eta = 61\%$, the dynamic range (Mok factor) $M/\# = 7.47$ and $S = 0.02\,\mathrm{cm\,J^{-1}}$ and sufficiently exceeded those in a $LiNbO_3$:0.06%Fe sample of the same size (36%, 3.88 and $0.01\,\mathrm{cm\,J^{-1}}$, respectively), whereas the optical erasure time in the former was noticeably larger. Interestingly, that in $LiNbO_3$:0.03%Fe:In the In threshold concentration manifests itself as a drastic reduction of $M/\#$ for 1 mol.% In, while S grows almost linearly with the In concentration.

The effect of codoping $LiNbO_3$:0.03%Fe with the new optical-damage-resistant impurity Hf ($[HfO_2]$ = 2–5 mol.%) was investigated by Li et al. [262]. Hf doping as well as doping with other ME ions sufficiently improves the recording parameters, the sensitivity of $LiNbO_3$:Fe:Hf for 532 nm is in the range of 4–5 $\mathrm{cm\,J^{-1}}$. The best results were achieved in slightly reduced crystals.

We now discuss the data on the UV-recording grating in highly-doped $LiNbO_3$:ME crystals. An interest in the UV recording is due to the following reasons. First, the response speed may be sufficiently improved. Second, the UV recording may provide a possibility of a nonerasable readout by a longer wavelength. At last, a UV readout of a grating is not accompanied by a change of refractive indices within the illuminated crystal area, which is characteristic for the readout by visible light and is related to the formation of the photovoltaic field in a nonshort-circuited state; this negative effect vanishes under UV radiation because of a negligible E_{pv}. The problem of the UV-induced photorefraction on a whole is not sufficiently investigated. The basics of the UV-induced (band-to-band) photorefraction are considered in a recent review [273]. Studies of the UV recording in $LiNbO_3$ were started by Laeri and co-workers [361, 418]. For two-beam recording in the UV at 351 nm in pure congruent $LiNbO_3$, a steady-state unidirectional energy transfer between recording beams was observed with a $\Gamma \approx 14\,\mathrm{cm^{-1}}$, which is evidence of a prevailing diffusion mechanism; the detected photovoltaic field was found to be $E_{pv} \approx 0.5\,\mathrm{kV\,cm^{-1}} < E_{dif}$. As the recording wavelength increased, the steady Γ decreased and completely vanished at 514 nm, where only the photovoltaic mechanism is involved. From the direction of the energy

Table 4.4. The photorefractive characteristics of $LiNbO_3$:Zn and $LiNbO_3$:In crystals for a recording at $\lambda = 351$ nm, taken from [420]. The grating period was $\Lambda = 0.5\,\mu m$, η was measured with $I_S = 121.7\,mW\,cm^{-2}$ and $I_R = 176.9\,mW\,cm^{-2}$, while for Γ the ratio I_R:I_S = 100:1 was used

Samples	CLN	5%Zn	7%Zn	9%Zn	1%In	3%In	5%In
σ_{ph} ($10^{-12}\,cm\,\Omega^{-1}\,W^{-1}$)	3.32	10.6	25.2	57.3	1.59	7.46	12.9
Diffraction efficiency η (%)	9.05	16.9	22.3	25.3	10.1	15.9	17.7
Response time τ_e (s)	12.4	1.97	1.01	0.88	13.9	3.06	1.58
Two-beam coupling Gain Γ (cm^{-1})	1.32	11.0	15.2	21.7	1.16	11.8	17
Sensitivity S ($cm\,J^{-1}$)	0.99	4.0	8.85	11.1	0.86	2.85	3.88
Dynamic range M/#	0.14	0.11	0.12	0.14	0.26	0.19	0.15

transfer, the photoholes were determined to be the dominant charge carriers at UV-wavelengths, whereas in the visible, the well-known charge carriers are photoelectrons. The UV-recording grating was performed in congruent $LiNbO_3$:Mg [419] and in $LiNbO_3$:Zn and $LiNbO_3$:In [420] in the range of Mg, Zn and In concentrations below and above the thresholds. The gratings were recorded at 351 nm. The summary results of these studies are as follows. The optical absorption at 351 nm in $LiNbO_3$:ME drastically decreased compared with $LiNbO_3$ in which $\alpha_{351} = 10\,cm^{-1}$, e.g., in $LiNbO_3$:9%Mg $\alpha_{351} = 1.2\,cm^{-1}$. In all $LiNbO_3$:ME crystals with Mg, Zn, and In concentrations close to the thresholds and above, a steady unidirectional energy transfer was observed with Γ growing with ME concentrations; the highest achieved values are $\Gamma = 15\,cm^{-1}$ in $LiNbO_3$:9%Mg, $\Gamma = 21.7\,cm^{-1}$ in $LiNbO_3$:9%Zn, and $\Gamma = 17\,cm^{-1}$ in $LiNbO_3$:5%In, see Table 4.4. So, the UV-recording in highly-doped crystals occurs dominantly by the diffusion mechanism as well as in undoped $LiNbO_3$ [361, 418]. The recording speeds and sensitivities are significantly enhanced for ME concentrations above the threshold, e.g., in $LiNbO_3$:9%Zn we have $S > 10\,cm\,J^{-1}$. An additional advantage of these crystals for the UV-recording is a relatively low light-induced scattering. Therefore, with regard to the UV-recording that is based on the diffusion mechanism, $LiNbO_3$:ME crystals offer very appropriate properties and by no means are optical-damage-resistant.

Interesting, but unclear results were obtained when determining the carrier sign from the direction of the energy transfer following the method of Orlowski and Krätzig [317]. At UV-wavelengths the photocarriers in highly-doped $LiNbO_3$:Mg are holes [419] similarly to undoped $LiNbO_3$ [361, 418]. On contrast, according to Qiao et al. [420] in highly-doped $LiNbO_3$:Zn and $LiNbO_3$:In, the photocarriers in the UV are electrons. At the same time, in $LiNbO_3$:Mg [327] and $LiNbO_3$:Zn [328] with Mg and Zn concentrations above the thresholds, the photocarriers in the visible determined by the same holographic method are holes. In principle, a different sign of the photocarriers in different spectral ranges for a given crystal is known in semiconductor physics and even could be discussed in the framework of a fundamental change of the charge transport scheme at ME concentrations above the thresholds. However, no speculations may account for the difference in the photocarrier sign in $LiNbO_3$:Mg on the one hand and $LiNbO_3$:Zn and $LiNbO_3$:In on the other hand.

5

Problem of a Non-Erasable Photorefractive Hologram

Summary. The chapter characterizes the methods providing a non-erasable optical readout of the photorefractive gratings. The first is the two-color holography which is based on recording a grating by a long wavelength (IR) possible due to a non-equilibrium population of shallow levels at the expense of a short-wavelength pumping-over of electrons from deeper levels. The two-step photoexcitation via short-living photoinduced centers ('gating') is described in $LiNbO_3$ and $LiTaO_3$ of different compositions. An emphasis is given to the involvement of the intrinsic defects, namely intermediate small polarons occurring due to the photodissociation of bipolarons in reduced crystals. A UV-gated recording in $LiNbO_3$ doped with optical-damage-resistant ions is described. An outline of data is given on certain combinations of impurities such as Fe and Mn in $LiNbO_3$ which under a short-wavelength photoexcitation form long-living metastable states resulting in slowing down the optical erasure. Another method providing a nonvolatile readout is the thermal fixation of gratings based on screening the recorded electronic space-charge by species of the opposite sign (ions or photoholes) activated when heating to certain temperatures. The thermal fixation and the current models of this process in $LiNbO_3$:Fe is described dwelling on the factors controlling the diffraction efficiency and storage times of a fixed grating. Particular attention is given to a discussion of protons as compensating charges in $LinbO_3$:Fe. A short outline of available data on thermal fixation in other photorefractive crystals closes this chapter.

The hologram storage should combine a high sensitivity and nonvolatility, that is a stability under the readout and in darkness. In photorefractive gratings the last problem still remains to be solved. When readout, a spatially modulated space-charge field is destructed on a homogenous illumination by a single Bragg beam, because of a photoexcitation and a spatial uniform redistribution of the trapped electrons. This effect is usually (but not always) undesirable and is the main factor limiting practical applications of photorefractive materials for holographic data storage. The rate of the erasure is proportional to $\sigma_{\rm ph} \approx I$ ($\sigma_{\rm ph} \gg \sigma_{\rm d}$), so even when reading with an attenuated beam, a recorded grating is erased rather quickly, e.g., in as-grown $LiNbO_3$:0.03%Fe we have $\sigma_{\rm ph} \approx 10^{-13}\,{\rm cm\,V^{-2}}I$, therefore with an intensity $I = 10\,{\rm mW\,cm^{-2}}$ a hologram is efficiently erased during less than 1 h of a continuous illumination. To overcome this obstacle, several methods were proposed, e.g., [421–424]. The most obvious way is using for the readout a wavelength longer than the recording one ($\lambda_{read} > \lambda_{rec}$) [421, 423], because the erasure is retarded owing to a lower σ_{ph} at longer wavelengths. This method actually suppress the erasure under readout, however, in the hologram multiplexing it results in an increased cross-talk and strong loss of the holographic efficiency due to the inability to achieve the exact Bragg phase matching. Additionally, the grating is anyhow slowly erased owing to a nonzero photoconductivity at λ_{read}. The techniques aimed to solve the problem of volatility of the photorefractive gratings are a two-color holography, or, what is the same, recording in the gating regime and various methods of fixation. These are the subjects of the present chapter.

5.1 Two-Color Holography

The general principle of the two-color (gated) recording and its difference to the usual (one-color) mode of operation is illustrated by Fig. 5.1.

5.1.1 Principles of the Two-Color Holography

At the traditional one-color recording (right side of Fig. 5.1) the formation of the space-charge field E_{sc} involves a direct photoexcitation of free carriers

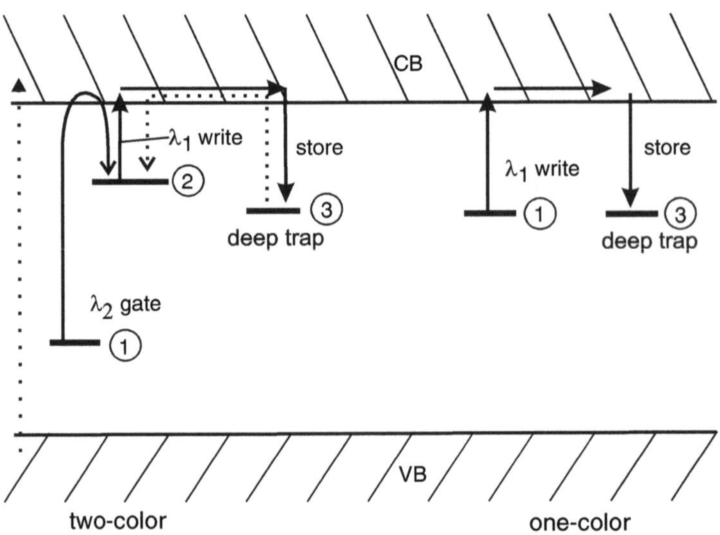

Fig. 5.1. Schematic diagram of the two-color and one-color photorefractive effects. CB, conduction band; VB, valence band, redrawn from [84]

from a donor level U_1, their subsequent drift or diffusion and capturing by deep traps U_3. In this case, the data medium is the recording wavelength λ_1, so the grating readout by λ_1 leads to an erasing E_{sc}, as in the charge transport scheme (3.2) underlying the photorefractive recording $U_1 \sim U_3$. To avoid the erasure, a two-step photoexcitation (left side of Fig. 5.1) may be used over an intermediate level $U_2 < U_1$. The free carriers are first pumped by an intensive illumination from level U_1 to U_2 ($\lambda_2 > \lambda_1$), then photoexcited from U_2 to the conduction band by a lower-energy wave λ_2 and after a drift or diffusion captured by U_3. The lifetime of the intermediate nonequilibrium state is dictated by the value of U_2. Provided that the photoexcitation leads to a nonequilibrium repopulation of relatively deep levels, then the lifetime of the intermediate state may be rather long, this case is sometimes referred to as a two-center recording. The deeper (feeding) level U_1 is denoted as a gating one U_g; the gating may occur via the interband photoexcitation, too. Usually, it is assumed that the formation of E_{sc} in the two-color case is due to a capturing of the photocarriers excited from U_2 by the same deep traps U_3 as in the one-color process. So, if a grating is recorded by λ_2, then the readout by this wavelength will not erase the grating, because the deep traps U_3 cannot be photoexcited. Obviously, a recording wave should be coherent, while for the gating wave this condition is not required and even sometimes is an obstructive factor. Besides a nonerasable readout, an additional advantage of the two-color recording is an improvement of the photorefractive sensitivity at longer wavelengths, thus the extension of the recording range towards the IR. Such an optically fixed grating may be erased by a uniform illumination with light of a short wavelength (U_1). The following qualitative expression characterizes the recording photorefraction in the two-color regime for the case of a short-living intermediate center; the subscripts "g" and "rec" correspond to the parameters of the gating and recording radiation, respectively [401],

$$\delta \Delta n_e = [\alpha(\lambda_g) I_g \tau] \, [\sigma(\lambda_{rec}) I_{rec} \, t] \times \text{constant} \,, \tag{5.1}$$

where α is the optical absorption coefficient, λ_g, λ_{rec}, I_g, I_{rec} are wavelengths and photon flux (intensities normalized to the photon energy), τ is the lifetime of the intermediate nonequilibrium state, σ is the photoexcitation cross-section for the intermediate level, t is the exposure time, and $t \gg \tau$ is required. The first and second square brackets characterize, respectively, the degree of a nonequilibrium population of the intermediate level and the photoexcitation rate from it. Therefore, a rather long τ is desirable to provide a sufficient population of the intermediate state and to reduce the intensities of the gating and recording lights. The recording may be performed either simultaneously with the gating, or with a delay for a time gap shorter than τ. We introduce some parameters used to characterize the two-color recording. The photorefractive sensitivity is defined in the usual way

$$S = \frac{1}{I_{rec} \, L} \frac{\mathrm{d}\sqrt{\eta}}{\mathrm{d}t} \Big|_{t=0} \,, \tag{5.2}$$

where L is the hologram thickness, η is the diffraction efficiency, I_{rec} is the intensity of the recording IR light. At the same time, in a two-color recording the sensitivity is varied during readout due to a partial erasure, therefore Adibi et al. [425] introduce another measure taking into account this partial erasure

$$S' = \beta S , \tag{5.3}$$

where β is the ratio of $\sqrt{\eta}$ after a long readout to $\sqrt{\eta}$ before a readout. As a rule, S increases with the gating light intensity in a relatively low range of I_g, whereupon it tends to a saturation. The authors of [84] propose to use as a gating standard the intensity $I_g = 1\,\mathrm{W\,cm^{-2}}$ in order to specify the gating conditions. To estimate the efficiency of the two-color recording, it is reasonable to correlate the achieved S to the usual empirical criterion, namely, to the photorefractive sensitivity of $LiNbO_3$:Fe observed usually under recording by an Ar-ion laser $S_{488\,\mathrm{nm}} \approx 0.05$–$0.3\,\mathrm{cm\,J^{-1}}$ and by an He–Ne laser $S_{633\,\mathrm{nm}} \approx 0.01$–$0.1\,\mathrm{cm\,J^{-1}}$ (both per incident energy). Sometimes the parameter *gating ratio* is used that is the ratio of the sensitivities in the presence of the gating light and without it.

The characteristics of photorefraction in a two-color regime differ from those in the one-step regime. This is due to the fact that in any case it is related to a nonequilibrium populated level, which leads to intensity and temporal dependencies of the optical absorption coefficient $\alpha(I,t)$, to nonlinearity of the lux–ampere characteristics $\sigma_{ph} \approx AI^{\gamma}$, etc. As a result, the photorefractive sensitivity $S \approx I^{\kappa}$. The well-known characteristic for the one-color regime between the recording τ_{rec} and erasure τ_{er} times $\tau_{rec}/\tau_{er} = I_{er}/I_{rec}$ is violated.

In what follows, one may see two methods used in $LiNbO_3$ for a two-step photoexcitation of the free carriers. First, it is a two-photon process, thus either an interband photoexcitation, or a two-step photoexcitation of an impurity center; in this case $I_{rec} = I_g$ and $S \approx I^2$. Second, it may be a non-equilibrium population of shallow traps at the expense of the photoexcitation of a donor level by means of a pumping by an intensive visible or UV light. It should be mentioned here that in the most cases it is tacitly assumed that Fe^{2+}/Fe^{3+} centers traditionally serve, respectively, as levels of photoexcitation (gating) and capturing electrons, even for an unintentional Fe concentration in the range of several ppm.

5.1.2 Two-Step Recording via Short-Living Centers (Gating Process)

For the first time the possibility of a two-step recording via a two-photon process was demonstrated in $LiNbO_3$ [94] and KTN [426,427] operating with $\lambda = 532\,\mathrm{nm}$ (second harmonic from an Nd-YAG laser). It was assigned to a two-step interband photoexcitation of the free electrons via a short-living virtual center. In $LiNbO_3$:Cr and $LiTaO_3$:Cr [95] the recording grating under identical conditions was attributed to a two-step ionization of a Cr^{3+}-ion, namely, the excitation from the ground state ${}^4A_2 \rightarrow {}^4T_2$ with a lifetime of ${}^4T_2$ at room temperature of about 500 ns and subsequent photoexcitation of the free charge carriers from ${}^4T_2$. A readout in a linear regime at relatively low intensities at $\lambda = 532$ nm did not erase the gratings.

It is necessary to emphasize the pioneering role of these first investigations, which underlay the subsequent development of this area. The results had a fundamental significance, but no practical matter, because the required peak power for recording was of about 10^7–10^8 W cm^{-2}. In later works, similar results were obtained in $LiTaO_3$:Fe and $LiNbO_3$:Cr when recording with $\lambda_{rec} = 694$ nm in a two-photon regime with a peak power of 6×10^6 W cm^{-2} [356, 428]. Later on, Buse et al. [295, 358, 429, 430] succeeded in recording a grating in $LiNbO_3$:Fe and $LiNbO_3$:Cu at $\lambda_{rec} = 1,064$ nm by means of an intensive pumping with $\lambda_g = 532$ nm. The lifetime of the photoinduced IR absorption band was in the range of ms which permitted to perform a two-step process, thus to delay the IR recording pulse with regard to the gating one. A disadvantage was again the necessity of high intensities. The photoinduced band at about 1,000 nm forming the base for the IR recording was interpreted as a result of electron capturing by Nb antisites. Therefore, for the first time the Nb antisites were regarded as an intermediate level participating in a two-color recording. This assumption was in agreement with the observation of a nonstable absorption band induced in $LiNbO_3$:Fe by an intensive visible light [357, 358]. Similar results were obtained when pumping $LiTaO_3$:Fe by an intensive UV laser pulses $\lambda_g = 355$ nm, $I_g \approx 3$ kW cm^{-2} and simultaneous recording with $\lambda_{rec} = 1,064$ nm and $I_{rec} \approx 10$ kW cm^{-2} [431]. A nonerasable readout was achieved, and UV-induced band in the infrared by analogy with $LiNbO_3$ was assigned to hypothetical Ta_{Li}.

A new era in the development of the two-color holography started in 1997, when several research groups practically simultaneously reported on the two-color recording gated by rather low intensities of a cw-light in SBN:Pr [432], and $LiNbO_3$ [89,89–91,433]. The role of a slight Pr doping was first seemed to be a decisive factor [90, 91, 432, 433], but in more recent researches it was not corroborated. These early results may be exemplified as follows. A recorded grating in $LiNbO_3$ at $\lambda_{rec} = 810$ nm from a Ti–sapphire laser with $I_{rec} = 10$ W cm^{-2} and a cw-gating with $\lambda_g = 515$ nm and $I_g = 10$ W cm^{-2} provided a gating ratio of 1,000 and a nonvolatile readout during 20 h; the hologram was stored in dark not less than during 200 h. Bai and Kachru [433] assigned this process to the existence of nonequilibrium centers with lifetimes in the range of tens of milliseconds and assumed these centers to be Nb_{Li}^{4+}. In $LiNbO_3$ doped with 0.1% Pr (810 nm, 10 W cm^{-2}) with gating at 515 nm provided a diffraction efficiency as high as 50% [89], that is five orders of magnitude higher than without pumping. In $LiNbO_3$ doped with 2% Pr the grating recording was performed at 810 and 852 nm with a gating between 400 and 500 nm from an Xe-lamp [90]; the recording sensitivity significantly increased at rather low I_g and comes to saturation at ≈ 2 W cm^{-2}. Lande et al. [91] reported on a digital 256 Kbit recording in Li-enriched $LiNbO_3$:Pr with $\lambda_{rec} = 800$ nm and

$\lambda_g = 476$ nm. So, the results were encouraging, because the two-step recording provided rather high S, diffraction efficiency and a nonvolatile memory with long storage times.

These basic works intuitively related an intermediate short-living level to Nb_{Li} and sometimes remarked on an improvement of the two-color recording parameters in Li-enriched crystals. This stimulated further playing with the intrinsic defects following the concept of bipolarons in reduced $LiNbO_3$ crystals described in details in Sect. 2.2. Recall that the illumination of a reduced $LiNbO_3$ by a visible light leads to a reconstruction of the optical absorption band in the visible (Fig. 2.6) and to the appearance of a band at 1.6 eV interpreted as formation of small polarons Nb_{Li}^{4+} whose photoexcitation by an appropriate IR light provides free electrons. The two-step process looks as follows

$$[Nb_{Li} - Nb_{Nb}]^{2-(h\nu=2.5\,\mathrm{eV})} \leftrightarrow \left[Nb_{Li}^{4\bullet}\right]^{-(h\nu=1.6\,\mathrm{eV})} \leftrightarrow Nb_{Li}^{4\bullet} + e^- . \tag{5.4}$$

Guenther et al. [84, 90] applied this scheme for the two-color recording in reduced $LiNbO_3$ crystals (Fig. 5.2) using $\lambda_g = 488$ nm for gating and $\lambda_{rec} = 852$ nm from a diode laser for recording. Studies were performed in

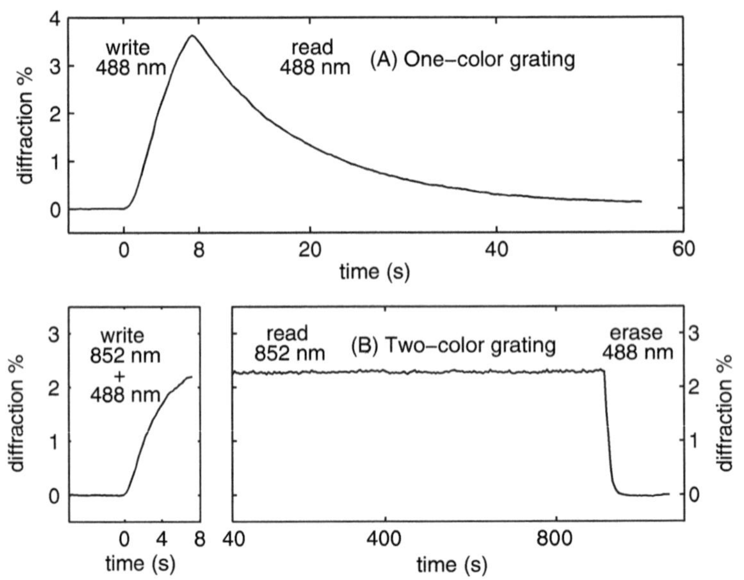

Fig. 5.2. Comparison of the one-color grating (**a**) and the two-color grating (**b**) schemes. For the one-color grating a laser with $\lambda = 488$ nm and $1\,\mathrm{W\,cm^{-2}}$ was used for both writing (two beams) and reading (one beam). In the second case the light source for writing was a distributed Bragg reflector laser diode with 852 nm and $4\,\mathrm{W\,cm^{-2}}$ (two beams), the gating was done with a second laser (488 nm; $1\,\mathrm{W\,cm^{-2}}$). Nondestructive reading was carried out with one of the writing beams ($2\,\mathrm{W\,cm^{-2}}$) and erasing with the gating light; redrawn from [84]

Li-enriched $LiNbO_3$ with different Li content. Figure 5.2a shows the recording and readout kinetics in a reduced crystal by visible light (488 nm); the grating is erased under readout. Figure 5.2b presents the IR recording in the same crystal (852 nm) with simultaneous uniform illumination with 488 nm. The readout of the recorded gratings at 852 nm was nondestructive and the gating ratio at 852 nm was in the range of 10^3–10^4. The saturation values of the sensitivity achieved at $I_g > 1$–$2\,\mathrm{W\,cm^{-2}}$ did not exceed $S = 0.02$–$0.03\,\mathrm{cm\,J^{-1}}$ per incident energy, being lower than the above-mentioned criterion. At the same time the sensitivity per absorbed energy was very close to that obtained in $LiNbO_3$:0.02%Fe when recording with 488 nm, which is due to the fact that the optical absorption at 800–900 nm in reduced crystals is very low. An important conclusion was a significant increase of the decaying time of the metastable band at 852 nm, i.e., of the Nb_{Li}^{4+} lifetime, with increasing Li content. In the crystal with 49.4%Li the decay in the dark was characterized by $\tau_d \approx 100$–200 ms, whereas in the crystal with 49.7%Li $\tau_d = 3.5$ s was found. Effects of Fe doping were found, namely doping with 60–100 ppm noticeably enhanced the sensitivity and reduced τ_d. This confirmed an a priori assumption of the Fe^{2+} level as a source of the electron photoexcitation. At the same time, an increase of the Li content or doping with Fe along with a sensitization of the crystals led to an undesirable decrease of the storage time in darkness, obviously due to an enhanced dark conductivity.

A relation of the photoinduced metastable IR bands to Nb antisites found in Li-enriched crystals stimulated studies of the two-color recording and photoinduced optical absorption in $LiNbO_3$ crystals doped with optical-damage-resistant impurities, because these ions are regarded as controllers of the Nb_{Li} content. The investigations of the two-color recording in as-grown (nonreduced) near-stoichiometric $LiNbO_3$:Mg crystals [88] found a fundamental difference in the behavior of crystals doped with below- and above-threshold Mg concentrations. It gives an insight into the specific of the charge transport in optical-damage-resistant crystals. The gratings were recorded at 852 nm with gating at 488 nm, the experimental conditions were the same as in [84]. The characteristics of recording in crystals doped with Mg below the threshold (0.6%) were very similar to those obtained in reduced SLN [84] (Fig. 5.2), so the process could be interpreted in terms of the reaction (5.4). In crystals doped with Mg above the threshold in this material (1.8% and 3.6%) the temporal characteristics of the photorefraction were fundamentally different. After turning off the gating light, the diffraction efficiency after passing over a slight maximum was quickly decreasing during about 10 min. A detailed analysis permitted the authors to suggest that this decay in darkness is caused by arising of a compensating grating, formed by the charge carriers of the opposite sign. In other words, an after-effect of gating is the formation of a hole-grating spatially modulated by an electron-grating recorded by the IR light. Winnacker et al. [88] qualitatively account for this effect by the appearance of an electron trap other than Fe^{3+}. This assumption is in agreement with the conclusion made by others [216, 218, 325] and discussed in

Sect. 4.2 of an alteration of the electron traps in $LiNbO_3$ crystals doped with above-threshold concentrations of optical-damage-resistant ions. A fundamental change of the charge-transport scheme is due to a partial incorporation of Fe^{3+} onto Nb sites and to a drastic decrease of the Li vacancy concentration. The assumption of a formation of a hole-grating is supported by observations of the hole-photoconductivity in $LiNbO_3$:Mg and $LiNbO_3$:Zn with concentrations above the thresholds [327, 328]. The idea of coexisting gratings formed by capturing the charge carriers of the opposite signs is in line with similar observations of mutually compensated gratings which accounts for a hologram fixation in some crystals having a bipolar photoconductivity, see Sect. 5.2.

The principle of the two-color recording in reduced crystals based on the photoexcitation of the small bound polarons, is very attractive from the fundamental view point, but exhibits drawbacks impeding its practical applications. The achieved sensitivities in the range of 10^{-3}–10^{-2} [84, 87] are far behind those in the one-color process. In addition to the above-mentioned decrease of the storage times because of increased dark-conductivity, a slight, but noticeable erasure of the grating during IR readout occurs due to a reverse bipolaron formation according to the reaction (5.4). Rather long lifetime of polarons achieved in reduced crystals may also be regarded as a negative factor, because it may lead to a dark decay of the grating instantaneously after recording [84]. A fundamental technological drawback of this method is an impossibility to achieve the identical starting states in reduced crystals, so one may a priori expect a nonreproducibility of results.

Later on, in developing the two-color recording methods, the interest turned towards UV-induced effects. This is due to the fact that, as discussed in Sect. 2.2, the UV irradiation induces in some $LiNbO_3$ and $LiTaO_3$ compositions unstable broad absorption bands, whose tail in the infra-red was often assigned, rather qualitatively, to the formation of Nb_{Li}^{4+}. This hypothesis was reasonable, because these effects again depend on the crystal stoichiometry and manifest pronouncedly in Li-enriched $LiNbO_3$ or in congruent $LiNbO_3$ doped with optical-damage-resistant dopants. Independently of the microscopic origin, this UV-induced coloration provides a sensitization of the crystal to an IR photorefractive recording, so can be used as the base for a two-color holography. It was observed in Tb-doped congruent and Li-enriched $LiNbO_3$ [92, 434–437], where the UV-induced band extends at least from the near infrared to the absorption edge, see Fig. 5.3.

Under gating with 313 nm from a mercury lamp, in $LiNbO_3$:0.01%Tb and $LiNbO_3$:Fe:0.01%Tb a nonvolatile grating was recorded at 852 nm [92, 435] (Fig. 5.4).

Additionally, a pre-exposure to the UV light significantly improved the sensitivity of recording by the visible light [436, 437]. The authors recorded by $\lambda = 532$ nm 50 plane-wave angle-multiplexed holograms in a $LiNbO_3$:Fe:Mg crystal pre-exposed to a UV irradiation from a Hg–Xe-lamp. When recording of each sequential grating the value of S is decreasing due to a step-by-step

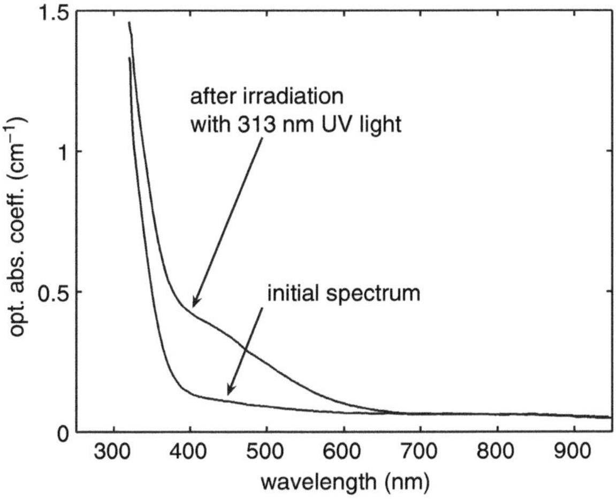

Fig. 5.3. Optical absorption of Tb-doped (100 ppm) $LiNbO_3$ with a [Li]/[Nb] ratio of 49.7/50.3 before and after illumination with UV light of 313 nm for 1 min with $50\,W\,cm^{-2}$ at room temperature; redrawn from [434]

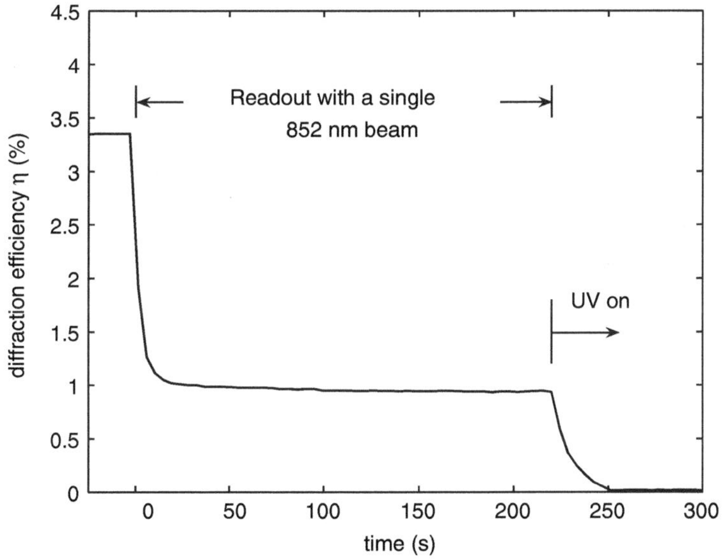

Fig. 5.4. Nonvolatile readout with a single 852 nm beam in Tb:NSLN. The total writing intensity was $300\,mW\,cm^{-2}$ at 852 nm, and the gating intensity was $30\,mW^{-2}$ at 313 nm; redrawn from [434]

optical bleaching of the photoinduced absorption, so an appropriate schedule of multiplexing should be elaborated for this special case [437].

Similar recording was performed in near-stoichiometric LiNbO:0.05% Er [438] and near-stoichiometric $LiNbO_3$:0.05%Tm [439]. The role of Ln^{3+} ions remains unknown yet. Lee et al. [440] tried to analyze a contribution from Tb ions into the UV-induced coloration in $LiNbO_3$:Tb and $LiNbO_3$:Tb:Fe by EPR measurements at low temperatures. A decrease of the Fe^{3+} concentration was observed due to electron capturing $Fe^{3+} + e^- \rightarrow Fe^{2+}$, but no evidences for an involvement of Tb was detected.

Usually Fe^{2+} ions, either unintentionally or specially introduced, serve as a source of photoexcited electrons. The authors of [441–443] probed Mn^{2+} ions for UV gating. The experiments on the UV-assisted two-color recording were performed in a near-stoichiometric $LiNbO_3$ crystal doped with slight Mn concentrations of 8–10 ppm. The UV irradiation of the crystal induced a broad optical absorption band expanding from the bandedge to the near IR with a diffuse peak at about 800 nm, decaying in the dark with characteristic times of 0.5–1.5 s depending on the crystal stoichiometry. The shape and kinetic characteristics of this band permit to attribute it to a formation of the bound small polaron. The two-color recording was performed at $\lambda_{rec} = 778$ nm with gating by $\lambda_g = 350$ nm and I_g in the range 0.2–1.5 W cm^{-2}. The write–read–erasure cycle reminds that observed in other crystals (Fig. 5.2). The recording parameters were promising, e.g., with $I_g = 1.5$ W cm^{-2} and $I_{rec} = 20.5$ W cm^{-2} a value of $S = 0.21$ cm J^{-1} was achieved.

A UV-assisted two-color recording was investigated in $LiNbO_3$ crystals doped with optical-damage-resistant impurities. The experiments in $LiNbO_3$ crystals highly doped with Mg [395, 396, 444] as well as the above-mentioned data in reduced $LiNbO_3$:Mg [88] revealed a specific of the transport process in optical-damage-resistant crystals. In the congruent and near-stoichiometric $LiNbO_3$ crystals doped with an above-threshold concentration of 5%MgO a noncoherent irradiation with 320–410 nm induces a band extending from the bandedge to the near infrared and a rather fast decaying in darkness; the same was observed by Volk and Rubinina [325], see Fig. 4.13.

Later on, a similar photochromic effect was observed in $LiNbO_3$ doped with 9% Zn [445]. In view of missing Nb antisites in the above-threshold $LiNbO_3$:Mg and $LiNbO_3$:Zn compositions, the UV-induced band cannot be attributed to Nb_{Li}^{4+}. From analyzing the kinetics of the decay, several authors [395, 396, 444, 445] assign it to the formation of a small bound polaron O^-. Their consideration is based on the old works [63, 64, 94], in which a similar broad band induced in $LiNbO_3$ under the two-photon photoexcitation was interpreted as a small hole polaron O^-. The two-color recording (780 nm) with the UV-gating was performed in these above-threshold crystals [395, 396, 444]; the authors attributed the process to the excitation of photoholes from O^-.

A UV-assisted two-color recording was obtained in Li-enriched $LiTaO_3$ crystals [446–448] with $\lambda_{rec} = 780$ nm, $\lambda_g = 350$ nm and I_g as low as 0.02–0.03 W cm^{-2}. The following data may exemplify an improvement of the sensitivity in the infrared achieved in the above-listed crystals when gating with UV light. Initially the photorefractive sensitivities at $\lambda_{rec} \approx 800\,nm$ are practically zero; for the two-color conditions the values 1.1×10^{-2} and $9 \times 10^{-4}\,cm\,J^{-1}$ were achieved in NSLN:Tb [92, 434, 435], and NSLN:Er [438], respectively. The sensitivity in Mg-doped CLN and NSLN was in saturation $0.08\,cm\,J^{-1}$ ($\lambda_{rec} = 780\,nm, \lambda_g = 350\,nm, I_g = 0.4\,W\,cm^{-2}$). In Li-enriched $LiTaO_3$ the value of S for 788 nm was in the range 0.02–0.08 cm^{-1} depending on the Li content, thus being comparable to the above-mentioned criterion in $LiNbO_3$:Fe. The two-color recorded grating in $LiTaO_3$ was extremely stable under a readout during several hours and showed no degradation in darkness, especially in dehydrated $LiTaO_3$ samples [446–448].

Note that the experiments on the two-step recording in $LiTaO_3$ [447] with varied Li/Ta concentration ratio provided indirect evidences for the relation of the UV-induced absorption band to small polarons, that is to hypothetical Ta_{Li} acting as shallow electron traps. When approaching the stoichiometric composition, the values of S and $\delta\Delta n$ with increasing Li content are growing up to a certain limit, whereupon they start to decrease. Such an optimal Li content may be interpreted as certain optimal conditions for a nonequilibrium population of Ta_{Li} following the reversible reaction (5.4).

Regardless of an ambiguity in the interpretation, the UV-assisted two-color recording in nontreated crystals exhibits certain practical advantage over that in reduced crystals, because it is not accompanied by an increase of the dark conductivity reducing the storage time.

A set of experiments on the two-color recording was performed in congruent $LiNbO_3$:In doped with a below-threshold concentration of about 0.6 mol.% In_2O_3 in the crystal using for gating $\lambda_g = 488\,nm$ and 514 nm with $I_{rec} = 1$–$4\,W\,cm^{-2}$ and recording $\lambda_{rec} = 780\,nm$ [449–452]. In as-grown (nonreduced) $LiNbO_3$:In the two-color recording provided a sensitivity in the range 3–$5 \times 10^{-4}\,cm\,J^{-1}$ and a nonerasable readout [449], a reduction of crystals led to an increase of the sensitivity $S = 3 \times 10^{-3}\,cm\,J^{-1}$ [450, 451]. The lifetime of the photoinduced IR band of about 4 ms was by orders of magnitude lower than in-reduced NSLN, since by this and other reasons an interpretation in the framework of the polaron concept is questionable yet.

Summarizing these data on the two-color recording in reduced crystals gated by visible light and in untreated crystals gated by UV, one may see that they bear witness to the participation of Nb_{Li}^{4+} in the two-step electron photoexcitation. This is consistent with the more recent data on the small polaron formation under pumping by an intensive visible light cited in Sect. 2.2. A pronounced dependence of the lifetime of the small polaron Nb_{Li}^{4+} on the crystal stoichiometry is clearly seen from all these data, namely milliseconds at room temperature in congruent $LiNbO_3$:Fe [86], tens or hundreds of milliseconds in reduced stoichiometric crystals [84, 87] and even tens of seconds in as-grown Li-enriched $LiNbO_3$:Mg crystals with Mg concentrations below the threshold [88].

5.1.3 Two-Color Recording in Doubly-Doped Crystals

Another method of a two-color recording assisted by UV light may be applied to doubly (or multiply) doped crystals, where UV irradiation results in a nonequilibrium redistribution of photocarriers between relatively deep levels, thus an intermediate metastable state is rather long living. This method requires finding an appropriate combination of dopants, which could provide an efficient refilling of levels. An interest to this type of recording flamed up after reporting on the UV-pumped two-color recording in $LiNbO_3$:Fe:Mn [453]. Later on this research group investigated this effect in $LiNbO_3$:Fe:Mn in detail [425,454–456]. The two-color recoding is based on the photochromic effect found formerly in these crystals [457]. A UV irradiation enhances the initial band at 480 nm, which evidences an increased Fe^{2+} concentration. As manganese and iron occur in $LiNbO_3$ in the valence states Mn^{2+}/Mn^{3+} and Fe^{2+}/Fe^{3+}, respectively, so this band was interpreted [457] as a metastable population of the Fe^{3+} level at the expense of the electron photogeneration from Mn^{2+}. Schematically the recording and readout process in this case is presented in Fig. 5.5.

Recording with an He–Ne laser after exposure to the UV light or simultaneously with the UV-pumping occurs with a diffraction efficiency significantly

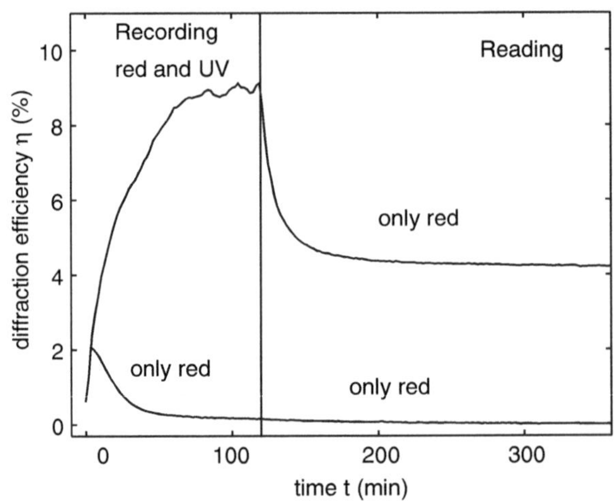

Fig. 5.5. Holographic recording and read-out curves. The diffraction efficiency η, the ratio between the intensity of the diffracted and of the total incident read-out light, is shown vs. time. *Lower curve*: the crystal is pre-exposed to the ultraviolet light (365 nm, 20 mW cm^{-2}, 120 min), then the hologram is recorded (633 nm, ordinary polarization, 600 mW cm^{-2}, 120 min) and finally it is read by one of the recording beams (633 nm, 300 mW cm^{-2}, 240 min). *Upper curve*: same procedure, but the UV light is also present during recording. Simultaneous presence of ultraviolet and red light yields higher efficiencies and in the final stage, nonvolatile read-out; redrawn from [453]

exceeding that achieved when one-color recording is used with the same I_{rec}. When reading out with 633 nm, a part of the recorded hologram is erased, and a residual part of it is stable under several hours of a permanent He–Ne illumination. This fixed residual η may be erased by UV. As an example, in $LiNbO_3$:0.075%Fe:0.01%Mn the recording with $I_{rec} = 35\,mW\,cm^{-2}$ and a simultaneous gating with nonpolarized light from a mercury lamp (365 nm, $20\,mW\,cm^{-2}$) resulted in an enhancement of η by a factor 4–5; under the readout a residual part of η was 32% and 4% after recording by the extraordinary and ordinary beams, respectively [453]. The process on a whole is qualitatively explained in the following way. A UV irradiation excites the electrons to the conduction band from both donor levels Mn^{2+} and Fe^{2+}, whereupon the electrons are redistributed between Fe^{3+} and Mn^{3+} traps. A balance between this nonequilibrium repopulated levels and thus the concentration of metastable extra Fe^{2+} depends on many parameters, particularly, on the initial population of the traps and on the relation between the recombination coefficients of Fe^{3+} and Mn^{3+}. Under readout by a longer wavelength, the electrons are dominantly excited from a shallower level Fe^{2+}, whereupon are again redistributed between the trap levels up to certain quasi-equilibrium balance. This accounts for a decrease of η in the beginning of the readout. A part of metastable nonequilibrium Fe^{2+} is preserved and provides an increased photorefractive sensitivity and a nonerasing readout. The photorefractive characteristics of such a scheme are controlled mainly by a shallower (nonequilibrium populated) level, therefore in the discussed case by Fe.

An optimal ratio between initial Fe^{2+} and Mn^{2+} concentrations may be achieved by variation of the doping levels [456] and an oxidation-reduction treatment [454, 456]. The best parameters are observed when the Fe (trap) concentration sufficiently exceeds (by a factor of 5–15) the Mn (donor) concentration. For example, at 0.075 wt% Fe_2O_3 the highest sensitivity and preserved value of η are observed for 0.005 wt% MnO and significantly decrease with increasing MnO concentration to 0.025%. A nonvolatile readout with reasonable values of the diffraction efficiency is achieved at oxidation conditions completely transforming Fe^{2+} to Fe^{3+} and only partially oxidizing Mn^{2+}, whereas in strongly reduced crystals the sensitivity is much higher, but the readout erases the grating. The achieved sensitivity (Fig. 5.5) is anyhow rather low $S_{633} \approx 0.0033\,cm\,J^{-1}$, and may be increased by using for recording an Ar-ion laser up to $S_{514} \approx 0.07\,cm\,J^{-1}$ [425], but in this case the grating is efficiently erased when readout.

The kinetics of the photorefraction in the two-color regime always differs from that of a one-color recording due to a decreasing of the photoinduced optical absorption when readout. In the case of the two-color recording in $LiNbO_3$:Fe:Mn the recording kinetics may be satisfactory approximated by a single-exponential function analog to (3.18), whereas an optical erasure (Fig. 5.5) is described by a two exponentials

$$\sqrt{\eta(t)} = A \exp\frac{-t}{\tau_{e1}} + B \exp\frac{-t}{\tau_{e2}}. \qquad (5.5)$$

In spite of a relatively low sensitivity, the discussed two-color recording in $LiNbO_3$:Fe:Mn has the advantage of a very low energy required for recording. The recording energy per 1% of the diffraction efficiency was $W_{\eta=1\%} = 3.5\,J\,cm^{-2}$ [453] which is comparable to $W_{\eta=1\%} = 1\,J\,cm^{-2}$ for a one-color recording with an Ar-ion laser in $LiNbO_3$:Fe. The practical potential of a two-color recording in $LiNbO_3$:Fe:Mn crystals was analyzed in [455], where the authors performed the hologram multiplexing using for recording an He–Ne-laser with simultaneous UV-pumping. The recording schedule was elaborated with the account for the specific of the erasure kinetics (5.5), and the authors succeeded in multiplexing 50 plane-wave holograms in $LiNbO_3$:0.075%Fe:0.01%Mn.

Several models were proposed to describe the two-center recording in $LiNbO_3$. The model [456], refined in [458] is applied for LiNbO3:Fe:Mn with account for experimental parameters. The model proposed by Liu and Liu [459] is devoted mainly to a discussion of the photochromic effect in doubly doped $LiNbO_3$ crystals on a whole. The model [456, 458] is similar to the two-center charge-transport model introduced by Jermann and Otten [98], see (3.38a)–(3.38e) and Fig. 3.7. It calculates the time evolution of the metastable Fe^{2+} concentration (or, which is the same, of the electron concentration on the trap Fe^{3+}) and the space-charge field when recording a grating by red light under simultaneous pumping by UV light. The set of equations reminds of the set (3.38a)–(3.38e) except for the fact that now the deeper (donor) level is Mn^{2+} and the shallower (trap) one is Fe^{3+}. An analytical expression is obtained for temporal evolution of the number of Fe^{2+} centers as a function of the population of Mn and Fe levels, the recombination coefficients γ of Fe^{3+} and Mn^{3+} and the absorption cross-sections S of Mn^{2+} and Fe^{2+} under UV and red light. The numerical calculations of $E_{sc}(t)$ required experimental data for γ of Mn^{3+} and of S in the UV for both species. The full current expression takes into account the photovoltaic current contributed from both dopants, so the missing Glass constants for Mn in the UV range was needed, too. The missing parameters were measured and reported [391]. The calculations of $E_{sc}(t)$ under recording and readout, made with certain assumptions, gave a satisfactory agreement with the experimental curves shown in Fig. 5.5. The important results of this work are estimations of the optimal concentration ratio $[Mn^{2+}]/[Fe^{3+}]$ providing the highest value for η, which agrees with results on the oxidizing/reducing treatment in these crystal. The optimal ratio for the recording and gating intensities is deduced to be $I_{rec}/I_g = 25$–30.

A search for other couples providing a UV-pumped two-color recording found several potential candidates – $LiNbO_3$:Cu:Ce [460–462], $LiNbO_3$:Mn:Ce [463] [464], and $LiNbO_3$:Fe:Tb [437]. In the first two cases, the deep levels Cu and Mn are the pumping donors and Ce is a shallow nonequilibrium pumped trap. Although a nonvolatile two-color recording was sometimes achieved in the above-listed crystals after appropriate oxygen/reduction treatment, however, all of them seem yet to stay behind $LiNbO_3$:Fe:Mn.

If commenting on the two-color recording from the fundamental viewpoint, the following remark may be done. For a two-center case the best parameters of recording are provided by the photovoltaic mechanism of recording, so a nonequilibrium populated level is desirably be related to a photovoltaic center. For example, in spite of a pronounced UV-induced coloration in $LiNbO_3$:Mn:Ce [464], the two-colour recording parameters are rather poor compared to $LiNbO_3$:Mn:Fe, mainly due to a much lower glass constant k_G of Ce in $LiNbO_3$:Ce [464]. However, for a two-color recording based on the photoexcitation of short-living intermediate Nb_{Li}^{4+} (or hypothetical O^-) the question of the recording mechanism is still open. The inspection of the literature data does not permit to conclude unambiguously the recording mechanism. The recording kinetics and the intensity dependencies of the parameters are too involved to separate the mechanisms. So, a fundamental question whether the intrinsic defects are photovoltaically active being nonequilibrium populated, is unclear yet.

5.2 Thermal Hologram Fixation

An alternative way of realization of a nondestructive readout is provided by methods of so-called hologram fixing. We shortly dwell on this problem, although it is not directly related to the main target of this monograph, the intrinsic defects, but is governed by the most prominent extrinsic defects, protons. The underlying principle of fixing is based on screening the recorded space-charge field $E_{sc}(z)$ by charge carriers having the sign opposite to those forming $E_{sc}(z)$. A complementary screening charge pattern provides a stability of $E_{sc}(z)$ under the read-out. As will be seen from the following material, a screening of the photorefractive space-charge fields by thermally activated charge carriers of different sign at elevated temperatures is a common effect for photorefractive media on a whole. In $LiNbO_3$ and some other oxides the fixation is owing to screening $E_{sc}(z)$ by thermally excited protons. Another possibility may be realized in the case of bipolar photoconductivity when a space-charge field $E_{sc}(z)$ created, e.g., by electron capturing is mutually screened by photoholes. For example, it was demonstrated at room temperature in $Bi_{12}TiO_{20}$ (BTO) [465,466]. As mentioned in Sect. 5.1, when recording grating in a gating regime in $LiNbO_3$:Mg with Mg concentration above the threshold, Winnacker et al. [88] found indirect evidences of formation of two mutually compensating gratings related obviously to a bipolar photoconductivity.

In ferroelectric photorefractive compounds like SBN or $BaTiO_3$ another possibility of hologram fixing was found, by means of applying after recording an external dc-field E_{ext} of the order of the coercive field E_c [467–473]. It may be regarded as a method relative to the thermal fixing, because the field fixing is assigned to a partial P_s reversal under combined external and internal fields ($E_{ext} \pm E_{sc}$), or, in other words, to screening $E_{sc}(z)$ by the bound charge of the depolarizing field on the boundaries of arising domains. This effect is due to the fact that in these crystals the coercive field $E_c \approx 10^3\,\mathrm{V\,cm^{-1}}$ is comparable to the space-charge field E_{sc} of the recorded grating. So far an electric fixing could not be performed in usual (congruent) $LiNbO_3$, as the coercive fields $E_c \approx 210$–$220\,\mathrm{kV\,cm^{-1}}$ are by orders of magnitude higher than the achieved values of E_{sc}. However, recently, the field hologram fixing was demonstrated in near-stoichiometric $LiNbO_3$:Fe crystals [474, 475] which became possible due to a significant lowering of the coercive field to $E_c \approx 21$–$25\,\mathrm{kV\,cm^{-1}}$ [476]. The number of publications on thermal fixing of photorefractive holograms is vast. The available data are summarized in recent reviews [477–479].

5.2.1 Basics and the Theoretical Approach of Thermal Fixing

The thermal fixation of holograms was discovered in several crystals like Fe, Mo-doped $Ba_2NaNb_5O_{15}$ [480], undoped $LiNbO_3$ [299, 481] and $LiNbO_3$:Fe [482] and observed later in a variety of similar and dissimilar materials (see Table 5.1). The procedure of the thermal fixation in $LiNbO_3$ and related

Table 5.1. Thermal fixation of holograms in materials other than $LiNbO_3$

Crystal	Doping	Comments on the principles of the method and mechanism[a]	Reference
$Ba_2NaNb_5O_{15}$	Fe, Mo	Same as in $LiNbO_3$	[480]
$LiTaO_3$	Fe	Same as in $LiNbO_3$	[307]
$Sr_{0.75}Ba_{0.25}Nb_2O_{12}$		Cooling from the para- to the ferroelectric phase, a domain mechanism	[496]
$BaTiO_3$		Same as in $Sr_{0.75}Ba_{0.25}Nb_2O_{12}$	[497]
$BaTiO_3$		Thermal treatment in ferroelectric phase	[498]
$KNbO_3$	Fe	Thermal treatment in ferroelectric phase	[499]
$La_3Ga_5SiO_{14}$	Pr		[500]
$KTa_xNb_{1-x}O_3$ (KTN)		Cooling from the para- to the ferroelectric phase, a domain mechanism	[501] [502]
KTN:Li			[503]
$Bi_{12}SiO_{20}$ (BSO)		Room temperature, complementary electron and hole charge patterns	[465] [466]
BSO		Heating and cooling, a screening by ions similar to $LiNbO_3$	[504] [505]
BSO		Heating and cooling, a screening by protons applying external field	[506]
$Bi_{12}TiO_{20}$ (BTO)		Heating with a simultaneously applied ac-field	[507]

[a]Comments correspond to explanations proposed by the authors

materials ($LiTaO_3$, $Ba_2NaNb_5O_{15}$) in general looks as follows. A holographic grating is recorded with the use of a two-beam scheme, an achieved diffraction efficiency being denoted as η_0. Then the crystal is heated up to temperatures in the range $120°C < T_{fix} < 200°C$ and cooled down to room temperature. In the very beginning of the Bragg readout at room temperature the initial value is $\eta \approx 0$, whereupon it is growing to a saturation value $\eta_{fix} \leqslant \eta_0$. Qualitatively similar results are obtained, if recording is performed at temperatures in the range $120°C < T < 200°C$; after cooling to room temperature the behavior of η under the readout is the same, as described above. For brevity, the two temperature procedures of fixing are referred to as *post-fixing* and *simultaneous* fixing, respectively. The behavior of η under the readout resembles qualitatively the chemical development of the usual photographic film, so is denoted by the same term. The development may be achieved under illumination by a non-Bragg beam and under noncoherent visible light. The authors of [483] proposed for multiplexed holograms a method of a local thermal fixation by means of heating an elementary grating by a focused CO_2-laser beam. A fixed pattern may be erased either thermally by heating the crystal up to $T > 300°C$, or optically by an uniform illumination at temperatures $T > 120°C$. The thermal fixation with slightly modified ranges for T_{fix} (120–200°C) is the same in $LiNbO_3$, $LiTaO_3$, and $Ba_2NaNb_5O_{15}$ [299, 480, 481, 484, 485].

A qualitative general explanation of the fixing process proposed in [299, 480, 481, 484, 485] and underlying the models [486–495], is that at elevated temperatures certain mobile ions move in the field E_{sc} within the recorded grating to screen E_{sc}. A formed ionic pattern $N_i(z)$ is a replica of the initial $E_{sc}(z)$. On cooling, this ionic pattern becomes frozen, hence in the beginning of the development no spatially modulated internal field exists because of screening effects. Homogenous illumination on developing excites and spatially redistributes captured electrons, thus partially releasing the frozen backbone $N_i(z)$. After attaining a steady state, it is a nonscreened part of $N_i(z)$ that dictates the properties (efficiency and temporal stability) of the developed hologram pattern.

The assumption of an ionic compensation mechanism was based on the following reasons: the activation energies of the thermal fixing and of the dark conductivity σ_d in $LiNbO_3$ are close (about 1 eV) and the value of a high-temperature σ_d estimated from the temperature dependence of the fixing was in good agreement with σ_d obtained by direct measurements [298]. As the high-temperature σ_d in $LiNbO_3$ is predominantly ionic [298], Staebler and Amodei [299] concluded that thermally-excited movable ions play an important role. In the temperature range 80–600°C this ionic conductivity is owing to the hydrogen transport [109, 303]. Additionally, in the references [299, 480, 481, 484, 485], the fixing and developing processes were characterized on a whole, particularly, the lowest temperature for fixing in $LiNbO_3$ at $T \approx 100°C$ and the temperature of the thermal erasure of fixed gratings $T \approx 300°C$ were established. Interestingly, that these regularities and the qualitative explanation of the thermal fixing published in the pioneering works are valid until recently and have been supported by a multitude of subsequent experiments.

Some authors of a modified explanation of the thermal fixing [324,493,508] suggested that the formation of an ionic pattern $N_i(z)$ (which is the cornerstone of all models) at elevated temperatures results in a spatial redistribution of the electron donors Fe^{2+} and traps Fe^{3+} which is frozen on cooling. So, by this consideration, the origin of a nonerasable grating is related to a permanent spatially-modulated photovoltaic current, supported by a pattern of redistributed Fe^{2+}/Fe^{3+} ions.

The whole scenario of the thermal fixing may be separated into several stages illustrated in Fig. 5.6. The initial efficiency η_0 of the grating achieved during recording (phase 0) corresponds to the space-charge field $E_1(0)$. In the phase I (heating) E_{sc} of the recorded electronic grating causes the transport of the thermally excited ions. This leads to a compensated grating, which is represented by the net space-charge field amplitude $E_1(1)$; at a limiting complete compensation of the electronic charge by ions, $E_1(1) = 0$. Simultaneously with the screening process a partial decrease of the initial $E_1(0)$ occurs in the phase I because of a thermal excitation of captured electrons. In phase II, characterizing the storage time of the frozen pattern in dark, the electronic grating is slowly decaying due to an electron thermal detrapping, whereas the ionic compensating pattern adiabatically follows this decay. Under a readout (development) in the phase III the light causes a partial redistribution of trapped electrons due to the photoexcitation, culminating in a quasi-stable field $E_1(2)$.

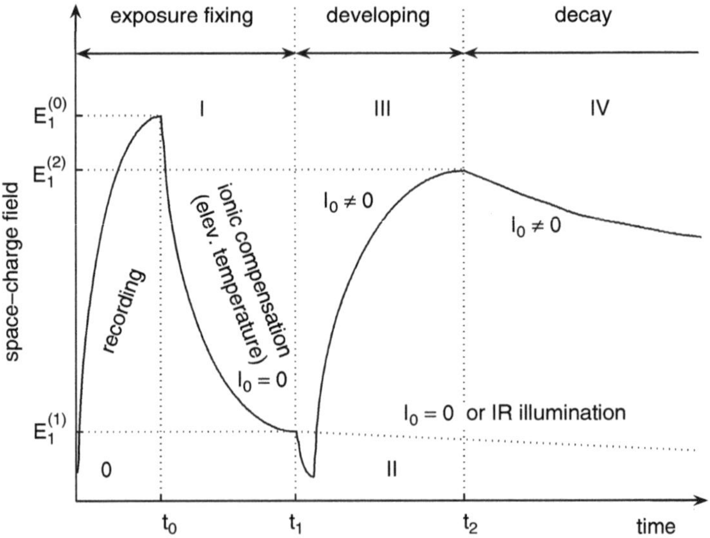

Fig. 5.6. Typical life history of a thermally fixed hologram for the case of postfixing. The roman literals are phase numbers, redrawn from [490]

The latter dictates the efficiency $\eta_{\rm fix}$ of the fixed hologram. Because of possible ionic transport, the field $E_1(2)$ may decay under illumination (phase IV).

There are three crucial parameters for practical applications of the thermal fixing: (1) the ratio of $\eta_{\rm fix}/\eta_0$ depending on $E_1^{(2)}$; (2) the storage time of the developed pattern in the dark; (3) the stability of the developed grating under continuous illumination, or, what is the same, under multiple cycles of the Bragg readout. In the framework of the qualitative model formulated above, one may a priori deduce an impact of some factors on these parameters. A necessary condition for a successful completion of phase I is that the screening of the space-charge field by movable ions occurs faster than the thermal excitation of the captured electrons, otherwise $E_1^{(0)}$ would disappear before being screened. This requirement gives an upper limit for the fixing temperatures $T_{\rm fix} \leqslant 200°{\rm C}$ in $LiNbO_3$, because at higher temperatures the thermal excitation of traps competes with a migration of ions [509]. In reduced $LiNbO_3$ crystals a temperature of de-trapping is much lower [296]. The low limit of $T_{\rm fix}$ is dictated by the requirement of a reasonably high mobility of the screening ions, for $LiNbO_3$:Fe it is $T_{\rm fix} \geqslant 120°{\rm C}$. Another requirement of the fixation is a reasonable concentration $N_{\rm i}$ of screening ions, which, on the other hand, should not be too large, because with increasing $N_{\rm i}$ the ionic conductivity $\sigma_{\rm i}$ would increase thus resulting in a decreased storage time. For practical applications the procedure of a simultaneous fixing gives higher $\eta_{\rm fix}$ than the post-fixing, because in the latter case a loss in η_0 occurs during heating. At the same time, recording at elevated temperatures provides additional experimental difficulties.

All the models developed are based on the qualitative picture described above. The first attempt of a model [492] assuming $E_{\rm pv} \gg E_{\rm dif}$ and a spatially uniform $E_{\rm pv}$ neglected the thermal decay of the space-charge, as well as the next, more rigorous charge transport model [486]. The further progress was achieved in the refined model [487] taking into account the thermal decay of the recorded pattern because of the electron thermal detrapping. Jeganathan and Hesselink proposed a description of the development for a nonphotovoltaic medium [510]. Later advanced models were developed [489–491, 494, 495, 511]. The most detailed description of the thermal fixing with account for all involved contributions, particularly concerning the recording at high temperatures one may find in [488] and the review [479]. The models start from a coupled set of material equations describing the one-dimensional electron transport and grating formation (5.6a–f) [281, 282] and incorporating additionally the continuity equation for ions and the ionic charge in the Poisson equation. The set of equations takes now the form

$$\frac{\partial n_{\rm e}}{\partial t} = \frac{\partial N_{\rm D}^{+}}{\partial t} - \frac{1}{e}\frac{\partial j_{\rm e}}{\partial z}, \tag{5.6a}$$

$$\frac{\partial N_{\rm D}^{+}}{\partial t} = \left(s_{\rm ph} I + s_{\rm T}\right)\left(N_{\rm D} - N_{\rm D}^{+}\right) - \gamma n_{\rm e} N_{\rm D}^{+}, \tag{5.6b}$$

$$\frac{\partial H}{\partial t} = -\frac{1}{e}\frac{\partial j_{\mathrm{h}}}{\partial z}, \tag{5.6c}$$

$$\epsilon\epsilon_0\frac{\partial E}{\partial z} = e\left(N_{\mathrm{D}}^{+} - N_{\mathrm{A}} + H - H_0 - n_{\mathrm{e}}\right), \tag{5.6d}$$

$$j_{\mathrm{e}} = e\mu_{\mathrm{e}}n_{\mathrm{e}}E + eD_{\mathrm{n}}\frac{\partial n_{\mathrm{e}}}{\partial z} + k_{\mathrm{G}}I\left(N_{\mathrm{D}} - N_{\mathrm{D}}^{+}\right)s_{\mathrm{ph}}, \tag{5.6e}$$

$$j_{\mathrm{h}} = e\mu_{\mathrm{h}}HE - eD_{\mathrm{p}}\frac{\partial H}{\partial z}. \tag{5.6f}$$

Here the mobile ions were assumed to be protons. The meaning of the variables is as follows: the dynamic variables n_{e}, N_{D}^{+} and H are the densities of free electrons, ionized donors and protons; N_{D} is the donor density and when applied to $LiNbO_3$:Fe, $(N_{\mathrm{D}} - N_{\mathrm{D}}^{+})$ and N_{D}^{+} are the densities of Fe^{2+} and Fe^{3+}, respectively; N_{A} is the density of a compensating passive acceptor, s_{ph} and s_{T} are the photoionization cross-section and the thermal ionization probability of the donor; γ is the recombination coefficient and I the averaged light intensity. The j_{e} and j_{h} describe the electron and proton current densities, respectively. H_0 is the average (initial) proton density, D_{n}, D_{p} and μ_{e}, μ_{h} are diffusion coefficients and mobilities, respectively, of electrons and protons; related by the Einstein relation $D = \mu k_{\mathrm{B}}T/e$ with k_{B} as the Boltzmann constant. In the expression for the electron current density (5.6e) the second and third terms are the diffusion and photovoltaic currents, respectively, E is the space-charge field. All the models for thermal fixing are based on the adiabatic approximation assuming a short free electron lifetime as compared to the relaxation times of the variables; additionally, small free electron density compared to the filled and empty trap densities is assumed.

Now following Yariv et al. [489,490], we briefly outline the analytic expressions characterizing the stages shown in Fig. 5.6. The deciding stage is phase I, where a compensation of the electronic space-charge by ionic transport occurs which governs the value of η_{fix}. If the screening of the electronic grating by movable ions is incomplete, a solution to the set (5.6) predicts the following expression for a residual steady-state field $E_1^{(1)}$ remaining in the end of phase I

$$E_1^{(1)} = \frac{D_{\mathrm{p}}K^2 + \mathrm{i}K\mu_{\mathrm{h}}E_0}{\omega_{\mathrm{h}} + D_{\mathrm{p}}K^2 + \mathrm{i}\mu_{\mathrm{h}}KE_0}, \tag{5.7}$$

where $\omega_{\mathrm{h}} = e\mu_{\mathrm{h}}H/\epsilon$ is the ionic dielectric relaxation frequency. The highest degree of compensation corresponds to $E_1^{(1)} \to 0$ (limiting complete compensation) which requires

$$H_0 \gg \frac{\epsilon k_{\mathrm{B}}K^2T}{e^2}. \tag{5.8}$$

This gives a lower limit for the ion concentration. A lower limit means that a further increasing of H_0 does not improve the fixing efficiency, but decreases the storage time because of an increasing ionic conductivity. Estimates of the proton density in as-grown $LiNbO_3$ show that it is always large enough to provide a complete compensation of the recorded gratings. The behavior of phase II, i.e., a decay in darkness of complementary screening ionic and electronic space-charges before development at room temperature depends mainly on a decay of the electronic charge, because the ionic space-charge is stable. As known, the E_{sc} decay is governed by the Maxwell relaxation time τ_M, that means by the dark electronic conductivity σ_d. However, in the presence of a screening ionic charge the decay is described by a longer time

$$\frac{1}{\tau_{dec}} = \frac{1}{\tau_M}\left(\frac{D_p K^2}{\omega_h + D_p K^2} + K^2 l_d^2\right), \tag{5.9}$$

where $l_d = (\epsilon\epsilon_0 k_B T/e^2 N_D^+)^{1/2}$ is the Debye screening length. Physically (5.9) means that at elevated temperatures the decay of an electronic space-charge in the dark is slowed down by screening, so requires a large enough ionic conductivity ($\sigma_h \gg \sigma_d$).

Another deciding stage is phase III at room temperature which describes both the development of a latent grating pattern under uniform illumination and its behavior (stability) under readout, that is under conditions $\sigma_{ph} \gg \sigma_h$. Electrons are uniformly photoexcited, move (drift and diffuse) both in the electric field of a present spatially-distributed ionic charge and under a uniform photovoltaic field E_{pv}, afterwards they are retrapped. So, even under a uniform illumination the equilibrium distribution of trapped electrons is not uniform because of a partial screening of $H(z)$. This screening is undesirable, because it decreases η_{fix} in comparison to the value expected from $H(z)$. The solutions to the maximum developed field $E_1^{(2)}$ (neglecting the ionic transport) and to the decay rate $1/\tau_1$ in phase III under short-circuit conditions are

$$E_1^{(2)} = \frac{ie}{\epsilon K} n_{i1}(t1) \frac{E_d - iE_{pv}\frac{N_a}{N_d}}{E_d + E_q - iE_{pv}\frac{N_a}{N_d}}, \tag{5.10}$$

$$\omega_1 = \frac{1}{\tau_1} = \frac{1}{\tau_M}\left(1 + K^2 l_d^2 - i\frac{E_{pv}}{E_q}\frac{N_a}{N_d}\right), \tag{5.11}$$

where

$$E_q = \frac{eN_a\left(1 - \frac{N_a}{N_d}\right)}{\epsilon\epsilon_0 K} \tag{5.12}$$

is a limiting space-charge field provided by a given concentrations of electron traps (see Sect. 3.2.2). Physically, (5.10) means that the value of η_{fix} controlled by $E_1^{(2)}$ may approximate η_0, if in the end of phase III the ionic pattern $H(z)$ is not screened or screened very slightly by trapped electrons. Accordingly to (5.10) this *naked* ionic framework may occur, when

$$\left|E_d - i\frac{N_a}{N_d}E_{pv}\right| \gg E_q. \tag{5.13}$$

The latter condition shows the ways to enhance $\eta_{\rm fix}$. The expression (5.13) means that the screening may be suppressed either by a strong diffusion field $E_{\rm dif} \gg E_{\rm q}$, or by a strong photovoltaic field $E_{\rm pv} \gg E_{\rm q}$. In as-grown $LiNbO_3$:Fe, where $[Fe^{2+}]/[Fe^{3+}] \approx 0.2$ and $E_{\rm q} \approx e[Fe^{2+}]/\epsilon K \approx 5\times10^5\,{\rm Vcm}^{-1}$, so the fulfillment of the inequality $E_{\rm dif} \gg E_{\rm q}$ would require nonrealistic hologram periods $\Lambda \ll l_{\rm d} \approx 0.1$–$0.2\,\mu{\rm m}$. The inequality $E_{\rm pv} \gg E_{\rm q}$ requires a strong enhancement of $E_{\rm pv}$. As $j_{\rm pv} \approx [Fe^{2+}]$ and $\sigma_{\rm ph} \approx [Fe^{2+}]/[Fe^{3+}]$ (see Sect. 3.2.2), we find $E_{\rm pv} \approx [Fe^{3+}]$ and this field may be essentially increased, for example, by a strong oxidation of $LiNbO_3$:Fe transforming $Fe^{2+} \rightarrow Fe^{3+}$ [116]. Therefore, one may expect higher values of $\eta_{\rm fix}$ in oxidized crystals, what is actually observed. In this relation we should remind on the results of R. Orlowski and E. Krätzig [317, 318] discussed in Sects. 4.2.2 and 3.2.2, accordingly to which an extremely strong oxidation may affect fundamentally the charge transport scheme. So, it may happen that the whole consideration of the thermal fixing would not be valid for a strongly oxidized $LiNbO_3$:Fe.

As mentioned earlier, (5.11) and (5.12) were derived for short-circuit conditions [489, 490]. Under open-circuit conditions ($j = 0$, $E_0 \neq 0$) the photovoltaic field acts analogously to the applied external field and a consideration comes to similar qualitative conclusions; except a lower diffraction efficiency as compare to the short-circuit ones [512]. Short-circuit conditions may be achieved, for example, by electroding the crystal edges side [513]. Taking into account the thermal excitation of electrons, thus an involvement of two types of species leads to existence in fixing phases of two kinetics conventionally referred to as fast and slow; for references see [479, 488, 490] and Fig. 5.7.

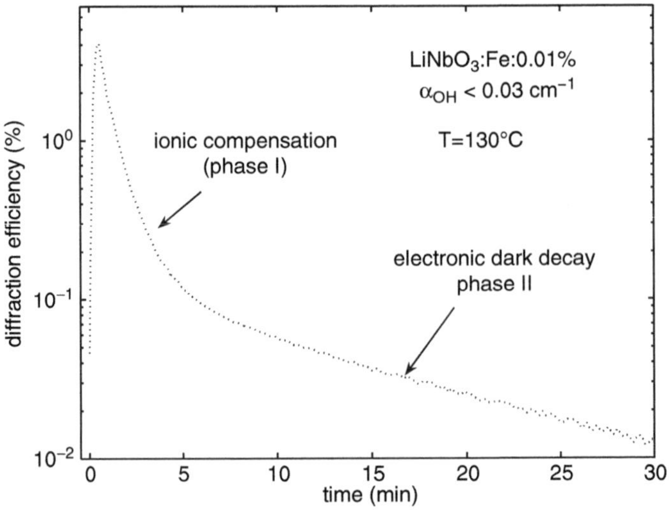

Fig. 5.7. An example of the temporal dependence of the diffraction efficiency at elevated temperature, redrawn from [490]

Note that two kinetics in grating recording at elevated temperatures were detected even at the earliest stage of studies in photorefraction [492]. In the model by Sturman et al. [488] taking as dynamic variables the space-charge field E, n_e, N_D^+ and H, a convenient presentation of a photorefractive two-stage kinetics was proposed. Accordingly to this model, the fast process is the dielectric relaxation depending on the conductivity, whereas the slow kinetics is caused by an interaction between charge carriers of opposite signs (photo- or thermally excited electrons and thermally activated positive ions). The slow kinetics may be imagined as a *slowing-down* of the Maxwell dielectric relaxation because of screening. Correspondingly, the recording (with light pattern) and the decay (without light pattern) of a space-charge field amplitude E_K are presented in a pictorial view

$$E_K(t) = E_f\left(1 - \exp\left(-\omega_f t\right)\right) + E_s\left(1 - \exp\left(-\omega_s t\right)\right), \qquad (5.14)$$

$$E_K(t) = E_f \exp\left(-\omega_f t\right) + E_s \exp\left(-\omega_s t\right), \qquad (5.15)$$

where ω_f and ω_s are the fast and slow relaxation rates, respectively. The recording presentation (5.14) means that the value $E_K = E_f$ is achieved at the end of the fast stage, and is then decreasing during the subsequent slow stage due to a slow build-up of the protonic grating. Obviously, at low-temperature recording, $\omega_s \to 0$, and expressions (5.14) and (5.15) are transformed to the usual one for the photorefractive recording. Presentation (5.15) describes the decay in the dark (phase II) or under IR illumination (phase IV); the result of the slow decay stage is the disappearance of any inhomogeneity of the charge distribution. The fast rate depending on the conductivity involves the protonic and electronic components

$$\omega_f = \frac{1}{\tau_{di}} = \omega_h + \omega_e, \qquad (5.16)$$

where $\omega_h = \sigma_h/\epsilon\epsilon_0 = H_0 e\mu_h/\epsilon\epsilon_0$, $\omega_e = \sigma_e/\epsilon\epsilon_0 = n_0\mu_e/\epsilon\epsilon_0$. In the dark and under illumination, σ_e is equal to σ_d and σ_{ph}, respectively. After certain re-arrangements, Sturman et al. [488] came to the following expressions for the fast and slow rates in $LiNbO_3$:Fe in the temperature range from 20°C to 300°C (with account for $\sigma_h \gg \sigma_d$ for as-grown $LiNbO_3$:Fe [300]. In darkness

$$\omega_f = \omega_h, \qquad (5.17)$$

$$\omega_s^e = \omega_e^d K^2 l_d^2 \left(1 + \frac{N_t}{H_0}\right), \qquad (5.18)$$

where $l_d = (\epsilon\epsilon_0 k_B T/e^2 N_t)^{1/2}$ is the appropriate Debye screening length and $N_t = (N_a N_d)/N$ is the corresponding effective trap concentration. Under illumination, $I \neq 0$, for the low-temperature limit ($T \leqslant 80$°C for $LiNbO_3$:Fe) the fast stage is controlled by the photoconductivity

$$\omega_f = \omega_e^{ph} \qquad (5.19)$$

and the slow stage neglecting the contribution from the photovoltaic field may be presented as

$$\omega_{\rm s}^{h} = \omega_{\rm h} K^2 l_{\rm d}^2 (1 + \frac{N_{\rm t}}{H_0}). \tag{5.20}$$

An important conclusion from these expressions is that the slow decay rate in the whole temperature range both in darkness (5.18) and under illumination (5.20) is proportional to the spatial frequency of the grating. The larger the grating period, the longer is the lifetime of the fixed grating. This fundamental relation to the diffusion mechanism was deduced in all models, e.g., (4.9) in [489,490], starting from the pioneering works [482,485]. The second important conclusion deduced from (5.18) and (5.20) is that a slow decay rate is strongly controlled by the ratio $H_0/N_{\rm t}$. The slowest decay may be achieved whenever decreasing this ratio. So, in $LiNbO_3$:Fe, where $N_{\rm t} = [Fe^{2+}]$ holds, the highest lifetime of a fixed grating would require $H_0 \ll [Fe^{2+}]$.

In the model developed in [487, 514] the emphasis was put on the role of the photovoltaic field $E_{\rm pv}$. The analysis has shown that the specificity of a photovoltaic crystal is a shift of the developed electronic grating with regard to the fixed ionic grating, which reminds of the effect of applying an external field. The phaseshift between the gratings is given by the expression

$$\varphi = \arctan\left(\frac{E_{\rm pv}^{*}}{(E_{\rm q} + E_{\rm D})}\right), \tag{5.21}$$

where $E_{\rm pv}^{*} = E_{\rm pv}[N_{\rm D}^{+}/(N_{\rm D}^{+} + N)]$. This conclusion was supported by the data of microphotometric investigations of the developing the holograms recorded in $LiNbO_3$:Cu at 180°C [471]. When approaching η to a saturation value, the grating phase is shifted almost by 2π, which gave an evidence for the mechanism of a mutual screening of two complementary gratings.

Later we exemplify experimental studies of the thermal fixing in $LiNbO_3$ crystals. It should be noted that in spite of a more than 30-years-old story of these experiments, the mechanism of fixing seems to be not fully understood. Particularly, there is no consensus regarding the origin of complex kinetics of decaying gratings at elevated temperatures. As mentioned earlier, Sturman et al. [488] attributed the slow and fast decay components to the dielectric relaxation and an interaction between charge carriers of opposite signs, whereas Yariv et al. [489,490] assign these kinetics to the fast ionic compensation and the much slower thermal decay of the electronic grating screened by the ions. The most uncertain question is the storage time at ambient conditions which cannot be obtained directly and is estimated either by direct extrapolating $\tau_{\rm s}(T)$ or by calculations from $\sigma_{\rm h}(T)$.

Note, that the thermal fixing in principle may be caused not only by a screening of the space-charge field $E_{\rm sc}$ by complementary charge carriers, but as well by a partial switching of $P_{\rm s}$ under $E_{\rm sc}$ as it takes place in SBN or $BaTiO_3$. Actually, the coercive field $E_{\rm c}$ in $LiNbO_3$ strongly falls down at $T > 120$°C, where a partial switching under external fields of about

$10^4\,\mathrm{V\,cm^{-1}}$ was detected in $LiNbO_3$:Fe [515]. However, so far no experimental evidences of a contribution from P_s switching to the thermal fixing have been obtained, although such a possibility was repeatedly discussed [471,492].

5.2.2 Experimental Studies in the Thermal Fixing

Summarizing the above qualitative consideration from the outlined models, one may conclude that the optimization of the parameters of fixed gratings, that is reaching longest storage times τ_s in darkness, slowest decay rates $1/\tau_{dec}$ under illumination and highest η_{fix} requires certain compromise between the concentrations of ions and electron traps. Their ratio should be rather low to give a high η_{fix} provided by a low degree of screening the ionic frame by electronic carriers, and at the same time large enough to provide a partial screening of the ionic frame to slow down its decay. Of course, slowing down the decay may be achieved by decreasing ionic conductivity, however, in this case the concentration of ions might become too deficient to provide a complete screening of the initial space-charge field. In other words, benefiting from τ_s, one loses partially η_{fix} and vice versa. In $LiNbO_3$:Fe this compromise may be achieved by picking up the ratio between concentrations of Fe^{3+}-like electron traps and protons.

Detailed studies of the grating behavior at elevated temperatures were performed [516–518]. It was found that a decay of holograms in the range from 50°C to 165°C in $LiNbO_3$:Fe may be fitted to a sum of several exponents [516]; the decay times show Arrhenius behavior $1/\tau = 1/\tau_0 \exp(-E_a/k_BT)$. A nonmonoexponential kinetics of the decay of η at elevated temperature was analyzed in several works [490, 517]. Taking into account the thermal excitation of electrons proposed in [487], Yariv et al. [490] interpreted the fast process as the compensation of the space-charge by thermo-excited ions and the slow one as the decay of the electronic space-charge because of the electron detrapping.

Arizmendi et al. [518] compared the decay kinetics in darkness and under illumination (phases II and IV) at elevated temperatures. In accordance with (5.18, 5.20), these kinetics are different. In the range from 90°C to 120°C a dark decay of η is due to the screening process, the ionic skeleton being almost preserved. This was proved by the argument that the developing procedure after η has come to zero could be performed repeatedly. In contrast, after a decrease of η under illumination the grating was not developed. Therefore, the light-induced decay is caused by a thermal erasure of the *released* ionic skeleton, whose thermal stability is no more supported by the mutually screening electronic space-charge. A decay under illumination may be fitted to a single exponent with $E_a = 1.03\,\mathrm{eV}$. Starting from this value one may estimate the decay rate of a fixed grating under readout at room temperature.

Studies of the decay rate vs. the grating period were performed at elevated temperatures in [490,517–519]. The decay rate is inversely proportional to Λ^2 consistent with the model of Carrascosa and Aguĺló-López [487] and later

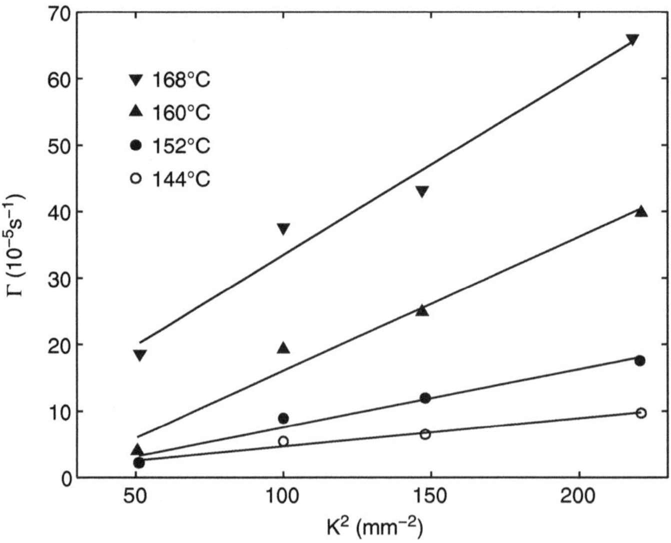

Fig. 5.8. Temperature dependence of the decay rate on the square of the spatial grating frequency in Fe-doped $LiNbO_3$, redrawn from [517]

models [488,490], for more details see (5.9) and (5.20) and Fig. 5.8. An example gives a crystal with a dependence $\tau_s \propto 5.7\Lambda^2$ [519]. The proportionality of storage times to Λ^2 is provided mainly by the slow stage of the decay (i.e., electron thermoexcitation), whereas the ionic decay, as expected, is practically independent of Λ. Consequently, the dependence $\tau_s \propto \Lambda^2$ is more pronounced in crystals with a higher Fe concentration (about 0.15%), in which a contribution from electronic charge is stronger.

Experimental studies of the developing process were performed by several authors [490, 511, 514, 520, 521]. Yariv et al. [490] and de Miguel et al. [514] investigated the dependence of η_{fix} on E_{pv} in $LiNbO_3$:Fe, particularly a predicted effect of enhancing E_{pv} by means of oxidation (transforming $Fe^{2+} \rightarrow Fe^{3+}$). To provide the slowest decay by means of reducing the ionic (protonic) conductivity these experiments were performed in dehydrated $LiNbO_3$:Fe [490]. The gratings were recorded at enhanced temperatures with a subsequent developing at 50°C. The value of η_{fix} in oxidized $LiNbO_3$:0.01%Fe was almost by an order of magnitude higher than in oxidized $LiNbO_3$:0.05%Fe ($\eta_{\text{fix}} \approx 0.8\eta_0$, and $\eta_{\text{fix}} \approx 0.1\eta_0$, respectively). Conversely, the decay time on developing in the low-doped crystal was by about an order of magnitude shorter than in the highly-doped one (about 2 and 20 h, respectively, at 90°C). The activation energy of protons of $E_a \approx 1.2\,\text{eV}$ predicted storage times of this *naked* backbone at room temperature in the range of 50–70 d only. To increase τ_{dec} and conserve simultaneously a reasonable value of η_{fix}, Yariv et al. [490] attempted to reduce the ionic conductivity. They tried either a very strong oxidation, which led to a significant reduction of the H content, or a preparation of a stoichiometric sample using VTE technique [30], which resulted in a decrease of the Li vacancy concentration. Both techniques caused an increase of the activation energy of the ionic conductivity to $E_a \approx 1.4\,\mathrm{eV}$ compared to $E_a \approx 1.2\,\mathrm{eV}$ for the protonic conductivity. This permitted the authors to speculate on an enhancement of τ_s at room temperature up to 2 years. Qualitatively similar results of oxidizing on the value of η_{fix} in $LiNbO_3$:Fe were obtained [514]. Additionally, the same authors [514] demonstrated an increase of η_{fix} with K which is specific for the photovoltaic crystals; this is especially pronounced in oxidized crystals due to a strong contribution from E_{pv} in this case.

An interesting example of studies on the microscopic origin of fixing with an emphasis on the role of the photovoltaic mechanism was demonstrated in [324]. The authors investigated a redistribution of Fe^{2+} and Cu^+ ions in $LiNbO_3$:Fe and $LiNbO_3$:Cu, resulting from illuminating the crystal by a narrow light strip normal to the polar axis during extremely long times (from 50 to more than 300 h) at 160°C. After developing at room temperature a scanning of the optical absorption along the polar axis revealed within the photorefractive area a nonuniform distribution of Fe^{2+} in $LiNbO_3$:Fe and Cu^+ in $LiNbO_3$:Cu, namely a moving of these ions towards +z borders of the damaged region. As in this case $E_D = 0$, so the observed spatial redistribution of trapped electrons is caused by the photovoltaic field. In the sequential work [305] a development without light in $LiNbO_3$:Fe was demonstrated by means of applying an external field of the order of several $\mathrm{kV\,cm^{-1}}$ to a crystal with a thermally fixed hologram. This effect was interpreted as a result of a nonuniformly distributed electronic dark conductivity $\sigma_d(z)$ following the redistribution of $[Fe^{2+}]$ proved in [324].

Of course, the above-listed set of experimental works does not claim to be exhaustive. We have chosen the publications demonstrating the most typical results in studying this problem and supporting the main conclusions of the models. It should be emphasized again that the situation with the storage times is uncertain. In Table 5.2, an attempt was made to summarize the experimental data on η_{fix} and experimental and calculated storage times τ_s of fixed holograms.

From the very beginning of studies on thermal fixing in $LiNbO_3$, hydrogen (in the form of protons or hydroxils) was assumed to be responsible for the screening of the electronic space-charge. The main argument was a closeness of the activation energies $E_a \approx 1\,\mathrm{eV}$ for the thermal fixing and the protonic conductivity. At the same time, as this value of E_a is not unique, alternative impurity ions, e.g., Si ions accidentally present in the crystal [523] were attempted to explain the thermal fixing. The suggestion of Si was ruled out by experiments in $LiNbO_3$ enriched with Si [492].

There is a lot of experiments aimed to prove the proton mechanism of the fixation by means of comparing the transport characteristics (E_a and D) obtained from the measurements of the conductivity to those calculated from the data on fixing. However, the ultimate conclusion was drawn after experiments

Table 5.2. Numerical characteristics of gratings thermally fixed in $LiNbO_3$ crystals. η_0 is the initial diffraction efficiency of the recorded hologram, η_{fix} is the diffraction efficiency of the fixed and developed hologram, Λ is the grating period, RT means room temperature, N_H is the hydrogen concentration, and α_{OH} is the optical absorption coefficient at 2.87 µm. In the column Comments extrapolation means estimates by extrapolating experimental curves from elevated temperatures to room temperature, calculations mean evaluations without experimental data

Impurity	η_{fix}	τ_s experim.	τ_s estimate	T	Comment	Ref.
Pure	$0.5\eta_0$			RT		[480]
0.01%Fe	Some %	5 m		RT	Slow decay	[481]
			5–15 a	0°C		
0.01%Fe	$(0.1–0.01)\eta_0$			RT		[485]
0.001%Fe	$0.3\eta_0$					
0.01–0.02%Fe	5–35%	$\gg$2 a	10^5 a	RT	Recorded and fixed simultaneously 500 gratings at 160°C τ_s grows with Λ	[482]
0.05% Cu				RT	Improved by oxidation	[522]
Fe, Cu			15 a	RT	Extrapolation from 165°C	[324]
0.1%Fe		1 d		$\approx$ 150°C	τ_s grows with Λ	[517]
0.1%Fe			3.7 a		Extrapolation from 145–165°C for $\Lambda = 0.7$ µm	[518]
0.1%Fe	$0.8\eta_0$				Recording at 170°C	[521]
0.1%Fe	$0.8\eta_0$		5.7 a	RT	Extrapolation from 150°C; $\Lambda = 1$ µm $N_H = 10^{19} cm^{-3}$	[519]
0.1%Fe	$0.8\eta_0$		200 a	RT	$\Lambda = 2$ µm $N_H = 10^{17} cm^{-3}$ $N_H = 10^{17} cm^{-3}$	[519]
0.01–0.02%Fe			10 a	RT		[489]
0.01%Fe	$0.25\eta_0$	1 h		90°C	$\alpha_{OH} = 0.035\,cm^{-1}$	[490]
0.05%Fe	$\ll\eta_0$	>10 h		90°C	$\alpha_{OH} = 0.01\,cm^{-1}$	
0.05%Fe			50–70 d	RT	Calc. from protonic conductivity	
			2 a		H-depleted (calc.); new type of ionic conductivity	
0.1%Fe	$0.7\eta_0$			RT		[514]

in $LiNbO_3$:Fe [324, 524] and $LiNbO_3$:Cu [324] directly proving the participation of hydrogen in this process. Taking into account the basic importance of these experiments, they are worthy of a more detailed description. The photorefraction was induced by a narrow light strip perpendicular to the polar c-axis. Vormann et al. [524] recorded the photorefraction at room temperature, whereupon the standard procedure of thermal fixing and developing was performed. Buse et al. [324], as mentioned above, recorded the photorefraction at elevated temperatures during extremely long times. After developing, the distribution of the optical absorption at $\lambda = 2.87\,\mu m$ was scanned along the z-axis. Recall that this band characterizes the hydrogen concentration, for details see Sect. 2.5.3. The results of both works were qualitatively similar, namely the distribution of the band intensity was nonuniform (Fig. 5.9) and showed anomalies (a maximum and a minimum) at the $+z$ and $-z$ borders, respectively, of the photorefractive region. Amplitudes of these anomalies increased with increasing concentration of hydrogen in the crystals [524]. Buse et al. [324] found a more pronounced effect, because the conditions of the

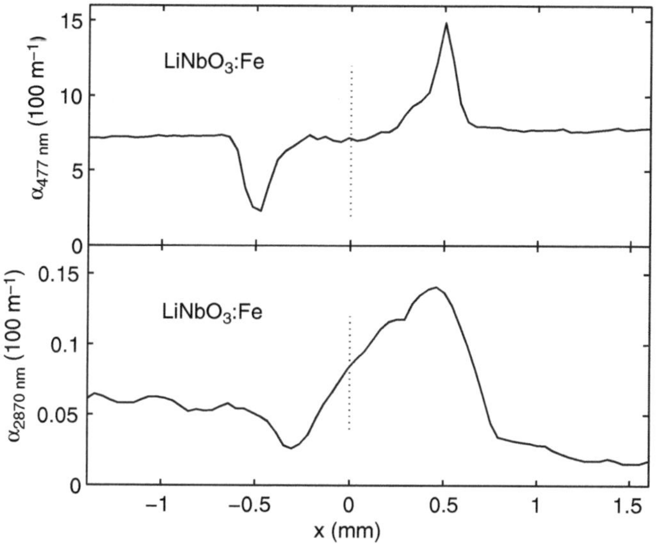

Fig. 5.9. Spatial distributions of the optical absorption coefficients at 477 and 2,870 nm along the polar z-axis after a prolonged recording of the photorefraction by a focused laser beam at 165°C in Fe-doped $LiNbO_3$. The origin at 0.0 corresponds to the center of the recording Gaussian beam. An increase of $\alpha_{477\,nm}$ (*upper curve*) and $\alpha_{2,870\,nm}$ (*lower curve*) at the *right* ($+z$) border of the photorefractive strip indicates the corresponding increases of the Fe^{2+} and H concentrations, respectively, in this region. The *left* ($-z$) border is Fe^{2+}-deficient, i.e., enriched by Fe^{3+}. The curves were obtained for ordinary polarization of the scanning beams. Redrawn with data taken from [324]

high-temperature recording provided a more efficient redistribution of hydrogen within the photorefractive region. From Fig. 5.9 one may see that the H concentration is increased in the region, where the concentration of Fe^{2+} (and of Cu^+ in the case of $LiNbO_3$:Cu) is increased, or, in other words in the regions of electron accumulation. This means that exactly protons are those mobile ions which screen the electronic space-charge. The results of the neutron diffraction on a thermally fixed grating supported this conclusion [525]. Since those experiments the protons are generally accepted as the main ions responsible for fixing in $LiNbO_3$. The conclusion of a decisive role of protons in the fixing process was supported as well by investigations of characteristics of fixed holograms in $LiNbO_3$ crystals with different contents of hydrogen [314, 324, 490]. As shown in Sect. 2.5.3 the H concentration estimated from the absorption band at 2.87μm is in as-grown or slightly oxidized $LiNbO_3$ crystals in the range of (0.2–5) 10^{-19}, so always satisfies the requirement (5.8) $N \gg 5 \times 10^{16}\,\mathrm{cm}^{-3}$ for the low limit of the ionic concentration providing the complete screening of E_{sc}.

There are evidences indicating that some ions other than protons may also provide a thermal fixing of holograms. For example, in strongly oxidized (hydrogen-free) $LiNbO_3$:Fe crystals despite the absence of hydrogen and an increase of E_a up to 1.4 eV, a thermal fixing of holograms occurs nevertheless [490].

It is worth noting that the range of the thermal fixing between 120°C and 150°C is peculiar for $LiNbO_3$, because various properties of these crystals exhibit certain anomalies within it. We may mention the thermal expansion coefficient, the dielectric permittivity and pyrocoefficient, some optical parameters, the dark conductivity, electron emission, the frequency of the Nb-nuclear quadrupole moment, the ratio of the lattice parameters a/c, and others [526–532]. These peculiarities impelled some authors to conclude about a phase transition in this temperature range, which might initiate an ionic conductivity on a whole.

5.2.3 Thermal Fixing of Holograms in Other Photorefractive Materials

We now mention very briefly the results on the thermal fixing of recorded gratings in crystals other than $LiNbO_3$. These are SBN [496], $KNbO_3$ [499, 533], $BaTiO_3$ [497, 498], KTN [501, 502], $Bi_{12}SiO_{20}$ (BSO) [505, 506] $Bi_{12}TiO_{20}$ (BTO), including applied ac-fields, [507], $La_3Ga_5SiO_{14}$:Pr (LGS) [500].

Note that in all these crystals the recording mechanism differs from that in $LiNbO_3$. In $BaTiO_3$, SBN, BSO, and BTO, it is purely diffusion, whereas in $KNbO_3$ it involves both photovoltaic and diffusion fields in a dependence of the impurity state [534]. At ambient conditions KTN is in the centrosymmetric paraelectric phase, so the photorefraction occurs via the quadratic electro-optic effect and the readout of a photorefractive grating requires applying an external bias field. An interesting example of the orientation dependence of the recording mechanism is observed in the piezoelectric crystal LGS, where the properties of the photovoltaic tensor dictate a dependence of the photovoltaic current j_{pv}, i.e., of the recording mechanism on the angle β between the light polarization vector and the polar axis x [535]. If $\beta = 0$, then $j_{pv} \neq 0$ and the hologram is recorded by the photovoltaic mechanism; for $\beta = 45$ we have $j_{pv} = 0$, so the recording proceeds by the diffusion mechanism.

Thermal fixing in these crystals has not been studied in detail due to their secondary role as storage media. The process of thermal fixing in them manifests itself as an unusual behavior of the diffraction efficiency $\eta(t)$ when recording at elevated temperatures ($T > 60$–$70°$C): the diffraction efficiency after coming to a saturation value η_{max}, starts to decrease and falls down to zero. When illuminating by a single Bragg beam at room temperature, a nonzero η appears increasing to a certain value $\eta_{fix} < \eta_{max}$. The fixing means that an optical erasure of this developed grating under the readout proceeds much slower than of the original grating recorded at room temperature. A principal distinction of all these materials from $LiNbO_3$ is that in all of them an optical erasure occurs anyhow, though relatively slow, whereas in $LiNbO_3$ a fixed grating is practically not optically erased at room temperature. The behavior of $\eta(t)$ at elevated temperatures is exemplified by $KNbO_3$, the caption to this Fig. 5.10 formulates the qualitative interpretation of the dynamics by the authors [499]. In this crystal $\eta_{fix} = (0.1$–$0.5)\eta_{max}$. A similar scenario was reported for KTN [502], $BaTiO_3$ [497], BSO [505], LGS [500]. Passing $\eta(t)$ over a maximum at elevated temperatures in all cases was attributed to the formation of a complementary grating, which screens the original (recorded) one. The existence of a complementary grating was sometimes proved experimentally. For example, in $KNbO_3$ simultaneous measurements of η and Γ during developing found a coexistence of two gratings (original and screening ones) mutually shifted by π [499]. In LGS the authors [500] found a dependence of the phase shift between the original and complementary gratings on the polarization vector of the off-Bragg developing light beam. This dependence was in accordance with the orientation dependence of the recording mechanism mentioned above.

In general, the interpretation of the kinetics of $\eta(t)$ at elevated temperatures presented in Fig. 5.10 as a manifestation of screening the photorefractive field E_{sc} by any charge carriers, does not raise doubts. However, the origin of the screening charges was determined in none of these crystals. The thermal fixing process in all crystals obeys the Arrhenius law with activation energies $E_a \approx 1$ eV in $KNbO_3$ [499], 1.44 eV in BSO [505], 0.89 eV in LGS [500]. Rather low E_a in LGS makes it possible to accomplish the fixing procedure at temperatures as low as 63°C. Several attempts were made to prove that fixing process is related to thermally excited ions rather than to thermally excited carriers of an *electronic nature* (electrons or holes). This attempt may be again exemplified by experiments in $KNbO_3$ [499], where in strongly reduced undoped crystals the photoconductivity σ_{ph} is electronic and E_{sc} is formed by captured photo-electrons, while in doped crystals $KNbO_3$:Fe and $KNbO_3$:Na:Li:Fe σ_{ph}

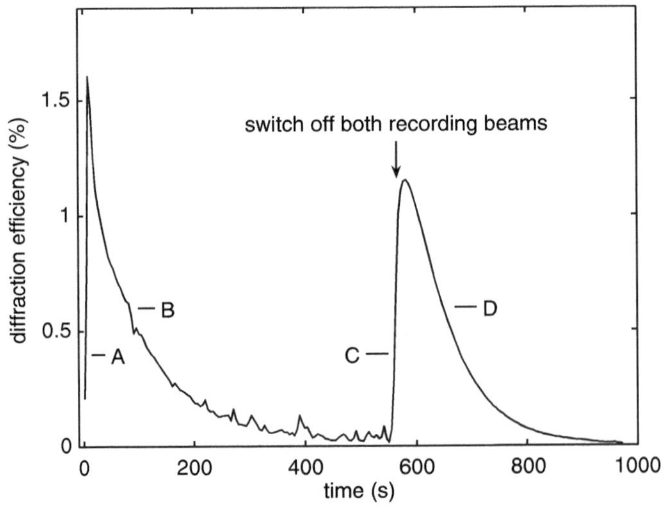

Fig. 5.10. Temporal dependencies of the diffraction efficiency during recording ($t < 550$ s) and after turning off the light ($t > 550$ s) in a $KNbO_3$:LiNaFe crystal at 95°C. The authors interpret the phases marked B, C and D as the compensation of the recorded grating by some screening (secondary) charges, the thermal decay of the photoinduced space-charge field and the thermal decay of the screening (secondary) grating, respectively. Redrawn with data taken from [499]

is of the p-type and E_{sc} is formed by captured photo-holes. Nevertheless, the activation energy of the thermal fixing in all these crystals is practically the same, which argued in favor of a common (ionic) origin of the screening carriers. As $E_a \approx 1\,\text{eV}$ for the thermal fixing in $KNbO_3$ is close to the activation energy of the protonic conductivity in oxides on a whole, so the thermally activated protons were suggested as a reason of fixing [499]. Interestingly that to account for fixing mechanisms in SBN [496], $BaTiO_3$ [497] and KTN [501] the authors engage the ferroelectric mechanism, because they perform fixing by means of cooling the crystals over the phase transition temperatures, thus involving possible domain redistribution under the space-charge field of the recorded grating.

Summarizing these isolated data, one can conclude that a screening of the photorefractive space-charge fields by thermally activated charge carriers (of different origin) at elevated temperatures is evidently a common effect for photorefractive media on a whole. However, only few cases of a real use of this effect for hologram fixing have been found so far. A principal difference of the thermal fixing characteristics in $LiNbO_3$ – like materials ($LiTaO_3$, doped $Ba_2NaNb_5O_{15}$) from those in other crystals is related obviously to the existence of the bulk photovoltaic effect.

5.2.4 Attempts of Practical Applications of Thermal Hologram Fixing in $LiNbO_3$

A first attempt of a practical application of thermal fixing has been performed in a pioneering work by Staebler et al. [482], where 500 holograms were recorded at 160°C in $LiNbO_3$:(0.01–0.02)%Fe and then developed at room temperature. The studies of applications of thermal fixation in the succeeding 30 years may be separated into two groups. First, thermal fixation is used in traditional applications of photorefractive crystals for constructing nonvolatile optical memory devices, which combine a high storage density with a fast random-access and high data transfer rate [287,332,511,536–546]. A more recent direction of applications is based on a high angular selectivity of recorded phase gratings, which may be used in devices requiring a narrow spectral bandwidth and a close spatial shift tolerance, like narrow-band filters, optical correlators and other related systems. We now briefly outline the available data on these two directions. For storage applications, multiplexing of many elementary holograms in the same photorefractive crystal is required. A high multiplexing capability in photorefractive media, particularly in $LiNbO_3$, was demonstrated; since 10 years the storage capability advanced from 500 [332,511,537,544] to 5,000 [538,541,542] and even 10,000–20,000 holograms [543, 545] with a reliable fidelity of a retrieval of an elementary hologram. The best techniques of multiplexing giving the largest number of recorded holograms M is the angular multiplexing. Its basics and the main obstacle for the multiplexing (a partial erasure of all prerecorded holograms on recording of each sequential one) were briefly outlined in Sect. 3.1. There were several attempts to fix thermal holograms multiplexed in $LiNbO_3$ [332,511,539,540,545]. The authors of [511] which achieved a capability of 1.2 MB during recording and fixing of 530 digital holograms, performed the first regular study of fixing conditions for multiplexing, particularly of the thermal regime of fixing. Recall that recording at elevated temperatures (producing a simultaneous fixing) gives higher η_{fix} and lower noise [482] than the postfixing. At the same time, the former procedure involves instabilities of the crystal parameters, an air convection and a consequent deterioration of the fixed hologram. To overcome this problem some authors [547–549] developed active stabilization systems. However, these methods may solve the problem only partially, because a deterioration of a fixed hologram at room temperature after recording at elevated temperatures is due also to fundamental reasons, such as the thermo-optical and piezo-optical effect (thermo- and strain-induced changes of the refractive indices) which cannot be avoided. So, the postfixing procedure is preferable for multiplexed holograms, although it leads to lowered η. To demonstrate a practical realization of thermal fixing of multiplexed holograms, we shortly dwell on one of these publications [545], in which the authors achieved thermal fixing of 10,000 holograms recorded in an $LiNbO_3$:0.015%Fe crystal $2 \times 1.5 \times 4\,\text{cm}^3$ in size. To record this number of gratings, five fractal rows in the crystal bulk were used with 2,000 angularly multiplexed holograms in each row. Holograms were recorded using the optimized schedule [287] providing an average (almost equal) diffraction efficiency of an elementary hologram of about 7×10^{-9}. Multiplexing was performed at room temperature with a post-heating to the fixing $T = 120°C$, whereupon the fixed system was developed at room temperature with the use of a mercury lamp. The authors attempted two regimes of fixation: a fixation of the recorded holograms altogether in one act of heating and an incremental procedure of fixing, namely, a step-by-step heating the sample after recording each successive thousand holograms. The latter procedure required 10 fixing (heating) acts and a sophisticated schedule of recording. The former procedure gave a ratio $\eta_{\rm fix}/\eta_0 \approx 0.26$, whereas the latter provided a $\eta_{\rm fix}/\eta_0 \approx 0.66$, the average efficiency of fixed holograms being about 4.6×10^{-9}. A surprising result is that in agreement with [482] a system of thermally fixed holograms shows a lower noise compared to the initial nonfixed one.

Let us briefly run through the more recent, alternative applications of photorefactive gratings [550–556] based on their angular selectivity, which is strictly constrained by the Bragg conditions. In the first turn, these are very-narrow-bandwidths interference filters required in a number of areas, such as solar astronomy, high-resolution spectroscopy, extra- or intra-cavity laser elements, wavelength multiplexers and demultiplexers for optical telecommunication systems, etc. First examples of these applications were demonstrated in [550, 551]. Rakuljic and Leyva [550] used a photorefractive grating as a Bragg-reflector which was not fixed. In the report [551] such a filter was constructed by means of recording and thermal fixing of a grating in a few millimeters thick $LiNbO_3$:Fe plate. For $\lambda = 514.5$ nm a bandwidth of about 50 pm, reflectivity, T of about 35% and an angular aperture of about 5° were achieved [551]. These parameters are somewhat worse than those of the most advanced narrow-band filters [557], however, the advantages are a much lower price and a possibility of a temperature and field tuning the peak wavelength, which is due, respectively, to the thermo-optic and electro-optic (and elasto-optic) effects. Tuning coefficients are $5\,{\rm pm}\,°{\rm C}^{-1}$ [551] and $6\,{\rm pm}\,{\rm kV}^{-1}$ [552]. In [556] a multi-channel Bragg reflector operating in the range of 670 nm was created by means of recording and thermal fixing of 16 gratings in an $LiNbO_3$:Fe plate. Other authors [553, 554, 558] tried the thermal fixing in wavelength-division multiplexers (WDM), which they develop on the base of superimposed phase gratings in a photorefractive medium. WDM serve to merge (multiplex) and separate (demultiplex) superimposed optical signals of different wavelengths, which are simultaneously transmitted by a single optical fiber in an optical communication network. A high Bragg-angle selectivity provides a necessary condition that a given photorefractive grating should diffract only a single light-wave thus being related to a definite single optical channel. Eight- and sixteen-channel multiplexers with bandwidths of 0.1 nm based on recording and thermally fixing gratings were demonstrated [558]. A multi-channel multiplexer was provided by recording with the use of an Ar-ion laser ($\lambda = 514.5$ nm) a corresponding number of gratings in an $LiNbO_3$:Fe plate; the orientation and the spatial period of a grating determines the wavelength and propagation direction of the corresponding diffracted beam. The multiplexing and demultiplexing was performed at the average wavelength $\lambda \approx 1.55\,\mu m$ in the reflection geometry. Fixed holgrams in $LiNbO_3$:Fe were also tried in an optical correlating system [555]. Development of real-time correlators in photorefractive media may be found in [559, 560]. Recall that an optical correlator serves to compare an unknown object to a known reference one. The principle is based on a correlation of the Fourier transform of an unknown object to the Fourier transform of the reference one. In the readout process the latter is incident on the correlator (in our case, a crystal with a thermally fixed reference image) at a Bragg angle, the diffracted beam intensity being proportional to the correlation function. In [555] a matched filter was created by means of recording a reference object in a sample $LiNbO_3$:0.1%Fe at 160°C (grating spacing $\Lambda = 1.35\,\mu m$). Finally, thermal fixing of holograms may find application in wave-guiding structures on $LiNbO_3$. A detailed bibliography one may find in recent reviews [561] and [479].

5.2.5 Conclusions for Thermal Hologram Fixation

By now the procedure of thermal hologram fixation has found a more or less adequate description and is examined sufficiently in $LiNbO_3$ crystals, especially doped with iron and in related materials such as $LiTaO_3$. At the same time there is a deficiency of these investigations in other photorefractive materials such as $BaTiO_3$, SBN, BSO, etc. Obviously, a screening of the photorefractive space-charge fields by thermally activated charge carriers of different sign at elevated temperatures is a common fundamental effect for photorefractive media on a whole and is worthy of further investigations. In $LiNbO_3$ crystals the microscopic origin of the thermal fixing, namely a screening of the electronic space-charge by protons thermally activated at temperatures $T > 120$–130°C, may be taken as settled. In line with the developed theoretical models the optimization of characteristics of thermally fixed holograms, that is an increase of η_{fix} and of the decay time τ_{dec} may be achieved in low-doped $LiNbO_3$:Fe crystals, particularly subjected to some treatments such as oxidation and dehydration. The maximum achieved experimental values for η_{fix} are 0.7–0.8 of the initial (nonfixed) η_0. In some experimental works no degradation of fixed holograms was revealed during a time as long as 2 years. The available theoretical description directs the ways to a further improvement of these parameters. The thermal fixing was successfully tested in nonvolatile optical storage systems based on hologram multiplexing and is coming into use in narrow-band devices requiring a high angular selectivity. The thermal fixation does not increase the noise of the multiplexed gratings compared to the initial nonfixed system. In spite of all these advantages, from the practical viewpoint this method seems to be technologically very complicated. Its fundamental drawbacks are the impossibility to perform the fixing in situ and to erase selectively and over-record an elementary hologram.

6

Properties to Characterize Undoped and Optical-Damage-Resistant $LiNbO_3$ Crystals

Summary. Many properties, relevant to the topics of this monograph depend on the composition of lithium niobate and tantalate. In an overview, we mention and evaluate, methods to determine the composition of the melt and the crystal. Emperical equations are given to estimate the crystal composition from that of the melt. Besides the measurement of the Curie temperature the position of the UV absorption edge can serve as a sensitive method for an accurate determination of the composition. The main topics of this section are the refractive index and birefringence including the modeling by various Sellmeier equations as a function of wavelength, composition, doping, and temperature. The phase-matching temperature and angle in the context of second-harmonic-generation are discussed as well as one other nonlinear feature, the electro-optic effect in lithium niobate doped with damage-resistant impurities.

Almost all properties of $LiNbO_3$ and $LiTaO_3$ are discussed in books and review articles mentioned in the introduction (Chap. 1). Therefore, we restrict this chapter to properties relevant to the topics of this monograph. We start with an overview of methods to characterize the composition of undoped crystals and those doped with damage-resistant impurities. Then, we continue with a more detailed description of methods, which are strongly related to photorefraction.

6.1 Methods to Measure the Composition and Homogeneity of Crystals

The composition of $LiNbO_3$ and $LiTaO_3$ can be determined with nonoptical and optical methods. The latter are usually nondestructive and more accurate, but need an individual calibration based on nonoptical methods. For undoped crystals many suitable methods are known and have been discussed and compared in 1996 [37]. The situation is less favorable for crystals doped with optical-damage-resistant impurities with concentrations around the threshold concentrations if the [Li]/[Nb] ratio is involved, because the threshold values strongly depend on this ratio. However, no complete knowledge of [Li]/[Nb] as a function of optical-damage-resistant impurities is of practical interest, because mainly stoichiometric compositions sufficiently doped with optical-damage-resistant impurities are needed for devices.

6.1.1 Melt Composition and Curie Temperature

For a quick estimate of the crystal composition, the following empirical equation

$$c_{\mathrm{Li}} = 48.5 + 0.3436(c_{\mathrm{Li,melt}} - 48.5) - 0.02897(c_{\mathrm{Li,melt}} - 48.5)^2 \qquad (6.1)$$

based on several reports [37], where c_{Li} and $c_{\mathrm{Li,melt}}$ denote the Li_2O content in mol.%, can be used in the range $45 < c_{\mathrm{Li,melt}} < 52$ mol.% with an accuracy of 0.3 mol.% (see Fig. 6.1).

The Curie temperature T_c changes by about 150°C in the solid solution range, thus providing a high sensitivity for composition measurements ([562], see Fig. 6.2). But the drawback is the closeness of T_c to the melting point. DTA measurements on four single crystals characterized by chemical analysis have been carried out and resulted in the fit

$$c_{\mathrm{Li}} = 0.02546\, T_c + 19.39 \quad \text{or} \quad T_c = 39.26\, c_{\mathrm{Li}} - 760.67, \qquad (6.2)$$

where T_c is given in centigrades [42]. Bordui et al. used five VTE equilibrated crystals for dielectric measurements [30] and found

$$c_{\mathrm{Li}} = 0.02557\, T_c + 19.149 \quad \text{or} \quad T_c = 39.064\, c_{\mathrm{Li}} - 746.73. \qquad (6.3)$$

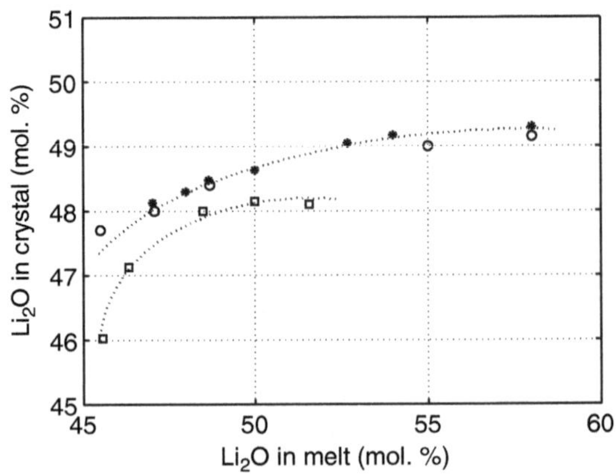

Fig. 6.1. Li concentration in the crystal vs. melt concentration. Data are taken for *squares* from [201], for *stars* from [562] and *circles* from [365]; *dotted lines* are guides for the eyes

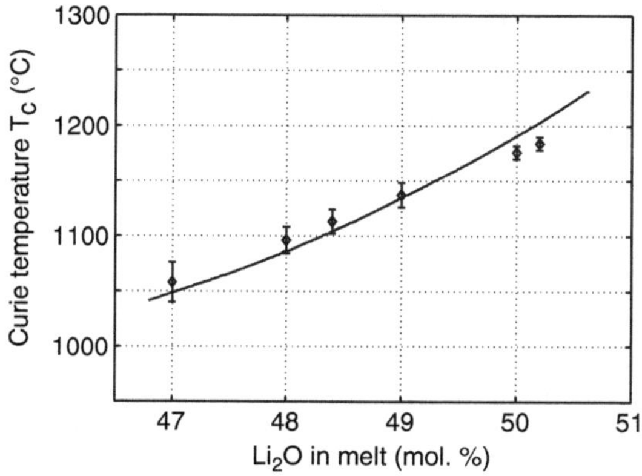

Fig. 6.2. Curie temperature of crystals as a function of the Li_2O content in the melt of $LiNbO_3$. The line was calculated using $T_c = 9,095.2 - 369.05\,c + 4.228\,c^2$ (for details see [6, 563])

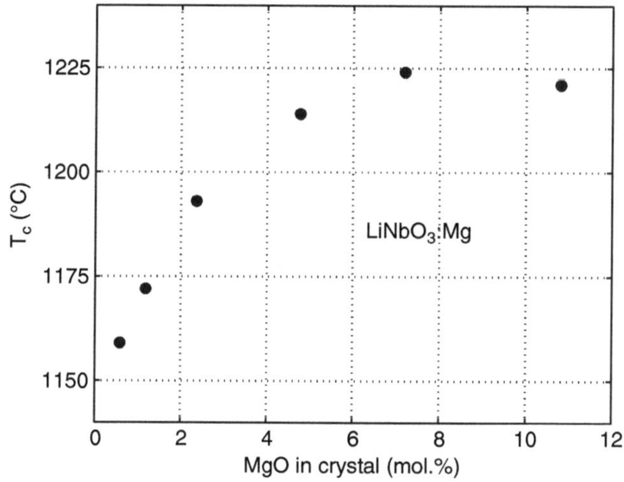

Fig. 6.3. Curie temperature of crystals as a function of the MgO content in the $LiNbO_3$ crystal, redrawn from [200]

Both fits differ only by 4°C or 0.1 mol.% and are frequently used. For $LiTaO_3$, a second-order polynomial of the form

$$T_c = 11,310 - 492.73\,c_{Li} + 5.6062\,c_{Li}^2 \tag{6.4}$$

was reported [564].

The Curie temperature as a function of the Mg doping increases as well, but has a different shape. Above the threshold, it starts to saturate or even diminishes (see Fig. 6.3).

6.1.2 UV Absorption Edge

The position of the fundamental absorption edge is very sensitive to the composition of $LiNbO_3$ and $LiTaO_3$. This feature is often used, because samples of 1 or 2 mm thickness with an orientation of about 5° are very suitable and a moderate optical quality with a density of scratches less than 1% of the total area is required. A reflection correction is necessary, shifting the edge by about one nm to the blue. Because there is no common definition of the bandedge, certain absorption values, suitable for a typical thickness of samples are to be investigated. For convenience of the reader, empirical equations are given for absorption values of 15 and 20 cm^{-1}, as well as a visualization in Fig. 6.4. Above 49.5 mol.% Li_2O the sensitivity is high, mapping a change of 0.1 mol.% to 2 nm. So even common spectrometers allow an accuracy of this magnitude

$$\lambda_{15} = 321.9 - 1.597\,x - 5.745\,x^2, \tag{6.5}$$

$$\lambda_{20} = 320.4 - 1.829\,x - 5.485\,x^2. \tag{6.6}$$

The variable x is the deviation from the congruent composition (48.38 mol.%) in mol.%. These equations are valid for light being polarized parallel and propagating perpendicular to the polar axis. Similar equations for other light polarizations are given in [565].

A nonlinear decrease of the bandedge wavelength with rising Li content was found in $LiTaO_3$ crystals, too, but described with an exponential fit function of the form

$$\lambda_{20} = 278.93 - 15.77\,\exp\left[(c_{\mathrm{Li}} - 50)\;1.75\right] \tag{6.7}$$

for the extraordinary polarization [38]. This behavior could be interpreted by assuming an appropriate overlay of an additional near UV absorption caused

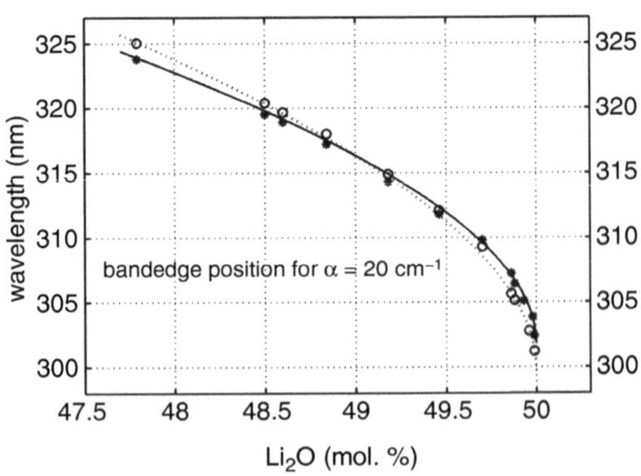

Fig. 6.4. Absorption edge of $LiNbO_3$ for $T = 22°C$ at $\alpha = 20\,cm^{-1}$ as a function of crystal composition. *Open circles* and *stars* indicate measured values for extraordinary and ordinary polarizations, respectively. *Lines* are calculated using an equation and parameters given in [565]

by Ta or Nb antisite ions to the base one. The concentration of antisites increases up to 50% and then stays constant causing the observed dependence of the bandedge on composition [38]. Finally, we mention that the bandedge position cannot be used for composition determinations, if an impurity like Fe causes some absorption close to the bandedge.

Other methods, sometimes used, depend on the linewidths of signals (NMR, EPR, Raman, OH stretch mode) that decrease with increasing stoichiometry, for details see the survey [37]. Recently, Dravecz and Kovács [110] proposed to determine the Li_2O content in $LiNbO_3$ from the intensity ratio $R = I_{3,480}/I_{3,465}$, where 3,480 and 3,465 cm^{-1} are two components of the OH-stretch mode. This criterion is put forward, because R linearly decreases with the Li_2O increase from 49.7% to 50%.

6.1.3 Homogeneity

Besides interferometric techniques, all optical methods working with spatially limited beam dimensions may be used to test the homogeneity. Specially the frequency doubling methods have been proposed. Measurements of the phase matching temperature T_{pm} can be applied, either working with a focused laser beam and moving the sample or illuminating it uniformly and using a one- or two-dimensional detector array to measure the intensity of the generated second harmonic light achieving a spatial resolution of 20 μm. Examples of the sensitivity of T_{pm} on variations of the content of damage-resistant dopants are given in Sect. 6.3. More methods and details are presented in [37].

6.2 Refractive Indices and Birefringence: A Generalized Sellmeier Equation

Doping with optical-damage-resistant impurities as well as variations of the [Li]/[Nb] ratio result in variations of the refractive indices and the birefringence. An extensive reference list on measurements and modeling of the data with various kinds of Sellmeier equations is given in [566,567]. Two differently motivated approaches can be suggested. Schlarb et al. [220, 222, 568–572] introduced a general approach for describing the dispersion of $n_e(\lambda)$, $n_o(\lambda)$ in $LiNbO_3$ of different composition or doped with optical-damage-resistant ions. The unique feature of this approach is the physical interpretation of each term of the Sellmeier equation with respect to the intrinsic defects. Jundt et al. [566, 567, 573] and Paul et al. [574] favor a description, which is valid over a large wavelength range (400–5,000 nm) but does not allow any special relation to the incorporation of optical-damage-resistant ions. Both models describe $n_e(\lambda)$ and $n_o(\lambda)$ very well for congruent and nearly stoichiometric as well as Mg-doped $LiNbO_3$ in the range 400–1,200 nm. We point out, that the fits of Schlarb et al. [220, 222, 572] are valid only in the range 400–1,200 nm and guarantee an accuracy of about 2×10^{-3}, while the empirical approach of

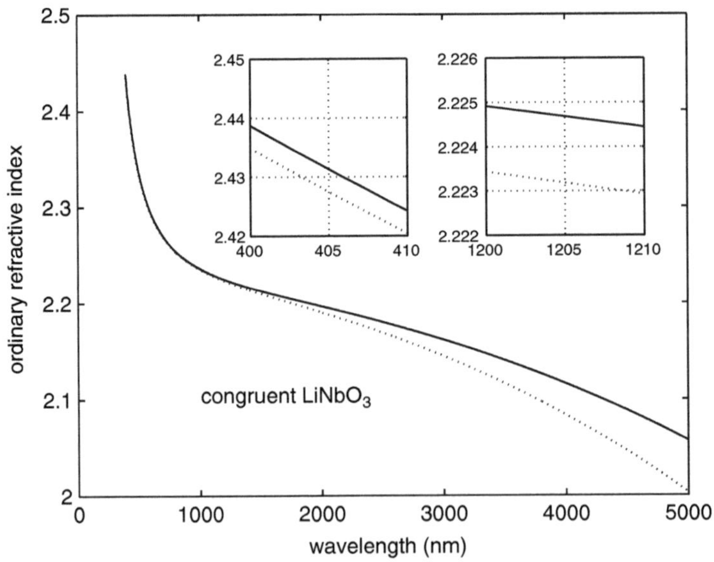

Fig. 6.5. Comparison of the calculated ordinary index of refraction n_o for congruent $LiNbO_3$ according to Jundt [567] (*solid line*) and Schlarb et al. [220,222,572] (*doted line*). The insets magnify the differences of the two Sellmeier fits in the blue and near infrared region

Jundt [567] extends up to 5 μm with a better accuracy of about 2×10^{-4}, but is limited to undoped and Mg-doped material. This conclusion is illustrated in Fig. 6.5, where the dotted curve denotes the ordinary refractive index according to Schlarb et al. [220, 222, 572], while the solid one to Jundt [567]. The curves are almost identical even in the blue region, if the uncertainty of 2×10^{-3} is taken into account, but strongly diverge for longer wavelength, as already pointed above. This is also true for the extraordinary index in congruent and nearly stoichiometric undoped material as well as in optical-damage-resistant $LiNbO_3$ grown from congruent melts. The insets in Fig. 6.5 illustrate the differences of both curves at about 400 and 1,200 nm.

We use the general fit to discuss the dependence on the optical-damage-resistant ions. Usually the index of refraction is described as a sum of several oscillators

$$n^2 = 1 + \sum_j \frac{A_j}{\lambda_j^{-2} - \lambda^{-2}}, \tag{6.8}$$

where λ_j^{-2} describes the resonance frequency of an oscillator and A_j is its strength proportional to the oscillator concentration (assuming their independence). Optical properties of $LiNbO_3$ in the near UV range depend on the single oscillator represented by an NbO_6 octahedron [575]. The approach is based on estimations of the change of the oscillator strength A_0 due to a substitution of a regular Nb_{Nb} atoms by defects like Nb_{Li}. Their concentrations depend on the Li content or on the concentration of the damage-resistant Mg ions controlling the content of Nb_{Li}. The consideration is based on the Li site vacancy model; a difference in the resonance energies of Nb_{Nb}, Nb_{Li}, and Mg_{Li}, (Zn_{Li}, In_{Li}) is neglected. This idea led to the following generalized Sellmeier equation for all optical-damage-resistant $LiNbO_3$ compositions

$$n^2 = \frac{A_0 + A_{Nb_{Li}}\, c_{Nb_{Li}} + A_{Mg_{Li}}\, c_{Mg_{Li}} + A_{Zn_{Li}}\, c_{Zn_{Li}} + A_{In_{Li}}\, c_{In_{Li}}}{\left(\lambda_0 + \mu_0 \left[f(T) - f(T_0)\right]\right)^{-2} - \lambda^{-2}} + A_{UV} - A_{IR}\,\lambda^2, \qquad (6.9)$$

where A_{UV} is the contribution from UV plasmons, $A_{IR}\,\lambda^2$ is the contribution from dipole-active phonon oscillations, and $f(T) = (T + 273)^2 + 4.023\,10^5\,(\coth[261.6/(T + 273)] - 1)$ takes into account the temperature dependence of the resonance frequency ($T_0 = 24.5°C$). $f(T)$ is an empirical parameter and has no clear relation to theoretical predictions derived in [576]. The concentration of Nb_{Li} as a function of the concentrations of optical-damage-resistant ions is given by

$$c_{Nb_{Li}} = \frac{2}{3}\left(50 - c_{Li}\right) - \frac{c_{Mg}}{\alpha_{Mg}} - \frac{c_{Zn}}{\alpha_{Zn}} - \frac{c_{In}}{\alpha_{In}}, \qquad (6.10)$$

where α_{Mg}, α_{Zn}, and α_{In} are fitting parameters corresponding to the experimental values of the threshold concentrations of Mg, Zn, In (5.0%, 6.5%, 1.5%), respectively. This presentation of the Nb_{Li} concentration is due to the fact that the authors assumed the threshold concentrations to correspond to the removal of Nb_{Li} and the following substitution of regular Li atoms by optical-damage-resistant ions. An incorporation of these ions onto Nb sites found later [215–217], was of course not considered. According to (6.10), a change of n_e, n_o at threshold concentrations means a vanishing oscillator strength for the resonator, and a linear increase of the strengths of Mg_{Li}, Zn_{Li}, and In_{Li} resonators.

Parameters of the Sellmeier equation (6.9) are shown for convenience, in Table 6.1. Despite some restrictions, particularly of a postulate on a relation between the disappearance of Nb_{Li} and the threshold concentrations, which lacked support from structure measurements [215–217], this type of Sellmeier equation (6.9) allows a discussion of the influence of optical-damage-resistant

Table 6.1. Parameters for the Sellmeier equation (6.9)

	λ_0	μ_0	A_0	A_{IR}	$A_{Nb_{Li}}$
n_o	223.219	1.1082×10^{-6}	4.5312×10^{-5}	3.6340×10^{-8}	-7.2320×10^{-8}
n_e	218.203	6.4047×10^{-6}	3.9466×10^{-5}	3.0998×10^{-8}	11.8635×10^{-7}
	A_{Mg}	A_{Zn}	A_{In}	A_{UV}	
n_o	-7.3648×10^{-8}	6.7963×10^{-8}	-2.4×10^{-7}	2.6613	
n_e	7.6243×10^{-8}	1.9221×10^{-7}	4.7×10^{-7}	2.6613	

ions on $n_e(\lambda)$, $n_o(\lambda)$ for Mg [572], Zn [577], and In [578] doping for wavelengths up to 1,200 nm and even for double-doping with (Zn+In) [579].

In the next step, we demonstrate the power of the Sellmeier fit by considering the influence of doping on $n_e(\lambda)$ and $n_o(\lambda)$. A rough inspection of Table 6.1 shows, that the parameters $A_{Nb_{Li}}$, A_{Mg}, and A_{In} have different signs for n_e, and n_o while they are equal for A_{Zn}. Therefore, we expect different slopes, this means an increase or decrease of the refractive index with increasing doping concentration. This is indeed observed and shown for the single laserline 633 nm (see Fig. 6.6). In addition, the threshold values for Mg, Zn, and In show up as a kink in the concentration dependence. Nevado et al. [580] found a clear singularity in the concentration dependence of n_o in $LiNbO_3$:Sc for a Sc^{3+} concentration of 3%.

Note that both $n_e(\lambda)$ and $n_o(\lambda)$ are growing with the Zn concentration in contrast to Mg doping which results in a decreasing $n_e(\lambda)$ [57, 572, 581]. This specific observation for Zn doping is of importance for practice (see below at the end of this section).

Although data on In and Sc are not included in Fig. 6.6, they behave quite similar [578, 580].

Recently data for Li-enriched (near stoichiometric) $LiNbO_3$:Mg were reported [582]. Refractive index data for NSLN are independent on the growth

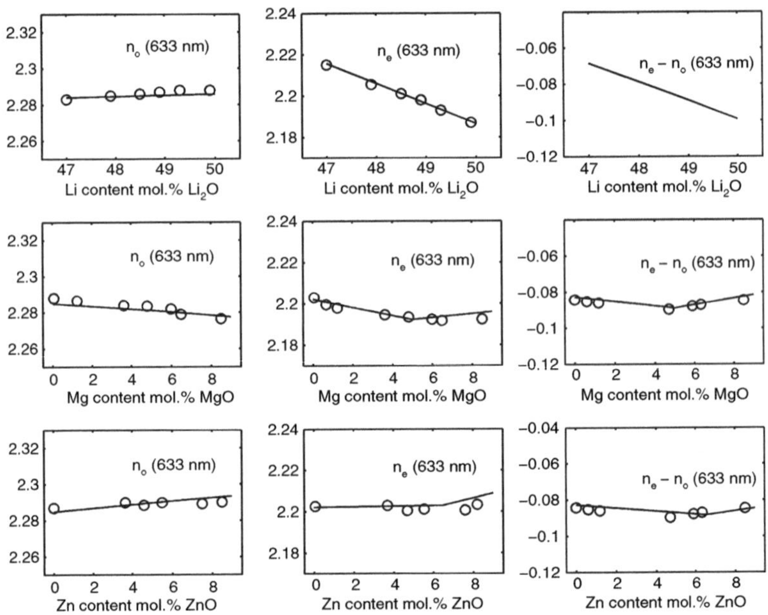

Fig. 6.6. Comparison of the calculated indices n_e, and n_o for $\lambda = 633$ nm for undoped $LiNbO_3$ of different composition and Mg-doped and Zn-doped grown from a congruent melt. The *solid lines* were calculated with (6.9) using the parameters of Table 6.1. Experimental data were taken from [220, 572]

method (TSSG or DCCZ). Although n_o decreases with increasing Mg concentration for both CLN:Mg and NSLN:Mg, the decrease of the latter is stronger. But the results differ strongly from those of Mg-doped congruent $LiNbO_3$ for the extraordinary index n_e, where the index decreases with rising Mg concentration and then slightly increases above 5 mol.% (Fig. 6.6). In NSLN, n_e is almost independent of Mg up to 4.5 mol.% MgO [582]. The different behaviors of NSLN and CLN can be partially explained by the fact that the Nb_{Li} antisite plays no more a dominant role. Changes in the electronic structure, due to modifications of the constituent chemical bonds induced by MgO or ZnO doping are probably responsible for the different influence of Mg and Zn. As a consequence, the Sellmeier approach used for (6.9) is no more applicable to NSLN, thus Nakamura et al. [582] derived Sellmeier parameters for each composition abandoning any physical interpretation. An exhaustive collection of Sellmeier equations including the parameters are published in the book by Nikogosjan [17].

A very similar situation is observed in $LiTaO_3$. The ordinary index n_o is almost equal for CLT and NSLT like in congruent and nearly stoichiometric $LiNbO_3$, while n_e is about 0.5% lower in NSLT than in CLT [28], but differ about 1% in NSLN and CLN (see upper part of Fig. 6.6). Mg doping of NSLT up to 1 mol.% MgO has no effect on n_o and n_e, but almost eliminates the photorefractive damage. Recently, a temperature-dependent Sellmeier equation has been reported for undoped NSLT and CLT with an accuracy of 0.0001 [26].

As mentioned earlier, a specific amount of Zn-doped $LiNbO_3$ among other optical-damage-resistant compositions is an increase of both the ordinary and extraordinary refractive indices with the Zn concentration (Fig. 6.6). Due to this effect, Young et al. [583, 584] proposed wave-guiding Zn-indiffused structures on $LiNbO_3$ substrates. As known, the most widely used methods to produce waveguides on $LiNbO_3$ are the proton-exchange or Ti-indiffusion. In spite of the great advantages of the H^+-indiffused waveguides (manufacturability owing to low temperature formation), their drawback is that they support extraordinary waves only. Additionally, protonated $LiNbO_3$ waveguides exhibit a noticeable photorefractive damage [193]. Ti-indiffused waveguides support both polarizations, but reveal a pronounced optical damage even on an $LiNbO_3$:Mg substrate [585]. Moreover, a high temperature during Ti-indiffusion $T > 1{,}000°C$ is accompanied by an Li-outdiffusion and a partial polarization reversal in the waveguiding layer, as shown in Chap. 7. Zn-indiffused waveguides on $LiNbO_3$ and $LiNbO_3$:Mg eliminate the drawbacks inherent to these types of waveguides, because they support both ordinary and extraordinary polarization modes and are extremely optical-damage-resistant. An additional advantage of Zn-indiffused waveguides is a rather high Zn diffusion coefficient exceeding by two orders of magnitude that of Ti, thus the required Zn diffusion temperature is considerably lower (800–900°C) [586]. Similar properties offer Zn-indiffused waveguides on $LiTaO_3$ [587]. Zn-indiffused waveguides on $LiNbO_3$:Mg were optically stable at an input power of 180 kW cm^{-2} for 515 nm radiation, whereas proton-exchanged waveguides on $LiNbO_3$:Mg were unstable from 7 kW cm^{-2} [583]. Encouraging results were obtained for proton-exchanged waveguides fabricated on a substrate of damage-resistant $LiNbO_3$:Zn with 6 mol.% ZnO [588]. Studies in various waveguiding structures on the base of $LiNbO_3$:Zn are in progress. For example, low-loss planar and channel waveguides for $\lambda = 1,500$ nm were obtained by diffusing ZnO in low pressure [589]. DiPaolo et al. [590] demonstrated a Zn-diffused $LiNbO_3$:Nd waveguide laser; channel waveguides formed on the z-cut of an $LiNbO_3$:Nd crystal provided the stable propagation of the pumping wave with 816 nm and of the π-polarized Nd^{3+}-emission wave at 1,085 nm. Kawaguchi et al. [591] reported on waveguiding properties of Zn-doped $LiNbO_3$ liquid-phase epitaxial thin films grown on $LiNbO_3$ substrates.

6.3 Phase-Matching Temperature and Angle

$LiNbO_3$ is one of the most attractive materials for optical frequency conversion due to its rather high NLO coefficients. The nonzero components of the nonlinear susceptibility tensor are $d_{31} = d_{32}$, d_{33}, $d_{22} = -d_{21} = -\frac{1}{2}d_{16}$, $d_{15} = d_{24}$ with values given in Table 1.2 for 1,064 nm light. A detailed investigation of nonlinear-optical coefficients of LN and LT are given in [593].

A great advantage of $LiNbO_3$ for the optical frequency conversion is the possibility to obtain a noncritical 90°-phase-matching by means of temperature tuning. Another possibility of controlling the phase-matching temperature T_{pm} and the angle Θ_{pm} is achieved by varying the Li content or/and doping with optical-damage-resistant ions (Fig. 6.7). These dependencies are caused by the compositional dependencies of the refractive indices discussed in Sect. 6.2. Particularly, a strong impact of the Li content of undoped $LiNbO_3$ on the birefringence results in a drastic growth of T_{pm} from −80°C in congruent $LiNbO_3$ crystals to 470°C in stoichiometric ones [7].

The threshold concentrations of the optical-damage-resistant impurities show up in the index of refraction as a smooth kink. Much more pronounced anomalies are predicted and indeed observed for properties depending on the dopant and temperature dependence of the refractive index. As an example, we discuss the phase-matching temperature T_{pm} for $\lambda = 1.06\,\mu$m (Nd-YAG laser radiation) for collinear noncritical SHG vs. the concentrations of optical-damage-resistant dopants (see Fig. 6.7). For both divalent and trivalent dopants, the maxima are found for concentrations corresponding to threshold values. The same is true for the phase-matching angle Θ_{pm} at a given temperature [220, 572, 578]. All these concentration dependencies are well described with (6.9).

Note that these nonmonotonous concentration dependencies of T_{pm} and Θ_{pm} present one of the most striking manifestations of the threshold. Studies on applications of $LiNbO_3$:Mg for optical-frequency conversion started immediately after finding the optical-damage resistance in this composition [405, 407]. For example, the second harmonic generation of an Nd-YAG

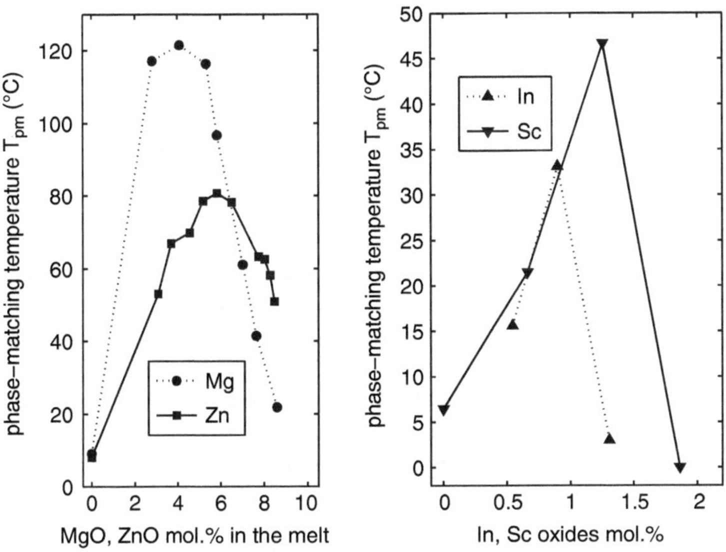

Fig. 6.7. Phase-matching temperatures for collinear noncritical SHG in $LiNbO_3$ for 1,064 nm vs. In, Sc (*right*) and Mg, Zn (*left*) concentrations. The data for Mg and Zn are taken from [56], for In from [592] and for Sc from [182], respectively

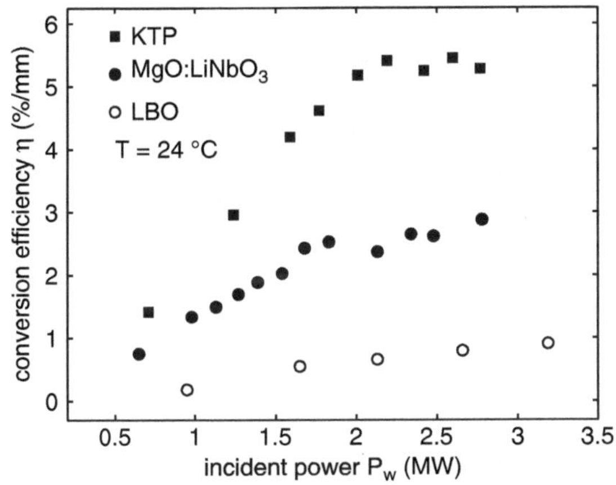

Fig. 6.8. Measured SHG efficiencies per mm crystal length in KTP, $LiNbO_3$:MgO, and LBO. Experimental data were taken from [407]

laser was investigated in $LiNbO_3$:7%Mg crystals that have $T_{pm} = 60°C$, and $\Theta_{pm} = 82.3$ at room temperature, so the walk-off angle is negligible $\approx 0.6°$) [407]. The authors estimated the type I SHG to be obtainable with this composition in a wavelength range from 1.05 to 3.4 µm. The SHG efficiency was compared to $KTiOPO_4$ (KTP) and LiB_3O_5 (LBO) (see Fig. 6.8).

6.4 Electro-Optic Properties

$LiNbO_3$ crystals are widely used as electro-optic modulators both in bulk and waveguiding structures (for references see the recent reviews [477, 594, 595]). The values of the nonzero electro-optic coefficients at 633 nm for static electric field (zero-strain or nonclamped conditions) and clamped conditions are listed in Table 1.2 for various compositions. These values oscillate within several percentage. As shown later, EO coefficients are modified like the NLO coefficients by a variation of the Li content or/and doping with optical-damage-resistant ions. In addition, they serve as a second example for a more pronounced anomaly induced by the dependence on an external parameter, like the electric field instead of the temperature, is the electro-optic coefficient (EO). An often used method to measure it, is the frequency doubling electro-optic modulation (FDEOM) introduced by the group of Aillerie and Fontana, where the phase shift between polarization components depends for $r_c = r_{33} - (n_o/n_e)r_{13}$ on the field-induced birefringence as follows:

$$\Gamma = \frac{2\pi L}{\lambda}\Delta n(E) = \frac{2\pi L}{\lambda} n_e^3 \, r_c \, E \tag{6.11}$$

with L as the crystal length along the x-axis and E along the z-axis. The often used transverse mode of operation, when the laser beam propagates along the z-axis and the field is normal to it, involves the coefficient r_{22}. Experimental results on r_{22} in undoped LN as a function of composition that slightly differ have been reported in [596, 597]. Starting at the congruent composition, the EO coefficient decreases up to about 48.9 mol.% and then increases. In a first attempt, the minimum was related to the behavior of the elastic constant [596] and then interpreted as due to the noncentrosymmetry in the electrostatic potential distribution around Nb–O_6 clusters [597]. An increase between 48 and 50 mol.% was reported for r_{33} [27], almost no difference for r_{13} in congruent and stoichiometric material [598]. A decrease of the EO coefficients with increasing Li content was observed in Fe-doped as well as in Cr-doped $LiNbO_3$, too, but in Fe-doped samples no increase was found for higher Li concentrations [599,600]. In both cases the EO coefficients depend almost linearly on the doping element. The most intriguing results are presented by Kondo et al. [27], where an increase of r_{33} in Ce- or Fe-doped near-stoichiometric $LiNbO_3$ by 30% was observed compared to CLN; the authors assign this anomalous effect to a possible variation of the crystal structure. We mention that the EO coefficients in congruent LT are much smaller than in LN [601]. Yonekura et al. [602] report on a r_{22} coefficient and a half-wave voltages $V_{\frac{\lambda}{2}}$ in congruent $LiNbO_3$:5%Mg and near-stoichiometric $LiNbO_3$:1.8%Mg in the spectral range 400–1,600 nm. The values and spectral dependencies of r_{22} in CLN:5%Mg are close to those in CLN, whereas r_{22} in NSLN:1.8%Mg are noticeably higher. For example, for 633 nm r_{22}: 6.54; 6.2 and 6.97 pm V^{-1} in CLN, CLN:5%Mg and NSLN:1.8%Mg, respectively, under constant stress (clamped) conditions.

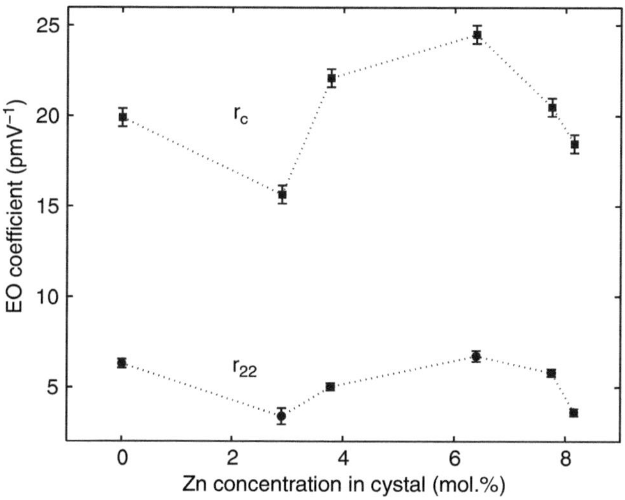

Fig. 6.9. Electro-optic coefficient r_c and r_{22} as a function of the Zn concentration in the crystals. The data are taken from [221]

In Zn-doped $LiNbO_3$, the varying Li vacancy concentration induces even a minimum and a maximum as seen in Fig. 6.9.

The dependencies of r_c and r_{22} vs. Zn [221] are nonmonotonic and additionally to maxima in the range of the threshold concentrations reveal anomalies in the low concentration ranges. Speculatively, in $LiNbO_3$:Zn it correlates with a disappearance of Nb_{Li} (Fig. 2.20). From a practical viewpoint, it is important that in an optical-damage-resistant concentration range, the values of r_{ij} are not decreased compared to an undoped congruent crystal.

7

Polarization Reversal and Ferroelectric Domains in $LiNbO_3$ Crystals

Summary. Numerous publications on the polarization reversal and domain dynamics in $LiNbO_3$ and $LiTaO_3$ are summarized. First, the basics of the ferroelectric switching are discussed starting from the Landau–Guinzburg–Devonshire phenomenology with an account for a contribution from domains and domain walls. The concept of the internal bias fields E_b in ferroelectrics related to an interaction of defects with the host is presented. The Miller–Weinreich model of the ferroelectric switching is formulated and various techniques of studies in the ferroelectric switching and domain dynamics are drafted with an emphasis on those applicable to $LiNbO_3$. The background part is closed by the basics of the optical frequency conversion using the ferroelectric domain structure. After a short historical review of the early results on the polarization reversal in $LiNbO_3$ and of the first attempts of the creation of periodically poled $LiNbO_3$ (PPLN), a specific of the ferroelectric P–E hysteresis loops in $LiNbO_3$ and $LiTaO_3$ is discussed. A drastic response of the coercive E_c and bias E_b fields and of the pulsed-switching parameters to an increase of the Li content or/and optical-damage-resistant impurity doping is described in detail. An interpretation of the polarization reversal specifics in terms of the pinning effects assuming the pinning centers to be the intrinsic defects is analyzed. In the next step an impact of a photoactive UV-light on the polarization reversal, particularly values of E_c and E_b and the switching velocity is presented which may be related to screening effects. In the end of the chapter, the experiments on recording regular domain patterns by means of electron and ion beams or under dc-voltages applied to AFM-tips are examined. These methods provide a possibility to obtain domain patterns on a nanoscopic level and to avoid the undesirable effects of back-switching and domain merging which are inherent for recording PPLNs by applying external fields to regular electrode patterns. A specific of the ferroelectric switching under an extremely spatially nonuniform field of an AFM-tip is characterized and current models are outlined.

7.1 Classical Description of the Ferroelectric Switching

To begin with, we recall the fundamentals of ferroelectricity with an emphasis on ferroelectric domains and the polarization reversal. This material presents a compilation from several monographs [3–5,297,603–607]. From a large amount of available material, we selected those details which are appropriate for discussing the polarization reversal in $LiNbO_3$. Therefore, we consider the case of uniaxial ferroelectrics with two antiparallel polarization directions $\pm P_s$. In this overview of the data for $LiNbO_3$ and $LiTaO_3$, we concentrate on the switching process. The results on the domain structure are presented as applied to the switching, whereas the domain structure of $LiNbO_3$ and $LiTaO_3$ by itself and the appropriate bibliography is discussed in recent reviews [608–610].

7.1.1 Phenomenological Description

The phenomenology of a ferroelectric phase transition is modeled with a thermodynamic potential, the elastic Gibbs function with the independent variables temperature, stress and the order-parameter. This order-parameter must reflect all possible changes (reduction of the number of symmetry elements) of the system. The order-parameter is defined to be zero in the high symmetry phase, i.e., for temperatures T above the phase transition temperature T_c, and nonzero else. In proper ferroelectrics the spontaneous polarisation P_s is used as order-parameter. Above T_c there is no P_s, thus no reduction of crystal symmetry elements occurs. On the other hand a $P_s \neq 0$ destroys all symmetry elements which do not keep P_s unchanged. Neglecting the strain, we have the differential $\mathrm{d}G = -S\,\mathrm{d}T + E\,\mathrm{d}P$ and a second-order phase transition is described by the Landau–Ginsburg $G(P,T)$ with a polynomial in P having at least terms up to the fourth order [4].

$$G(P,T) = G_0 + \frac{\alpha(T)}{2}P^2 + \frac{\beta(T)}{4}P^4. \tag{7.1}$$

Only terms with even power are used, because usually the high temperature phase has an inversion center or a mirror plane perpendicular to the principal axis.

We mention that this simple form is only valid if a uniform P_s distribution is assumed and the piezoelectric effect and electrostriction are neglected. For a first-order phase transition close to the second-order one an additional term $\frac{\gamma}{6}P^6$ is added to (7.1) and $\beta < 0$. In congruent $LiNbO_3$ the second-order phase transition occurs at $T_c \approx 1{,}190°\mathrm{C}$ from the centrosymmetric-paraelectric phase (space group $\mathrm{D}^6_{3\mathrm{d}}$) to the noncentrosymmetric phase (space group $\mathrm{C}^6_{3\mathrm{v}}$).

The appearance of P_s results in spontaneous elastic strains

$$u_i = d_{ij}P_j + Q_{ijk}P_jP_k \quad \text{and} \quad u_{jk} = d_{ijk}\,E_i, \tag{7.2}$$

where d_{ij} and Q_{ijk} are piezoelectric and electrostriction coefficients, respectively. Consequently, the free energy includes the elastic energy and may be presented in a generalized Landau–Ginzburg–Devonshire (LGD) form as

$$G_{\mathrm{LGD}}(P_i,P_j,T,u_k) = G_{\mathrm{LG}}(P_i,P_j,T) + G_{\mathrm{el}}(u_i) + G_{\mathrm{es}}(P_i,u_k) \tag{7.3}$$

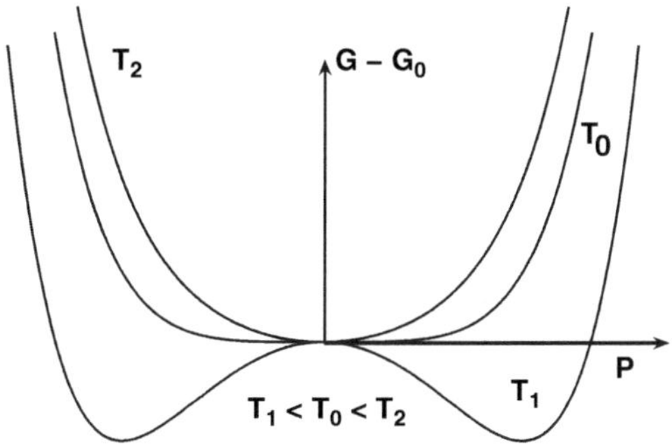

Fig. 7.1. Free energy $G - G_0$ as a function of the polarization P for three temperatures $T_1 < T_C = T_0 < T_2$

where G_{LG}, G_{el} and G_{es} are the LG-energy (7.1), the elastic and the electrostrictive energies, respectively.

In Fig. 7.1, G is depicted for temperatures below, at the phase transition and above. In the paraelectric phase only one minimum exists, while two show up in the ferroelectric state. They correspond to the $\pm P_s$ values, which can be derived from (7.1) solving $(\partial G/\partial P)_{P_s} = 0$, i.e., $P_s(\alpha + \beta P_s^2) = 0$, yielding

$$P_s = \pm\sqrt{\frac{-\alpha}{\beta}} \quad \text{for} \quad T < T_c$$
$$P_s = 0 \quad \text{for} \quad T > T_c. \tag{7.4}$$

The parameter α can be expressed in the paraelectric phase in terms of the Curie–Weiss law

$$\chi = \frac{1}{\alpha} = \frac{C}{T - T_c}, \tag{7.5}$$

where χ is the dielectric susceptibility. For a constant temperature T the above-mentioned differential reads $E = \mathrm{d}G/\mathrm{d}P = \alpha P + \beta P^3$ defining a hysteresis loop as shown in Fig. 7.2.

In the ferroelectric phase the curve between the two states A, C describes unstable states, thus the system jumps from $A \to B$ and $C \to D$. The corresponding field in A or C is known as the coercive field E_c and calculated from the roots of $\mathrm{d}E/\mathrm{d}P = \alpha + 3\beta P^2 = 0$, i.e., $P_c = \pm\sqrt{-\alpha/3\beta}$, by substituting P_c back into $\alpha P + \beta P^3$ we yield

$$E_c = \pm\frac{2}{3\sqrt{3}}\sqrt{\frac{\alpha^3}{\beta}} \approx 0.385\,\alpha\,P_s \tag{7.6}$$

using the solution (7.4) with $\beta P_s^2 = -\alpha$.

This theory describes adequately the anomalous properties in the phase transition regime and according to (7.4) and (7.6) predicts qualitatively a two-valued (hysteresis) dependence of $P(E)$. However, this concept is not

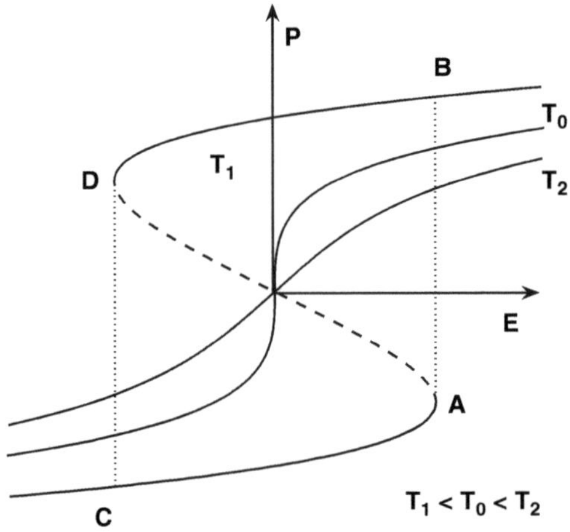

Fig. 7.2. Polarization P vs. electric field E for three temperatures $T_1 < T_C = T_0 < T_2$, using the same parameters as in Fig. 7.1

valid for a quantitative description of the observed hysteresis phenomena. Indeed, calculations using experimental values for χ and P_s give coercive fields orders of magnitude higher than the experimental ones. For example, in $LiNbO_3$ ($\epsilon_{33} \approx 30$, $P_s \approx 75 \times 10^{-6}\,\mathrm{C\,cm^{-2}}$) the relation (7.6) leads to $E_c \approx 5.4 \times 10^6\,\mathrm{V\,cm^{-1}}$ [611], whereas the experimental values in stoichiometric and congruent $LiNbO_3$ are 1.7×10^4 and $2.2 \times 10^5\,\mathrm{V\,cm^{-1}}$, respectively. As $LiNbO_3$ exhibits the highest E_c among the ferroelectrics, so for other crystals this discrepancy is even larger. This discrepancy is due to the fact that the P_s reversal occurs via the formation of ferroelectric domains, which are regions with different orientations of P_s. This mechanism is beyond the phenomenology, which was developed for a uniform infinite ferroelectric without account for discontinuities of P_s on the crystal surfaces. The domain formation is caused fundamentally by depolarizing fields ($\nabla P_s \neq 0$). The thermodynamic potential for a ferroelectric crystal with a nonuniform P_s distribution (i.e., one containing domains) takes into account the depolarizing energy W_E and the energy of the domain walls W_w

$$G(P,T) = G_0 + \int_V \left[\frac{\alpha}{2}P^2 + \frac{\beta}{4}P^4 + \frac{1}{2}\delta(\nabla P)^2\right] \mathrm{d}V + W_E + W_w . \tag{7.7}$$

The depolarizing energy depends on the crystal geometry and on the surface domain configuration. On the whole, the depolarizing field inside the crystal is

$$E = -\frac{LP}{\epsilon_0}, \tag{7.8}$$

where L is the depolarizing factor and ϵ_0 the dielectric constant. Thus, the depolarizing energy is

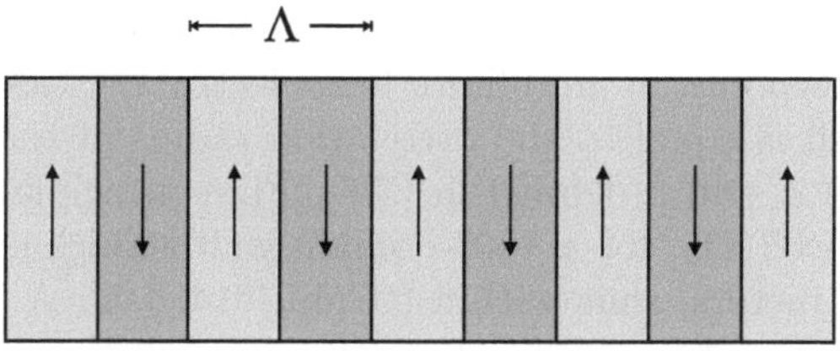

Fig. 7.3. A model periodic domain structure

$$W_{\mathrm{E}} = \frac{1}{2}\int_V PE\,\mathrm{d}V = \frac{1}{2}\int_V \frac{\epsilon}{\epsilon_0}L^2P^2\,\mathrm{d}V. \tag{7.9}$$

For a thin single-domain crystal with a uniform polarization, we have $L = 1$. There were many attempts to estimate W_{E} and W_{w}. The latter parameter requires the energy of the unit domain wall area, usually described by σ. For example, for a practically interesting case of a regular (periodically poled) domain structure (Fig. 7.3) the calculations give

$$W_{\mathrm{E}} = \frac{\epsilon^\star w P_0^2 V}{d} \tag{7.10}$$

$$W_{\mathrm{w}} = \frac{\sigma V}{w} \tag{7.11}$$

where d is the crystal thickness, w the domain width ($d \gg w$), V the crystal volume, and $\epsilon^\star$ a coefficient depending on the dielectric constants. Minimizing $W_{\mathrm{E}} + W_{\mathrm{w}}$ for this structure one obtains an equilibrium domain width

$$w = \sqrt{\frac{\sigma d}{\epsilon^\star P_0^2}} \tag{7.12}$$

The dependence $w \propto \sqrt{d}$ was actually observed in some crystals. The calculated values for σ are strongly scattered, e.g. for 180°-domains the estimates give $\approx 10^{-2}$ and $10\,\mathrm{erg\,cm^{-2}}$ for Rochelle salt and $BaTiO_3$, respectively.

In the equilibrium state the spontaneous strains (7.2) on either side of the wall in the adjacent domains should be equal to avoid large stresses at the domain boundary. This condition of the 'mechanical compatibility' based on the crystal symmetry and the values of d_{ij} and Q_{ijk} imposes limitations on the orientation of the domain walls with regard to the crystallographic axes. Their favorable orientations for some point groups were predicted and tabulated [612]. Recently a phenomenological approach to the domain walls was developed for uniaxial trigonal ferroelectrics, particularly for $LiNbO_3$ and $LiTaO_3$ crystals [610, 613]. According to these calculations the most energetically favorable are the domain walls lying in the yz symmetry plane, which was experimentally supported. Note, that the calculations were performed with no account for a contribution from defects, so are more appropriate for the stoichiometric crystals.

The domain wall thickness is under discussion for long. From the phenomenological viewpoint, in an infinite ferroelectric crystal the thickness of a single domain wall is equal to the correlation radius of the order parameter $r_c = \sqrt{\delta/2\alpha}$ with α and δ defined in (7.7). First-principle calculations give, when applied to $PbTiO_3$, for a 180°-domain wall a thickness of the order of the unit cell parameters, thus within 0.5 nm [614, 615]. A phenomenological consideration taking into account the elastic energy G_{el} and the nonuniformity of P_s, i.e., the existence of domains, was for the first time proposed by Zhirnov [616] by an example of a periodic domain structure like that shown in Fig. 7.3; recently this approach was developed for domains in trigonal crystals [613]. The polarization and elastic strain fields are functions of the coordinate x_n normal to the domain-wall plane, and minimizing the free energy by P gives the solution $P_s(x_n) \sim P_s \tanh(x_n/H)$, where H is the domain-wall thickness. The following expression linking H, the domain width w, and the crystal thickness d was derived by Zhirnov [616] for the equilibrium of the periodical structure (Fig. 7.3)

$$\frac{w}{Hd} = \frac{-K\alpha}{\epsilon} = \text{const} \tag{7.13}$$

where $K = 1.5-2$ and α is the coefficient of the first term in the LG-potential. The corresponding estimates for various crystals gave $H \approx 1$ nm. Unluckily, the accuracy of the most precise methods like TEM or PFM does not permit to estimate the thickness of an 'ideal' domain wall with a satisfying reliability.

A real domain wall thickness is, as a rule, far from these estimates. For example, measurements by the PFM method give for $LiNbO_3$ a domain wall thickness in the range from 65 to 150 nm [617–619]. Additionally, a polarization reversal is accompanied by an onset of nonequilibrium slow relaxing elastic strains adjacent to a domain wall. Another effect is a tilting of a domain wall with respect to the polar axis, which is accompanied by the formation of a space-charge field at the domain boundary. Both situations are characteristic for $LiNbO_3$ and $LiTaO_3$ as discussed in details in Sect. 7.2. These elastic strains and space-charge fields adjacent to the domain boundary result in a disturbed area, thus a domain wall in fact is as thick as hundred nanometers and even micrometers. It should be emphasized that the problem of the domain wall thickness is at present very far from a comprehensive solution.

Screening effects are most important for the mechanism of the polarization reversal and domain dynamics. First, the bound charge P_s is screened by free charges. Additionally, a screening of the depolarizing field accompanying a change of P_s (under external fields, temperature variations, etc.) may sufficiently affect the polarization kinetics and the resulting domain structure. An impact of a photoactive radiation on the domain dynamics, which usually manifests itself as an acceleration of a domain evolution received the name *photodomain effect* [606]. As will be shown later, this effect is characteristic for $LiNbO_3$ and $LiTaO_3$. According to Tagantsev et al. [620], two particular cases of screening exist, namely, the screening of charged domain walls and screening of bound charges on the polar interface between the crystal and the electrodes. The former process obviously contributes to the domain sideways movement, including the on-wall nucleation, whereas the latter is important for the on-surface nucleation. Note, that the latter process is of a special importance for ferroelectric films in view of the concept of so-called interfacial passive or silent surface layers. An important parameter of screening effects is the dielectric (Maxwell) relaxation time

$$\tau_{\mathrm{M}} = \frac{\epsilon\epsilon_0}{\sigma}, \tag{7.14}$$

where ϵ is the dielectric constant, $\sigma = \sigma_{\mathrm{d}} + \sigma_{\mathrm{ph}}$ with σ_{d} and σ_{ph} are the dark- and photoconductivity, respectively. Let us give some estimates. As mentioned in Sect. 3.2.1, in as-grown undoped or slightly doped $LiNbO_3$ we have $\sigma_{\mathrm{d}} \approx 10^{-16} - 10^{-18}\,\Omega^{-1}\,\mathrm{cm}^{-1}$ (by indirect estimates), corresponding to $\tau_{\mathrm{M}} \approx 5 - 500$ days. This means that screening by dark carriers plays no role in the polarization reversal, which occurs significantly faster (maximum length are tens of seconds). On the other hand, the photoexcited carriers may essentially affect the domain dynamics, especially at high intensities. As shown by Fridkin [606], these effects are very pronounced in photosensitive ferroelectrics like $BaTiO_3$ and SbSJ. In the case of $LiNbO_3$, they may be expected, e.g. in optical-damage-resistant crystals doped with Mg. In compositions like $LiNbO_3$:5%Mg σ_{ph} becomes at moderate light intensities $10^{-11}\,\Omega^{-1}\mathrm{cm}^{-1}$ (see Fig. 4.8), thus τ_{M} is of the order of tens of seconds or lower. In summary, a contribution from screening effects to the polarization and domain kinetics is determined by the balance between switching and screening velocities. The role of screening effects on the domain dynamics in $LiNbO_3$ and $LiTaO_3$ crystals is discussed in detail by Shur [609].

Polarization processes in ferroelectrics are strongly affected by features, combined in the general term *real crystal structure*, which involves various accidental imperfections, such as point defects and dislocations, growth inhomogeneities, etc. One of the manifestations of the real crystal structure impact is the internal field in ferroelectrics. As this factor is of special importance for $LiNbO_3$, we shortly dwell on it. P–E hysteresis loops are often shifted along the E-axis or constricted in one direction (this asymmetry may appear with time). A shift of the P–E loop is denoted as unipolarity and described in terms of a so-called *bias internal* field

$$E_{\mathrm{int}} = \frac{E_{\mathrm{c1}} - E_{\mathrm{c2}}}{2}, \tag{7.15}$$

an averaged $E_{\mathrm{c}} = E_{\mathrm{c1}} - E_{\mathrm{b}} = E_{\mathrm{c2}} + E_{\mathrm{b}}$. The origin of E_{int} is a basic problem of ferroelectricity, first of all because it is closely related to the ageing process, which particularly reveals oneself in a distortion of the P–E loop. As E_{int} means a preferred and stabilized orientation of P_{s}, and thus making a reversal difficult or even preventing it, the unipolarity is sometimes referred to as a stabilization effect.

Several qualitative microscopic mechanisms for E_b may be proposed [621].

- Bulk effects related to the alignment of dipole defects in the polar matrix.
- The wall effect in a polydomain crystal caused by a defect diffusion toward the domain walls and the subsequent pinning of the latter.
- Effects related to a volumetric nonuniformity of P_s (e.g., in ceramics or relaxor ferroelectrics), which results in a screening of the electric fields originating from $\nabla P \neq 0$ either by charged defects or free charges.
- Stress effects associated with the macroscopic mechanical stress produced by the domain structure itself.

Note that E_b may be due additionally to incidental reasons, for example such as a nonequivalency of polar surfaces (electrodes, surface layers), a prearranged impurity distribution over the crystal bulk, etc. According to the concept of Lambeck and Jonker [621], in a single-domain uniformly polarized ferroelectric the dominant origin of E_b is a volumetric alignment of polar defects. From the thermodynamical viewpoint the internal field may be discussed in the following simplified way [622]. A uniform alignment of dipole defects in consistence with the P_s direction results in a decrease of the free energy by a value $\Delta F = N\Delta W$, where N is the concentration of noninteracting defects, ΔW is the energy of the dipole field. Thus, the symmetric thermodynamic potential Fig. 7.1 transforms to an asymmetric two-minimum function similar to one in the presence of an external field

$$G(P,T) = G_0 + \frac{\alpha(T)}{2}P^2 + \frac{\beta(T)}{4}P^4 + \frac{\gamma(T)}{6}P^6 - E_b P, \tag{7.16}$$

where $E_b = N\Delta W/P_s$. So, the internal field is a formal equivalent of the energetic difference of the two poling directions.

The origin and properties of E_b were investigated in perovskites, mainly in $BaTiO_3$, and in triglycinsulfate (TGS). According to these results, two alternatives may occur, namely the E_b orientation (thus the direction of the defect dipoles) can or cannot be reversed following the P_s reversal. The first process was observed and investigated in $BaTiO_3$ crystals and ceramics and its relatives, for references see [623]. In this case E_b appears in TM (Mn, Fe)-doped crystals after a reduction (hydrogen-annealing), which produces oxygen vacancies [621]. The dipole defects are formed by the coupling of Mn (or Fe) substituting for Ti to the nearest V_O, so E_b is owing to aggregates like $Mn^{2+} - V_O$. Applying a switching external field to a poled $BaTiO_3$:Mn at elevated temperatures of about $T \approx 100°C < T_c = 120°C$ leads to a change of the sign of E_b [621]. Simultaneously, EPR measurements [623] found a realignment of the defect Mn^{2+}–V_O in line with the reversed P_s direction. Note that the dipole realignment which proceeds during seconds, is due to the P_s reversal rather than to a field action itself, because applying the same field at $T > T_c(P_s = 0)$ does not affect the dipole orientation [623]. The possibility of the realignment of Mn^{2+}–V_O dipoles is accounted for by thermally activated diffusive jumping of V_O via equivalent O-sites in oxygen octahedra. A quantitative model for the corresponding rotation of these defect dipoles was proposed [622]. Interestingly, that isolated defects (e.g. Ti^{3+} or Pb^{3+} in PLZT ceramics) also may be realigned by applying fields in the ferroelectric phase, which is attributed to a change of their off-center position in O-octahedra [623]. In contrast to $BaTiO_3$, in TGS crystals E_b cannot be realigned by applying a field in the ferroelectric phase.

7.1.2 Model Description of the Domain Dynamics and Ferroelectric Switching

In this section we do not discuss the dependence of the polarization reversal on the crystal thickness, because these effects manifest themselves only for a thickness of some tens of μm.

Similarities of the domain dynamics and the switching, mainly found in $BaTiO_3$, and partially in TGS, Rochelle salt, GASH and other soft ferroelectrics, formed the groundwork for the model description. Studies were performed by correlating optical observations with switching currents.

Experimental studies of the switching current found in the low-field range an exponential field dependence of the switching time t_s. By convention t_s is the time corresponding to 90% drop off of the switching current (Fig. 7.4), or, what is the same, to 90% of the switched charge) and is described by

$$t_s \propto \exp\left(\frac{\alpha}{E}\right), \tag{7.17}$$

where α is an activation field. This empirical activation law (7.17) is valid for almost all ferroelectrics, although the corresponding low-field range varies in different ferroelectric crystals. With increasing field $t_s(E)$ diverges from (7.17)

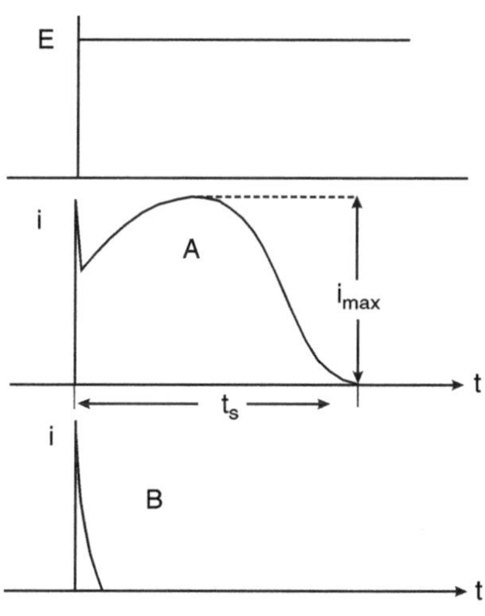

Fig. 7.4. Applied field E and switching current i vs. time in $BaTiO_3$. *Curve* A corresponds to switching pulses with E antiparallel to P, while *curve* B is obtained for E parallel P when no switchings occurs, redrawn from [2]

tending to a power law $t_s \propto E^{-n}$ with n depending on the crystal, e.g. $n = 1.5$ for $BaTiO_3$, 1 for TGS, and 5.2 for $LiH_3(SeO_3)_2$, etc.

Figures 7.5 and 7.6 illustrate the field dependencies of t_s for $BaTiO_3$ and TGS, respectively. An exponential behavior at low fields is observed in both crystals. Naturally, in a unipolar crystal the empirical dependence (7.17) for fields of opposite sign are described by different activation fields.

Optical studies detected four stages of the domain kinetics under external fields:

- the domain nucleation;
- the forward (transit) growth of nucleated domains along the field direction;
- the sideways domain growth via the domain wall lateral movement in the direction normal to the polar axis;
- the coalescence of domains, tending to a single-domain state.

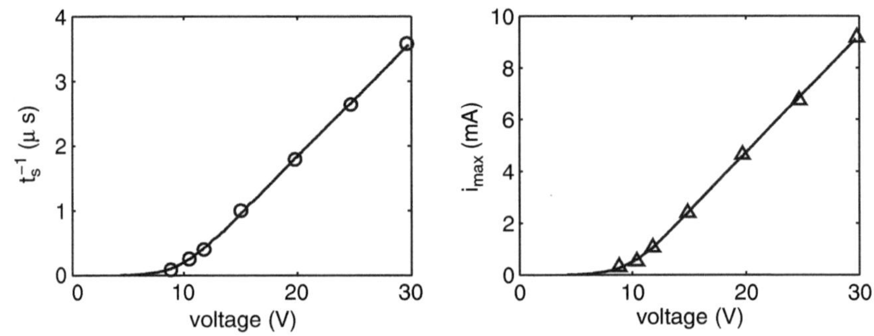

Fig. 7.5. Reciprocal switching time t_s and maximum current i_{max} vs. applied voltage in $BaTiO_3$, redrawn from [2]

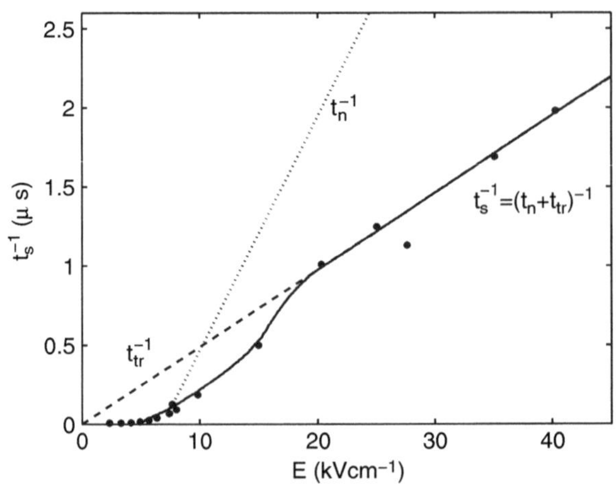

Fig. 7.6. Reciprocals of the switching time t_s, the transit time t_{tr} and the nucleation time t_n vs. applied field for TGS, redrawn from [2]

So, the main questions were, first, what is the mechanism in each stage and, second, which process dictates the switching time t_s.

Let us discuss the switching stages in the generally accepted qualitative language. The mechanism of the domain nucleation is a subject of numerous discussions and has not been clarified yet. Usually it is presented as a statistical process with the probability depending on the external field E directed oppositely to P_s

$$p = p_0 \exp\left(\frac{-\beta}{E}\right) \tag{7.18}$$

predicting qualitatively an exponential dependence of the nucleation rate t_n^{-1} on E. A change of the energy accompanying the nucleation of an antiparallel domain is

$$\Delta W = W_{\mathrm{E}} + W_{\mathrm{w}} - 2PEV \tag{7.19}$$

where the third term is the electrostatic energy of a nucleus of volume V under an external field E. Obviously, among the formed nuclei only a rather large one (exceeding certain critical dimension) can grow without collapse. In the classical model of the domain nucleation [624] an estimate of energies required for the formation of a critical nucleus (an isolated antiparallel nucleus in a polarized body) gives enormous values, for example ΔW 10^8 kT for $BaTiO_3$, which eliminates the possibility of a nucleation via thermal fluctuations. Evaluations of the activation field give $\beta \approx 10^5 - 10^6\,\mathrm{V\,cm^{-1}}$, which by orders of magnitude exceeds the experimentally observed fields of the domain nucleation. To remedy this problem it was reasonable to assume the existence of certain centers triggering the nucleation by analogy with the crystallization in under-cooled solutions. An example of the probable nucleation centers might be unregulated micro- or nano-domains located on the polar surfaces. Actually, the modern precise methods detect such microstructures in some ferroelectrics even in a nominally single-domain state. Particularly, those were noticed in nominally single-domain $LiNbO_3$ [625]. A spontaneous appearance of antiparallel domains on the polar face may be qualitatively accounted for by a specific of the surface layers on the polar faces, which may be related particularly to screening effects. For example, Janovec [626] assumed that high screening fields may lead to a partial polarization reversal within a surface layer. This might be a possible reason of the surface microdomains initiating the domain nucleation.

The sideways domain velocity v_{sw} obtained from optical observations in $BaTiO_3$ obeys an exponential law in the low-field range

$$v_{sw} = v_\infty \exp\left(-\delta/E\right) \tag{7.20}$$

where δ is the activation field for the sideways domain motion depending on many factors, such as electrodes, crystal thickness, etc. At higher fields the dependence transforms to a power law

$$v_{sw} \propto E^{1.4} \tag{7.21}$$

The calculations have found, that the real lateral domain wall movement under moderate fields is energetically unfavorable. To overcome this argument, a concept of an apparent sideways domain motion was suggested [627, 628]. This process may be imagined as a generation of antiparallel domain nuclei on an already existing domain wall, and their subsequent coalescence. It is energy-facilitated when compared to the formation of an isolated nucleus in a fully antiparallel body. Most energetically favorable is a triangle-shaped nucleus. According to estimates of Miller and Weinreich [628] the energy of formation of such a nucleus with several lattice parameters in thickness on a 180°-domain wall is of about $\delta \approx 10 - 12\,\mathrm{kT}$ at ambient conditions. Multiple nuclei generated on the domain wall form a pattern of ledges, which tend to merge making an impression of a sideways wall movement. This process of the on-wall domain nucleation was recently directly observed in $LiNbO_3$ crystals with the aid of a precise optical method [608]. This model accounts for an activation field dependence (7.20) of the sideways wall velocity at relatively low fields, because $v_{sw}(E)$ is proportional to the rate of the on-wall nucleation obeying an exponential field dependence analogous to (7.18). As seen from the comparison of $t_{\mathrm{s}}^{-1}(E)$ in $BaTiO_3$ and TGS (Figs. 7.5 and 7.6), an exponential law is fulfilled in the low-field ranges in both crystals and is assigned to the dominant role of the nucleation process. However, at higher fields the deviation of $v_{sw}(E)$ from the exponential law (7.20) has found no unambiguous explanations. In $BaTiO_3$ it was accounted for by either the dominant role of the real sideways domain wall movement obeying a power law, or by the fact that at high fields the domain nucleation itself follows a power law similar to (7.21). A final conclusion for $BaTiO_3$ was that in a low-field range an activation nucleation process is dominant, whereas at higher fields the sideways movement (both via the on-wall nucleation and a lateral movement) governs the switching time. As seen in Fig. 7.6, in TGS there exists an intermediate field range between the low-field (exponential) and high-field (linear) parts. It was attributed to a participation of the domain transit (frontal growth), which may be regarded as a viscous drag

$$t_{\mathrm{tr}}^{-1} = kE = \frac{\mu E}{d}, \tag{7.22}$$

where μ is the domain wall mobility and d is the crystal thickness. The total switching time is defined as a combination of t_{n} and t_{tr} and may be presented as

$$t_{\mathrm{s}}^{-1} = (t_{\mathrm{n}} + t_{\mathrm{tr}})^{-1} = \left[t_0 \exp\left(\frac{\beta}{E}\right) + \frac{d}{\mu E} \right]^{-1}. \tag{7.23}$$

Expression (7.23) describes adequately the experimental dependence of $t_{\mathrm{s}}^{-1}(E)$ in TGS, see Fig. 7.6, in a wide field range and means that the switching time is controlled by a slower process, namely, by the nucleation (including the on-wall nucleation) in a low-field range, and by a frontal domain movement at higher fields.

In summary, the outlined general model of the switching process based on the research of three groups [624, 628, 629], so-called Miller–Weinreich model, is commonly accepted up to the present. The main results of their studies are the existence of several stages of the domain evolution and the separation of their contributions to the overall switching velocity in the model crystals. Note again that the exponential law (7.17) is general for low fields and may owe either to the nucleation, or to the sideways wall movement via the on-wall nucleation. The behavior at higher fields is more individual.

Ishibashi and Takagi [630] made an attempt to describe the switching process in the framework of the general approach to the phase transformation developed independently by Kolmogorov [631] and Avrami [632, 633] for the description of the nucleation and growth kinetics of the crystallization process. In the concept referred to as the Kolmogorov–Avrami–Ishibashi (KAI) switching model, the switching kinetics is expressed in terms of the temporal dependence of the switched fraction $Q(t)$ of the crystal volume; the value $Q(t)$ depends on the domain-wall sideways motion after the nucleation. Following Avrami [632, 633], two fundamentally alternative cases are discussed resulting in different $Q(t)$ kinetics: (1) in the course of switching under an applied field the domain nuclei arise progressively with the constant rate (the heterogeneous or α-case); (2) only latent initial nuclei are involved and no new nuclei appear when switching (the homogeneous or β-case). For these two cases the KAI model leads to different expressions for the ratio $i_{\text{max}} t_{\text{max}} / P_{\text{s}}$, based on which Ishibashi and Takagi [630] tried to interpret in those times available data for $BaTiO_3$, TGS, and $NaNO_2$. However, the comparison with experiment seems to be not very successful, obviously because the KAI model was developed for an infinite crystal and assumed a uniform distribution of nuclei both in the α and the β case.

Nowadays, the Miller–Weinreich model is a subject of further developments when discussing the domain-wall motion in the language of pinning effects. A pinning means that the domain wall motion under an external field is impeded because of the wall coupling with a defect, i.e., because of formation of certain potential barrier. Overcoming this barrier, i.e., the wall depinning may be achieved by increasing the applied field or/and heating. As shown below, pinning effects pronouncedly manifest themselves in $LiNbO_3$ and $LiTaO_3$ and are evidently responsible for extremely high E_{c} and E_{int} in these crystals.

In addition to the regularities of switching discussed above, experimental studies in pulsed switching found several specific effects of practical importance. In the first instance these are fatigue (or ageing) effects manifesting themselves as decreasing values of the switched charge Q_{s} (or, which is the same, of the switched crystal volume) after multiple switching cycles (not exceeding 10^6 in $BaTiO_3$ or TGS). Just fatigue effects set limits on the most important applications of ferroelectrics.

An important factor controlling the polarization process is the shape of the voltage trains applied to the crystal, namely the pulse duration and the off-duty factor. An impact of these factors manifests particularly in the effect of the domain backswitching observed for the first time in $BaTiO_3$ [634]. The backswitching means a partial (or, sometimes, a total) depoling of the crystal right away after turn-off the switching field. A related effect is an influence of the pulse duration t_p on the switched charge Q_s which is characteristic for relatively low field amplitudes (close to E_c). If t_p is shorter than certain critical $t^\star$, then after applying a single pulse the value of the remnant polarization is negligible or even zero. In the first case, Q_s may be accumulated by means of applying a series of short pulses (up to 10^8) of the same polarity and duration. If $t_p > t^\star$, then several pulses or even a single one is enough to switch P_s completely. For soft ferroelectrics like TGS, $t^\star$ is in the range of few milliseconds and decreases with E and T. In rigid ferroelectrics like $LiNbO_3$, $t^\star$ is in the range of seconds, as shown later. The physical meaning of $t^\star$ is yet unclear. Sometimes it is regarded as a time required for the formation of a critical nucleus, in other cases, as in $LiNbO_3$ it is accounted for by the backswitching process. At present, the backswitching is sometimes believed to be related to pinning effects, namely to the existence of certain critical $t*$ required to depin the wall.

We now dwell shortly on surface layers on the polar faces of ferroelectrics, because such a layer has an influence on the domain nucleation. First, specific skins do exist on the surfaces of as-grown crystals, or mechanically treated (e.g. polished) samples. The properties of these skins, such as the dielectric constant, conductivity, pyroelectric coefficients, etc., differ from the bulk parameters, which manifests particularly in the thickness dependencies of these values. Moreover, the structure of the surface layer may be distorted, which is exemplified by $LiNbO_3$, where a surface layer after a mechanical treatment contains micro-inclusions of another phase ($LiNb_3O_8$). These skins and their impact in some cases can be avoided, for example, by etching in $BaTiO_3$, cleavage in TGS, or by a special heat treatment in $LiNbO_3$ [635]. More important is a fundamental specific of the polar surface layer in ferroelectrics related to the screening of P_s by charge carriers. The properties of a formed interface depend on the screening conditions. Particularly, at the interface *insulating ferroelectric–conductive electrode* several specific effects are expected, for example, the depletion effects characteristic for wide-gap compensated semiconductors. An estimate for a reasonable carrier density close to the surface of about 10^{19}–10^{21} cm^{-3}, forming a Schottky barrier, gives a surface field in the range of 10^4–10^5 V cm^{-1} [4]. Another estimate of the field developed in a layer with a thickness of about 0.5 μm and a concentration of about 10^{19} cm^{-3} of single-ionized donors gives about 2×10^4 V cm^{-1} [620] (in the recent bibliography on the screening effects in ferroelectrics one may find e.g. in [620]). These fields might be an substantial factor for the domain nucleation. A well-known predominant domain nucleation by electrode edges is evidently related to the edge anomalies of these fields.

Different screening conditions cause an influence of the electrode materials on the switching process. For example, an essential dissimilarity is observed in the switching characteristics under metal and liquid electrodes. It may be exemplified by the switching experiments in SBN and $Ba_2NaNb_5O_{15}$ crystals [636] where on using liquid electrodes (LiCl water solution) the values of E_c were twice as much lesser than for metal ones.

7.1.3 Experimental Methods for Domain Studies and Polarization Kinetics

Observations of the Ferroelectric Switching

The integral $P(E)$ characteristics are obtained either by measuring the switched charge Q_s or the switching current $i_s = dQs/dt$ (Fig. 7.4) accompanying the polarization reversal. The switched charge is often measured with the aid of the Sawyer–Tower method (Fig. 7.7); the switching current is observed with the Merz method by measuring the voltage drop at the calibrated resistor placed in series with the sample. In the original version the Merz method used square (Π-) field pulses, so the whole model description and terminology was developed for the pulsed switching. An appropriateness of Π-pulsed fields is the affine shape of the current (i.e., its independence on the Π-parameters). Particularly, for Π-pulses the field dependencies $t_s^{-1}(E)$ and $i_{s,max}(E)$ are identical. Under Π-pulses of alternating polarity the switching current exhibits the well-known form presented in Fig. 7.4, provided that the pulse rise-time is much shortened than t_s. The switched charge $Q_s = \int_0^{t_s} i_s dt$ at the limit tends to $2P_s/A$, where A is the area under the electrodes, at $E \geqslant E_c$. It should be emphasized that the term *switching time* may be used provided that the P_s reversal occurs completely; otherwise, if Q_s is lesser, or, which is the same not the whole crystal volume is involved in the switching, as it takes place e.g. in relaxor ferroelectrics even at $E \gg E_c$ [637]), then this term loses its meaning. Applying abruptly rising high fields with alternating polarity is poorly suitable for ferroelectrics with a strong piezoelectric effect, $LiNbO_3$ among them, because such fields promote a crystal cracking.

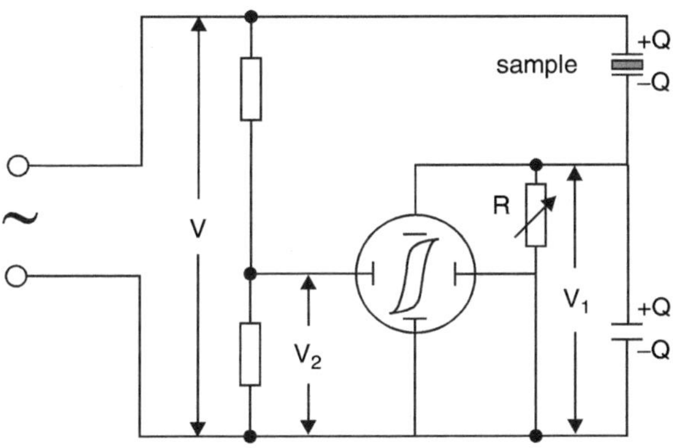

Fig. 7.7. Schematic circuit of a Sawyer–Tower bridge for observing P–E relations in ferroelectrics

A method often used for $LiNbO_3$ and similar objects is the application of ramping (saw-tooth) fields.

A special case of switching currents are Barkhausen jumps, which are characteristic for switching under relatively low, slowly increasing (ramping) or step-by-step rising fields, and look like a series of the noise-spikes superimposed onto the switching current. Their overall charge does not exceed 1% of the total switched charge. Barkhausen currents are evidence of a discrete polarization process. As shown for $BaTiO_3$, a positive Barkhausen pulse (that is the spike of the same polarity as the switching current) accompanies either the formation of a nucleus with a size exceeding the critical value, or a step-like sideways domain jump. A negative Barkhausen pulse seems to correspond to a backswitching act. At the same time, the microscopic origin of the Barkhausen pulses on the whole has not been clarified yet.

Domain Observations

In a detailed bibliography on the methods of the domain structure observation one may find particularly in the recent review by Soergel [638], so we mention very briefly those methods, which found or may find application in $LiNbO_3$. These methods may be divided into those for the surface and the volume. All the surface methods are based on descrying a distribution of the surface charge (or of a surface mechanical relief following the charge distribution), which replicates the near-surface domain distribution. These methods include the chemical etching, powder deposition and colloidal decoration methods, the scanning force microscopy (SFM) and scanning electron microscopy (SEM). In addition to the SEM method, the transmission electron microscopy (TEM) method can be used for these aims, but it requires very thin samples. The etching method is universal for all ferroelectrics and is based on a different etching rate for the positive and negative ends of the domains by an individual etchant. In $LiNbO_3$ the −c end of domains is etched much faster than the +c end [639, 640]. In contrast, in $BaTiO_3$ the positive end is etched faster, although at first sight these two crystals seem to be relatives.

The etching forms the basis for the so-called Miller–Savage method of the domain dynamics observations under low fields [641]. The switching field is applied step-by-step to the crystal, whereupon, subsequently to each step the crystal is etched and a shift L of a domain wall is estimated to give the wall velocity $v_s = L/T$, where T is the duration of the field step. This widely used method gives an average velocity for a given field step, because v_s reaches its maximum instantaneously by applying the pulse and further decreases to a saturation value.

SEM and scanning probe microscopy (SPM) methods are very informative and provide extremely high resolution up to nanometers and even better. The SPM group includes several methods, from which the most convenient for ferroelectric domain studies are scanning force microscopy (SFM) techniques. The latter may be applied in various modes of operation, such as contact and noncontact modes using either insulating or conductive tips, etc. The most often used mode for the studies in the domain structure is the contact mode with a conducting tip being under an ac-voltage. This method of using the converse piezoelectric effect $u_{jk} = d_{ijk}E_i$, see (7.2), is based on scanning the distribution of oscillations of the sample surface induced by the tip ac-voltage. According to this widespread (but not unique) explanation, the domains are imaged via the phase contrast of the local piezoresponse, which is due to the fact that the components d_{ijk} are sensitive to the sign of P_s. This method as a rule is referred to as the piezoresponse force microscopy (PFM).

The bibliography on domain observations in $LiNbO_3$ by SPM methods one may find for example in [617] and in a recent review [642]. As an associated surface method of the domain visualization one may regard the exoelectronic emission arising from the polar cuts of a pyroelectric crystal when changing the temperature. This emission which is obviously due to noncompensated strong pyroelectric fields, is different for $+P$ and $-P$ cuts. Most of the surface methods are appropriate mainly for observations of the static domain structure. The main drawback of all these methods is that they characterize the domain structure near the surface rather than in the bulk. At the same time, in many ferroelectrics the surface and bulk domain densities are completely different. Additionally, some of the surface methods (SFM, SEM) are not completely noninvasive, because both the electron beam and SFM-tips may affect the domain structure. As shown later, these instruments may be used for a creation of domain patterns. Recently, with $LiNbO_3$ as an example it has been shown that even the chemical etching may affect the domain pattern, too [643].

The bulk methods are first of all the structure studies. Domain observation by X-ray diffraction methods are based on the anomalous dispersion of the X-rays resulting from a difference of the reflexion intensities from positive and negative domains. For $LiNbO_3$ this technique was applied for the first time by Wallace [644]. Another possibility is based on the observations of the domain walls rather than the domains themselves, owing to the existence of a lattice distortion in the vicinity of a nonideal domain wall, which is due to various reasons. Particularly, in $LiNbO_3$ the measurements by means of a high-resolution synchrotron X-ray diffraction method [645] and high-resolution X-ray topography [646] detected extremely high mechanical strains up to 10^{-4} in the vicinity of the domain walls. Thick domain walls formed in a head-to-head (tail-to-tail) grown PPLN were observed by means of X-ray topography [647]. Basics of the X-ray diffraction observation of periodically poled ferroelectric structures, particularly PPLN, and the appropriate bibliography one may find in the following references [648–651].

Optical methods belong to the volumetric ones and permit to probe the switching process and the domain distribution over the crystal bulk. The eldest is the polarizing-optical microscopy applied by Merz for pioneering investigations of domains in $BaTiO_3$. This crystal is optimal for this method, because it exhibits both antiparallel (180°)$-c$-domains and two equivalent

(90°)-a-domains, where the P_s vectors in the adjacent domains are orthogonal, and a domain boundary is inclined at 45° to P_s. When a polarized light-wave incidents normally to the polar c-axis, the 90°-domains are visualized. This method, in principle is not appropriate for optically uniaxial crystals, because the 180°-domains are optically indiscernible, i.e., the refractive indices $n_i^+ = n_i^-$. For these crystals the polarizing-optical method may be refined utilizing the electro-optic effect. If applied to a crystal, an external electric field $E_j < E_c$, then the refractive indices in the adjacent domains are changed

$$\Delta n_i = -\frac{1}{2} n_i^3 r_{ij} E_j \tag{7.24}$$

giving an optical contrast $\Delta I \propto \Delta\Phi^2$, between light waves transmitting through the opposite domains; with $\Delta\Phi = (2\pi/\lambda) d\Delta n_i$, and the crystal thickness d.

An additional possibility of the domain observations is provided again by nonideal (thick) domain walls. If strong mechanical strains occur in the vicinity of domain walls, they affect the refractive indices via the elasto-optic effect,

$$\Delta B_{ij} = p_{ijkl} u_{kl} \tag{7.25}$$

where p_{ijkj} and u_{kl} are the elasto-optic coefficients and mechanical strains, respectively. A combination of the conditions (7.24) and (7.25), namely, appearance of the optical contrast between the antiparallel domains under external fields and additional contrast caused by mechanical strains at the domain walls permitted Gopalan et al. [652] to elaborate the electro-optic imaging microscope for the observation of slow domain dynamics in $LiTaO_3$ and $LiNbO_3$. If a domain wall is tilted with respect to the polar axis, it results in a formation of a space-charge field directed normally to the wall. This field changes the refractive indices via the linear EO-effect following (7.24). In both the cases, the thick domain walls may be observed by the polarizing-optical method. As seen later, these cases are realized in $LiNbO_3$, especially for a nonequilibrium (freshly formed) domain structure.

The components r_{ij} of the linear electro-optic (EO) effect and those of nonlinear optical coefficients are related in oxygen octahedra ferroelectrics in the ferroelectric phase by

$$\begin{aligned} r_{13} &= 2g_{13} P_s \epsilon_{33} \epsilon_0 \\ r_{33} &= 2g_{33} P_s \epsilon_{33} \epsilon_0, \end{aligned} \tag{7.26}$$

using the contracted notation [653]. The g_{13} and g_{33} are quadratic electro-optic coefficients in the centrosymmetric paraelectric phase.

Further, the electro-optic components r_{ij} can be determined from the nonlinear susceptibility components d_{ij} by interchanging rows and columns [654]

$$d_{ji} = -\frac{n^4}{4\pi} r_{ij}. \tag{7.27}$$

These relations form the basis for several optical methods for observations of the ferroelectric switching and of domains. The fast and nondestructive EO method permits a real-time observation of the switching kinetics by means of measuring the induced phase-shift $\Delta\Phi(t)$ under applied E. The $r_{ij}(E)$ are governed by the P–E hysteresis, so the EO method permits to probe the distribution of the P–E loops over the crystal bulk [655]. A scanning electro-optic microscopy method was developed, a mapping of the linear electro-optic effect [656]. To visualize the 180°-domains the contrast is controlled by the direction of the modulating external field applied to the crystal through semi-transparent electrodes. Additionally, a spatial distribution of the EO-coefficients may characterize qualitatively a spatial distribution of the domain density, which is seen from the following simple consideration. Let an elementary volume of the crystal be v and the corresponding volumes of positive and negative domains are v^+ and v^-, respectively; then, we have $v^+ + v^- = v$ provided that the domain wall thickness is much smaller than the domain size. For example, a positive domain density in a negative matrix is $D^+ = v^+/v$, thus a linear EO-coefficient is related to the degree of the total volume polarization as $r/r_0 = (v^- - v^+)/v$, where r_0 corresponds to the single-domain state. This leads to the following relation between the EO-coefficient and the domain density.

$$\frac{r}{r_0} = 1 - 2D^+. \tag{7.28}$$

So, EO-coefficients linearly decrease with D^+ (on the contrary, the half-wave voltage $V_{\lambda/2}$ increases) approaching zero in a polydomain crystal, where by definition $D = 0.5$. Therefore, in a given crystal, a spatial nonuniformity of r_{ij} or $V_{\lambda/2}$ may be interpreted in terms of the spatial distribution of the domain density D.

The relations (7.26) provide a group of methods, specific for domain observations in photorefractive crystals, because in photorefractive holographic gratings the diffraction efficiency η (3.11) and the two-beam coupling gain Γ (3.14) depend on r_{ij}, thus being controlled by P_s.

In accordance with (7.26), in antiparallel domains the recorded gratings are shifted by π with respect to each other, which is revealed on read-out [657]. A relatively new digital holography interferometry technique [658, 659] of an in situ observation of the domain dynamics is based on the same principle. The dependence of Γ on the sign of P_s permitted to develop a method of three-dimensional (3D) mapping of the antiparallel domains hidden within the crystal bulk [660]. The ferroelectric switching may be investigated by means of measuring Γ on the applied field E [661] or of the hysteresis dependence of the photoinduced light-scattering (fanning) [662].

The NL-O methods utilizing a fundamental relation of the nonlinear susceptibility to P_s were developed already in early studies in ferroelectrics; an example is the classical express search method for a ferroelectric phase transition via the generation of second harmonic light [663].

The relation of the NL-O parameters to P_s underlie several modern NL-O methods. These methods are especially appropriate for $LiNbO_3$ and $LiTaO_3$ crystals due to rather high nonlinear susceptibilities. Recall that the polarization term responsible for the SHG in the general form is presented as

$$P_i(2\omega) = d_{ijk} E_j(\omega) E_k(\omega), \tag{7.29}$$

where $E_j(\omega)$ and $E_k(\omega)$ are the field components of the fundamental harmonic (FH). The SH-intensity is proportional to the square of P_i and controlled by the nonuniformity of d_{ijk}. One of the methods based on these relations is the nonlinear-optical microscopy (SHG microscope) [664,665] aimed to obtain a 2D-image of a sample with a nonuniform d_{ijk} distribution, particularly caused by the ferroelectric domain structure. At first, SHG microscopy was successfully tested by an example of the 90°-domain structure in $BaTiO_3$ [664], and then was applied for visualization of the 180°-domains in $LiNbO_3$ and $LiTaO_3$, particularly of a PP-structure [665]. As the values of d_{ijk} and refractive indices in 180°-domains are equal, so no spatial variation of the SH-intensity may occur in this case. The possibility for a visualization is due to the fact that the SH waves generated by opposite domains are out of phase according to (7.26), therefore a domain visualization may be realized by means of detecting this phase shift. An additional possibility of the domain imaging is provided by thick domain walls, as was demonstrated by an example of PP-KTP [666]. The SHG method combined with a scanning confocal microscope, [667] permitted to observe domains at the microscopic level, e.g. in PPLN. A relatively new volumetric method analyzing the spatial profile of the quadratic susceptibility $\chi^{(2)}$ (a nonlinear tomography) has been developed by Kitaeva et al. [668,669]. It is based on a specific three-wave process of the spontaneous parametric light scattering in media with spatially nonuniform $\chi^{(2)}$ and is dealt briefly in Sect. 7.1.4

Note that a certain drawback of the nonlinear optical methods is that they cannot always be regarded as fully noninvasive, because the SH-light of the green-blue spectral range can affect the domain characteristics, first of all due to the screening effects.

Thick domain walls provide some new optical techniques of measuring the domain dynamics, such as the light deflection on the domain walls in $LiNbO_3$ and $LiTaO_3$ [670] and the Rayleigh-like light scattering of the transmitted light at the domain walls [671].

The main drawback of almost all available optical methods is a relatively low spatial resolution, which due to the diffraction limitations is of the order of the wavelength, thus being strongly behind the resolution of the surface methods. Some ways of improving the resolution are using the scanning confocal microscopy [667] or the near-field scanning microscopy (NSOM) [672,673].

7.1.4 Background of the Optical Frequency Conversion in the Quasi-phasematching (QPM) Regime

In this section, we present briefly the background of the optical frequency conversion on the ferroelectric domain structure. Waves with frequency ω traveling through nonlinear optical media produce a polarization $P(2\omega)$ according to (7.29), which acts as a source term in the wave equation and consequently produces radiation at the second harmonic frequency. $E(2\omega, z)$ depends on $\Delta k = k(2\omega) - 2k(\omega)$ and gives, neglecting depletion of $I(\omega)$, an intensity $I(2\omega)$

$$I(2\omega, z = L) \propto \frac{d_{\text{eff}}^2 \, L^2 \, I(\omega)^2}{n(\omega)^2 \, n(2\omega)} \left(\frac{\sin\left(\Delta k \, L/2\right)}{\Delta k \, L/2} \right). \tag{7.30}$$

The general dependence $I(2\omega)$ is presented in Fig. 7.8 by the lower (light-gray) curve, which shows an interference quenching at positions which are multiples of a unit length. The parameter defining this periodicity is the coherence length $L_c = \pi/\Delta k$. To provide an amplification, energy and momentum conservation should be fulfilled; $2\omega_1 = \omega_2$ and $2k_1 = k_2$. This may be achieved due to birefringence, i.e., the identity of the phase velocities of the fundamental and second harmonic. In this case the intensity is presented in Fig. 7.8 by the dotted parabola.

A different approach is the compensation of the mismatch between the ω_1 and ω_2 by an appropriate phase shift, a quasi-phase-matching (QPM). This means that every time when the optical path increases by L_c, the phaseshift between the two waves is changed by $\pi/2$ thus eliminating the mutual interference quenching. Taking into account the expression $\Delta k = (4\pi/\lambda)(n(2\omega) - n(\omega))$, the required period is given by

$$\Lambda = m \, 2 \, L_c = \frac{m\lambda}{2(n(2\omega) - n(\omega))} \tag{7.31}$$

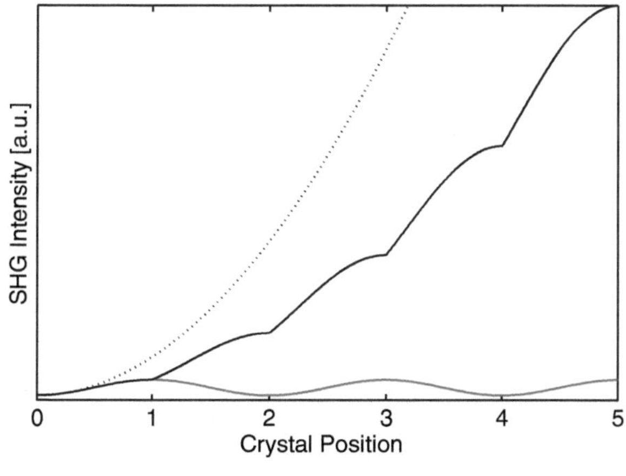

Fig. 7.8. Comparison of the various types of phase-matching. Intensities for phase-matched SHG (*dotted parabola*), quasi-phase-matched (*black solid line*) and nonphase-matched SHG (*light gray*)

with $m = 1, 3, 5, \ldots$. Because of $r_{ij} \propto P_s$ (7.26), an appropriate candidate for such a QPM is a ferroelectric domain structure, where P_s can cause a sign change. On the basis of this simplified consideration a concept of the optical frequency conversion on a regular domain pattern, a periodically poled structure with alternated P_s in each building unit was developed. An extensive literature on the quasi-phase-matching, particularly with the use of the regular domain structure in various ferroelectric crystals one may find in several reviews, the most recent is [674].

The required periods of such domain gratings are in the range of microns, for example for $\lambda \approx 1\,\mu m$ the first-order QPM ($m = 1$) requires $\Lambda \approx 7\,\mu m$ in $LiNbO_3$. Figure 7.9 presents the PPLN periods needed for SHG in the QPM regime in a wide commonly used spectral range.

Figure 7.8 may suggest that the phase-matched signal is always larger than the QPM one. However, this assumed drawback is not true in general. For example, in the QPM regime the maximum nonlinear coefficients can be used which are not necessarily involved in the conventional phase-matching. This is the main advantage of QPM!

This will be exemplified by $LiNbO_3$. Here, the nonlinear coefficients are $d_{33} = 34.4\,pm\,V^{-1}$, $d_{31} = d_{32} = 5.95\,pm\,V^{-1}$; and $d_{22} = 2.4\,pm\,V^{-1}$. The relevant coefficient for the SHG in the phase-matching regime is d_{31}, resulting in $d_{eff} = 5.7\,pm\,V^{-1}$ for SHG generated by $\lambda \approx 1\,\mu m$. On the other hand, for QPM there is no limitation for the polarization of the fundamental wave and because of $d_{eff} = (2/m\pi)d_{ij}$ one may use the large value of d_{33} to obtain for $m = 1$ a $d_{eff} = 22\,pm\,V^{-1}$, thus a 15 times larger intensity with the QPM method.

One more advantage of QPM is a possibility to realize SHG under conditions, when the conventional phase-matching does not occur, therefore it offers the chance to involve new materials (e.g., highly nonlinear $LiTaO_3$ or

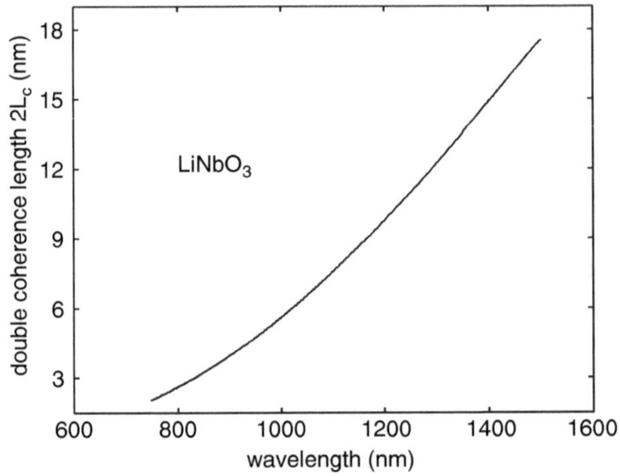

Fig. 7.9. Calculated double coherence length of $LiNbO_3$, redrawn from [675]

SBN) and to expand the spectral ranges to shorter wavelengths in $LiNbO_3$ or KTP. Note that a low birefringence, often a reason for missing conventional phase-matching, is an advantage in the case of QPM, because according to (7.31) larger periods for the domain pattern are required, offering an easier production. In addition, in QPM the problem of the walk-off angle is eliminated, because the fundamental and frequency-doubled waves are collinear.

Note, that in general the momentum conservation depends on the angle between k_1 and k_2, extending the SHG to noncollinear configurations. This means that in the case of an arbitrary domain structure, whose geometrical parameters (domain diameters and spacing) are in the range of microns, one may observe noncollinear SHG with an angular intensity distribution dictated by a given domain structure. This underlies, for example, the observations caused by arbitrary needle-like domains in SBN crystals [676–678].

The aforementioned method of the nonlinear tomography is based on the QPM conditions, too [668,669]. In this case, a three-wave mixing on a spatially modulated quadratic susceptibility is considered, where the phase-mismatch corresponds to a parametric process. The method uses the general relations between the Fourier harmonics of $\chi^{(2)}$ and the lineshape of the parametric signal.

A further development of the QPM frequency conversion is the concept of nonlinear photonic crystals proposed in the seminal publication by Berger [679]. Recall that the term *photonic crystal* implies a regular spatially modulated pattern of the linear susceptibility $\chi^{(1)}$. A nonlinear photonic crystal means a regular spatial array of the nonlinear susceptibility $\chi^{(2)}$) [679], the linear $\chi^{(1)}$ being spatially constant. From this viewpoint, a QPM structure, i.e., a PP ferroelectric structure, presents a 1D photonic crystal. Therefore, a regular 2D pattern of antiparallel domains, in other words a 2D QPM structure, is a 2D nonlinear photonic crystal. The optical frequency conversion in the QPM regime on such a 2D structure promises interesting properties, particularly, the possibility of multiple wavelength generation by SHG. Experimentally, a 2D nonlinear photonic crystal in $LiNbO_3$ was for the first time realized by Broderick et al. [680].

At the end of this section, we note that the PP structures may be used also as electro-optic modulators or deflectors. As r_{ij} is sensitive to the conversion of the P_s direction (7.26), so under an external field the sign of the refractive index change Δn in antiparallel domains is opposite, thus a PP structure transforms to a grating. For a wavevector k parallel to domain walls, a Raman–Nutt diffraction occurs, whose diffraction intensity of order m is

$$I_m = I_0 J_m^2[(2\pi/\lambda)d\Delta n],$$

where I_0 is the incident intensity, d the crystal thickness in the k direction, and J_m is the Bessel function for the order m. Applying an external field may overpump the light intensity to the diffraction maxima of higher orders, whereas the zero-order intensity decreases. For example, placing such a deflecting element into a laser cavity provides a Q-switching with the frequency of the field E. An advantage of this device over the conventional electro-optic modulators is that it may control the intensity of a nonpolarized light.

7.2 Ferroelectric Switching of $LiNbO_3$ and $LiTaO_3$ Crystals

In this section, one will see that the model scenario of the polarization reversal is more or less appropriate for $LiNbO_3$ and $LiTaO_3$. At the same time, the use of precise modern techniques permitted to detect those details of the domain dynamics, which have never been observed in other ferroelectrics but may be expected in principle in a wide group of ferroelectrics. This concerns, in the first instance, the effect of domain wall pinning by point defects. The outline of this section is as follows. After a short review of the initial steps, we concentrate on the investigations giving an insight into the mechanism of the polarization process in $LiNbO_3$ and $LiTaO_3$. Studies of the domain structure itself and particularly of periodically poled (PP) structures will be dealt with, only in relation to the switching mechanism, because the available enormous material is beyond our aims and possibilities. One may find a detailed description about the domain dynamics in $LiNbO_3$ and $LiTaO_3$ and the appropriate bibliography in the recent reviews [608–610].

Although in $LiNbO_3$ the ferroelectric phase transition and domains were found long ago, it was regarded for a long time as a *frozen* ferroelectric following Megaw [681]. The early results on the switching in $LiNbO_3$ are rather scattered and incomplete. Camlibel [636] was the first who observed the pulsed switching in $LiNbO_3$. The experiments were performed with the aid of a modified Sawyer–Tower method in extremely thin samples (30 μm) under field pulses of 1,500 V with a duration of 30 ms. The total switching of the crystal was observed with $P_s = 0.71\,\mathrm{C}\,m^{-2}$, which is in excellent agreement with the later measurements. As the switching field observed in $LiNbO_3$ at room temperature was too high and the dielectric breakdown occurred often, the attempts were made to operate at elevated temperatures. For example, Ballman and Brown [682] tried to pole $LiNbO_3$ and its relative $LiTaO_3$ at elevated temperatures and managed to pole $LiTaO_3$ by moderate fields of $1\,\mathrm{kV\,cm}^{-1}$ at 350°C; however, their attempts failed in $LiNbO_3$, which was assigned to a high ionic conductivity. Kovalevich et al. [515] with the use of the Barkhausen current method also found in $LiNbO_3$:Fe a significant decrease of E_c with T, not less than an order of magnitude at 120°C. Subsequent studies in $LiNbO_3$ and $LiTaO_3$ corroborated the conclusion of a lowering of E_c with T. When heating $LiNbO_3$ samples by intense laser pulses up to temperatures of about 800–900°C, antiparallel domains appeared under fields as low as $200\,\mathrm{V\,cm}^{-1}$ [683]. Later, in agreement with former results a decrease of E_c by a factor of four at 250°C was reported in $LiTaO_3$ [684]; E_c at 500°C fell down to $60\,\mathrm{V\,cm}^{-1}$ [685]. Haycock and Townsend proposed a method of poling $LiNbO_3$ and $LiTaO_3$ far below T_c by means of a joint action of an electron beam and external field [686]. At 600°C an $LiNbO_3$ sample was switched by a field as low as $10\,V\,cm^{-1}$ under simultaneous irradiation by the electron beam with an energy of 1.8 MeV. This mysterious result was explained as follows: an electron-bombardment produces short-living oxygen vacancies providing a pathway for Li ions, which in their turn are ordered by the external field to give the poled (single-domain) state.

7.2.1 First Steps of Regular Domain Pattern Creation in $LiNbO_3$ and $LiTaO_3$ Crystals

The switching of $LiNbO_3$ attracted attention after the pioneering work of Feng et al. [687], who were the first to observe the SHG in QPM mode operation on a regular domain structure in $LiNbO_3$ produced by the crystal growth. The Y_2O_3-doped $LiNbO_3$ crystals were grown by the Czochralski-technique along the a-axis (so-called 90° – crystals), the yttrium-concentration being sinusoidally varied along the a-axis by means of controlling the temperature field and the pulling and rotating speeds. It resulted in a creation of a domain-structure in form of inserted cylinders aligned parallel to the c-axis with a period following the regular yttrium distribution. Nonpolar cuts from such a PPLN-boule provide plates with a regular laminar domain pattern. A similar structure using the identical growth method was obtained in $LiNbO_3$:Cr [688]. The periods of the grown PPLN in these works were 7 and 8 μm, respectively, which is close to $2L_c$ for $\lambda = 1.06\,\mu m$ with L_c as the coherence length. Results of SHG generation from an Nd-YAG laser were encouraging and initiated the PPLN era. For a long time the efforts were aimed to obtain PPLN by growth methods, namely by using the techniques similar to those described earlier for nonlinear-optical applications and by growing the oppositely oriented domain layers (head-to-head, tail-to-tail domains) for acoustical applications. However, the periodicity and regularity of such PPLN were insufficient for SHG, so a search started for external actions providing a satisfying optical quality of PP-structures. An important contribution to the development of this subject was given by the results of Miyazawa [689], who noticed a partial polarization reversal in Ti-indiffused waveguides formed on $LiNbO_3$ polar z-cuts. This effect (very negative with respect to the waveguide operation) occurred only on the +z-cut and appeared after a Ti indiffusion procedure performed at temperatures between 950 and 1,020°C; the degree of the polarization reversal depending on the annealing time and reached 100% (the total P_s reversal). Later observations have found a partial or total polarization reversal in the surface layers on the z-cuts after various heat treatments at rather high temperatures. A reverted layer appeared after heating plates of $LiNbO_3$ with no electrodes to $T \approx 1{,}110$°C in air or Ar-atmosphere (obviously accompanied by an Li_2O out-diffusion) [690], the depth of this layer reached one half of the plates thickness at a prolonged heating. Interestingly, this effect was not observed in plates cut normally to nonpolar x or y axes and arose in plates cut even at small angles to P_s. A polarization reversal was observed in a surface layer produced by cladding a polar face by a dielectric SiO_2 layer [691]. In all cases, the switching occurred on the +z-face only. A possibility of controlling the polarization state of the surface layer by a heat treatment was used as the basis for the first attempts of creating regular domain gratings in optical waveguides, particularly in Ti-indiffused [692–694] and Li_2O-outdiffused ones [695]. The domain patterns produced by these methods looked like shallow triangles, so demonstrated only very low parameters of the frequency conversion. The reason of the surface polarization reversal has not been analyzed in detail yet. Miyazawa [689] explained it with the existence of a ΔP_s, i.e., a depolarization field at the interface bulk – waveguide, because of a change of P_s in the Ti-indiffused layer; the change of P_s was assigned to a hypothetical decrease of the [Li]/[Nb] ratio during Ti-incorporation. The same explanation can be expanded to the case of the Li_2O outdiffusion [690]. Actually, a change of P_s in a surface layer even by only 1% would give $E \approx 10^4\,\mathrm{V\,cm^{-1}}$, which is sufficient for switching $LiNbO_3$ at high temperatures. At the same time, as there is no experimental technique to measure P_s with such an accuracy, so this explanation cannot be verified. The main result obtained by Miyazawa [689] and other works of this type was the conclusion of a dominant nucleation of antiparallel domains on the $+z$-face.

The next step in creating PP-structures was the recording by an electron beam. In [696] a PPLN was produced at elevated temperatures in a waveguiding layer with the use of the technique proposed by Haycock and Townsend [686]. The crystal was irradiated by an electron beam through a periodic metal mask during 1 h at 580°C under $E = 10\,\mathrm{V\,cm^{-1}}$. The penetration depth of about 1 μm of the electrons was greater than the waveguide thickness, being approximately 0.7 μm. Several groups succeeded in obtaining regular domain patterns at room temperature by means of recording by electron beams without electric fields [697–699]. Very similar results were obtained in $LiTaO_3$ [700]. In all these cases, the depth of the recorded domain gratings along the polar axis, revealed with the aid of chemical etching, was in the range of 0.5 mm, or even up to 1 mm [698], thus exceeding by orders of magnitude, the electron penetration depth being of the order of micrometers. The possibility of creating volumetric PP-structures by this method and first results on SHG resulted in optimism.

The first successful preparation of PPLN by means of a direct polarization reversal under an external field was reported somewhat later [701]. A high negative field pulse of $24\,\mathrm{kV\,mm^{-1}}$ with a duration of 100 μs was applied to the planar electrode deposited onto the −c-cut, whereas a regular Al-electrode pattern deposited onto the +c-cut was grounded. The subsequent etching of the nonpolar y-cuts revealed a laminar domain pattern with a period of 2.8 μm and a thickness along the c-axis within several tens of micrometers. To evaluate the quality of the PPLN, an H-indiffused waveguide was then formed and tested with the aid of a Ti-sapphire laser. For a power of about 200 mW for the fundamental ($\lambda = 851.7\,\mathrm{nm}$), the SHG power was about 20 mW.

From this time on, the PPLN preparation methods were developed following these two routes, namely, either by a direct polarization reversal under an external field applied to a regular electrode pattern, or by surface methods, the electron-beam among them, which imply recording of a precursory domain pattern on a polar face of the crystal. In what follows, we dwell first on the domain patterns produced by applying external fields. The most attention will be concentrated on the investigations, informative with respect to the polarization reversal mechanism. At the end of this chapter, we shortly review the results on the domain patterns created with the aid of surface methods (an electron-beam, scanning force microscopy, etc.).

In the beginning of 1990s, several research groups elaborated empirical methods for producing PPLN by applying fields. These experiments were aimed to improve the parameters of PPLNs by trial-and-error methods, e.g. by varying electrodes, field regimes, temperatures, etc. It is appropriate to mention here an important characteristic of a PP-structure, the commonly used term *duty cycle*, which means the ratio of the domain width (or, what should be the same, of the electrode width) to the domain period. The optimum duty cycle for QPM is 0.5. Methods for a minimization of the PPLN period have been elaborated in 1994/1995 [702–704]. The periodicity of the structure was provided by using regular patterns of liquid electrodes separated by isolating photoresist stripes. The experiments were performed in thin samples 0.2 mm thick having a length of 6 mm by applying field pulses of 4.5–4.7 kV with a duration of several hundred milliseconds. The researchers experienced the effect of a domain merging with respect to the electrode, that means that the domain width as a rule exceeds the electrode width. In the course of a laborious activity aimed to suppress the domain merging by means of varying the duty cycle of the electrode patterns, the groups made progress from a PPLN period of 9 μm (third-order QPM for $\lambda = 832.5$ nm from a Ti-sapphire laser) [702] to a period of 4.5 μm (first-order QPM for $\lambda = 946$ nm from a diode-pumped Nd-YAG laser) [704]. Thus, d_{eff} was improved from 2 to 19 pm V^{-1} approaching the theoretical limit $d_{\text{eff}} = 22\,\text{pm}\,V^{-1}$ for d_{33}. It resulted in rather good features for the optical frequency conversion, for example, with a fundamental of 1.07 W a second harmonic as high as 49 mW was obtained [704], hence giving a conversion efficiency of about 4.6% for an interaction length of 6 mm. At the same time the authors faced the problem of photorefractive damage in the PPLN obtained, but the data were somewhat ambiguous. For example, Pruneri et al. ascribed an optical instability observed at a power of 1.07 W ($\lambda = 946$ nm) to the photorefractive damage [704] whereas Webjörn et al. [703] detected no optical degradation for the same wavelength up to 160 MW cm^{-2}, corresponding to a second harmonic intensity in the range of kilowatt.

The investigations of Myers et al. ([705] and references therein) aimed to the construction of optical parametric oscillators (OPO's) on the base of PPLN, discovered important features of the polarization reversal in $LiNbO_3$. First of all, a domain backswitching was detected, when the pulse duration was too short. For a field of 25 kV mm^{-1} the critical time $t^\star$ was not less than

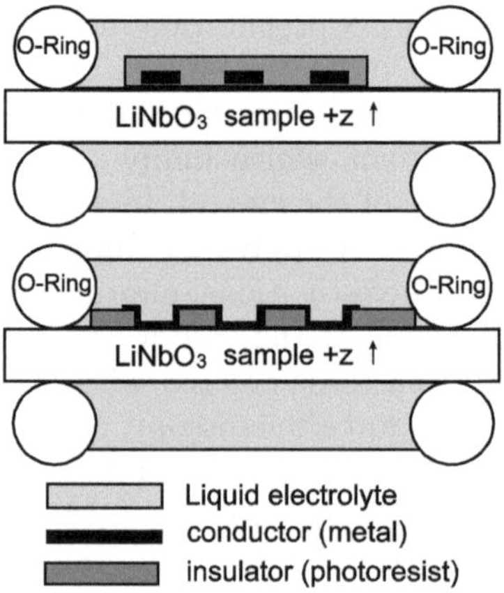

Fig. 7.10. Electrode configurations for electric-field poling. The *upper figure* shows a variant with insulator over conductor; the *lower figure* shows conductor over insulator. In both cases, the covering electrolyte connects the sample to the circuit and forms an isopotential surface that helps to control fringing fields around the pattern, redrawn from [705]

50 ms. To suppress the backswitching, the following method was proposed: after turning-off a switching pulse, the sample was held for a while under a lower field of about $E_c/2$ of the same sign. Additionally, it was shown that using ramping (saw-tooth shaped) fields prevents from the backswitching. An additional important result was the discovery of a high thermal stability of PPLN structures, which were not destroyed even after heating up to 800°C. Myers et al. [705] introduced an appropriate system of combined periodical electrodes, see Fig. 7.10, used later by other research groups, too. The electrodes were a system of Al stripes 200 nm thick either covered by an insulating photoresist (insulator over conductor), the upper part of Fig. 7.10 or spaced by photoresist stripes, in turn covered with a conductor (conductor over insulator), the lower part of the same figure. In both cases, the regular electrodes were deposited onto +z-cut and connected to the circuit via a liquid electrode (LiCl water solution); the plane electrode on the −z-cut was also an electrolyte. In 0.5 mm thick samples, the developed regimes permitted to obtain rather regular PPLNs with various periods and to realize several OPO schemes. Unluckily, a photorefractive damage was detected which manifested itself as a beam fanning. To avoid this undesirable effect, first steps were made in obtaining PP-structures in the optical-damage-resistant $LiNbO_3$:Mg crystals [706]. Therefore, since those times the researchers faced two crucial problems when creating PP structures: the domain merging with respect to the electrodes and the backswitching. Coexistence of these two obstacles requires searching for sophisticated field regimes, since from one hand the field applying should be long enough to avoid the backswitching and provide a stable reversed structure, and from the other hand not too long to prevent domain merging.

7.2.2 Fundamental Role of the Composition in Ferroelectric Switching

Essential progress of the insight into the polarization process in $LiNbO_3$ and $LiTaO_3$ was achieved by several research group [24, 476, 610, 625, 652, 684, 707–711]. These authors developed a specific terminology for the description of polarization features. The polarization reversal from the virgin single-domain state is referred to as a *forward* poling, and the switching from this polarization-reversed state to the initial state as *reverse* poling; the corresponding subscripts in notations are 'f' and 'r', respectively. However, below we try, whenever possible, to use the traditional terms.

In these studies the switching process was registered by measuring the switching current under either pulsed or ramping ($15 - 20\,\mathrm{V\,s^{-1}}$) fields. The occurring of a polarization reversal was supplemented by emerging antiparallel domains observed optically. Additional to the aforementioned Miller–Savage method requiring a chemical etching, $LiNbO_3$ and $LiTaO_3$ crystals provide a new specific possibility of domain visualization [708]. The fresh domains arising on applying a switching field are visualized due to a contrast appearing both in the polarized and unpolarized light when observing along the polar axis. The contrast is owing to a birefringence close to domain walls, which in congruent crystals is as high as $\Delta n = 10^{-5} - 10^{-4}$ [712,713] and is significantly lower in stoichiometric crystals, for example, approximately 10^{-6} in near stoichiometric $LiTaO_3$. This specific of the domain walls was investigated with the use of precise optical methods having a resolution in the range of microns, such as electro-optic imaging microscopy (EOIM) [652,711] and near-field scanning optical microscopy (NSOM) [712,713]. The birefringence was shown to be due to nonequilibrium mechanical strains according to (7.25) [645,646,712] or/and a space-charge field normal to the domain wall [712]. This specific of the domain properties in congruent $LiNbO_3$ and $LiTaO_3$ and using the transparent liquid electrodes made it possible to observe the domain dynamics in situ.

Unipolarity of $LiNbO_3$ and $LiTaO_3$ Crystals

The first fundamental result of this group was the discovery of a strong unipolarity manifesting as an asymmetry of the P–E loop in both crystals, for an example see Fig. 7.11. The values obtained in $LiNbO_3$ were $E_{\mathrm{int}} = 33.5\,\mathrm{kV\,cm^{-1}}$, $P_{\mathrm{s}} = 71 \times 10^{-6}\,\mathrm{C\,cm^{-2}}$ and in $LiTaO_3$ $E_{\mathrm{int}} = 50\,\mathrm{kV\,cm^{-1}}$, $P_{\mathrm{s}} = 50 \times 10^{-6}\mathrm{C\,cm^{-2}}$. The switching fields from the virgin state (E_{f} in the terminology of the authors) were of about 210–220 $\mathrm{kV\,cm^{-1}}$ in agreement with old data. Later the data for E_{int} were refined, namely reported as 22.2 and 41.2 $\mathrm{kV\,cm^{-1}}$ for $LiNbO_3$ and $LiTaO_3$, respectively [710]. The value of E_{int} does not depend on the crystal thickness (checked for 0.5 and 0.25 mm), therefore it was assigned to the volume rather than to an influence of the surface. With increasing temperature both E_{c} and E_{int} decrease in agreement with the old data, e.g., in $LiTaO_3$. E_{int} was close to zero

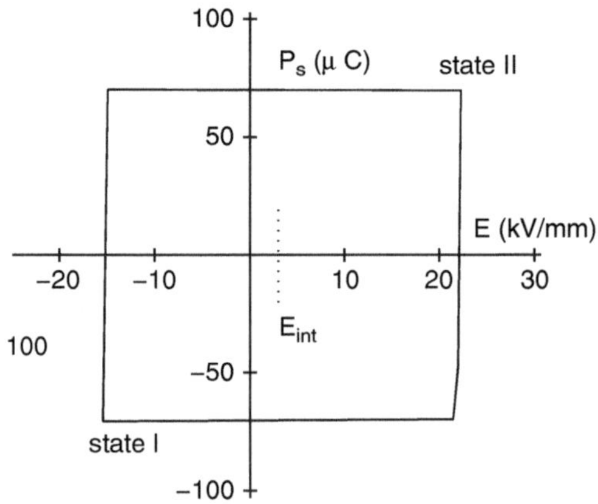

Fig. 7.11. Example for an asymmetric P–E loop for z-cut $LiNbO_3$ at room temperature, redrawn from [709]

at $T > 200°C$. Under field cycling at room temperature, the shape of P–E loops, i.e., the sign and value of E_{int}, are constant. This stable unipolarity is the reason, why the field required for the reverse switching (that is from the polarization-reversed state to the state with the initial P_s direction) is lower than for the forward switching in the course of multiple switching cycles at room temperature. However, E_{int} may be realigned at elevated temperatures in consistence with the P_s direction. If after a polarization reversal the crystal is quickly heated up to $T > 100$–$150°C$, then the sign of E_{int} is reversed. In other words, if a P–E loop in a virgin crystal is initially shifted, say, to the right, then after applying a reversing field and a subsequent heating to certain temperatures it will be shifted to the left. In $LiTaO_3$ this reversed loop is a full mirroring of the initial one, whereas in $LiNbO_3$ it is slightly different. The thermal reversal of E_{int} reveals a temperature threshold of about 100°C for $LiNbO_3$ and 150°C for $LiTaO_3$. This is illustrated in Fig. 7.12. As seen, at room temperature the amount and sign of E_{int} is practically unchanged even during 10 h of exposure to a switching field, whereas it is fully reversed after heating to $T \geqslant 150°C$ during minutes. This behavior qualitatively resembles the characteristics of E_{int} in the model crystal $BaTiO_3$:Mn [622,623] described in Sect. 7.1.2.

Gopalan and Gupta [707,708] predicted a fundamental impact of the crystal stoichiometry on the polarization characteristics, which was experimentally proved for $LiNbO_3$ [476] and $LiTaO_3$ [24].

Figure 7.13 presents the P–E hysteresis loops for the congruent and near-stoichiometric $LiTaO_3$. In both cases, the Li content was determined from the phase transition temperature T_c following [714]. According to these pioneering results, E_c in stoichiometric $LiNbO_3$ (SLN) is decreased by a factor of four to five as compared to the congruent crystals (CLN) and in SLT more than by an order of magnitude compared to CLT. Simultaneously, E_{int} tends both in SLN

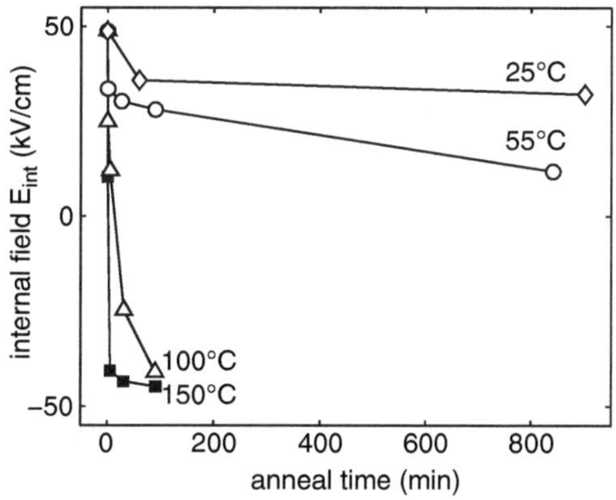

Fig. 7.12. Internal field in a domain reversed z-cut $LiTaO_3$ at room temperature as a function of annealing time at fixed annealing temperatures, data taken from [707]

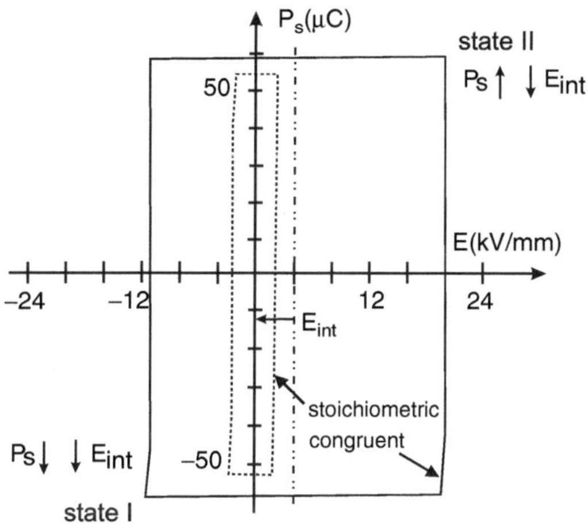

Fig. 7.13. Comparison of hysteresis loops in stoichiometric and congruent $LiNbO_3$, redrawn from [24]

and SLT to zero. A great number of subsequent studies confirmed and refined these conclusions. Bermúdez et al. [715] estimated E_c vs. smoothly increasing Li content, which was achieved in $LiNbO_3$ crystals by means of increasing step-by-step the K_2O concentration in the melt, see Fig. 7.14. In a recent work, the measurements were performed in VTE treated crystals; however, in this case E_c was not a linear function of the smoothly growing Li content and pronouncedly dropped down when approaching the near-stoichiometric composition [716]. Kim et al. [717] confirmed their former results on E_c and E_{int} in SLT with measurements of the domain kinetics. Here in near-stoichiometric $LiTaO_3$ grown by the double-crucible Czochralsky method, the switching fields for forward and reverse switching were 17 and 15 kV cm^{-1}, respectively, and

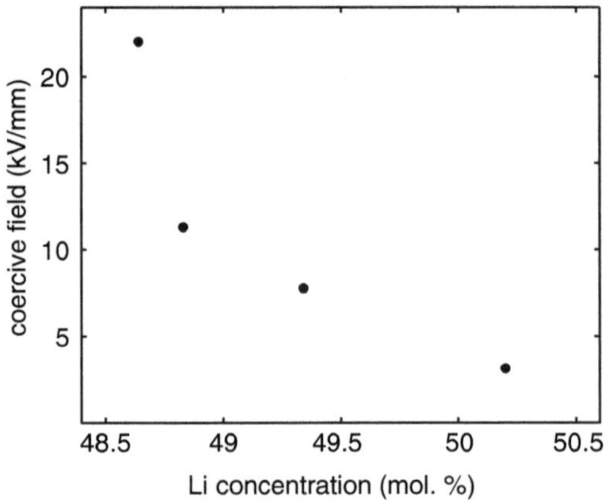

Fig. 7.14. Coercive field vs. Li-concentration in $LiNbO_3$, redrawn from [715]

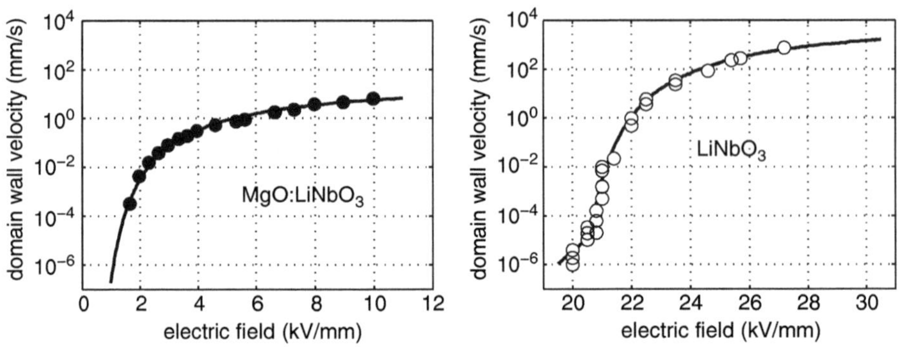

Fig. 7.15. Domain wall velocity as a function of electric field. The *squares* are the experimental results for MgLN and LN. The *solid lines* are the fitting *curves*. Note that the *curves* have different abscissae, data taken from [719]

$E_{int} = 1\,kV\,cm^{-1}$. A decrease of the switching field in the near-stoichiometric $LiNbO_3$ by a factor of three was found as compared to CLN [22]. In this work, the Li content was determined from the widths of characteristic Raman lines.

Doping with the optical-damage-resistant impurities gives results analogous to an increasing Li content. Kuroda et al. [718] studied the pulse switching in 0.5 mm thick wafers of congruent $LiNbO_3$:5%MgO and observed typical switching currents under fields of $44.5\,kV\,cm^{-1}$, which is five times lower than in undoped crystals. Mizuuchi et al. [706] succeeded in obtaining a PP-structure in a 0.3 mm thick sample of $LiNbO_3$:4.8%MgO under pulse fields as low as $25\,kV\,cm^{-1}$. Later studies confirmed a decrease of E_c and E_{int} in highly doped $LiNbO_3$:Mg [719–721].

In Fig. 7.15 a comparison of the field dependencies of the domain wall velocities in undoped and Mg-doped $LiNbO_3$ crystals measured by the Miller–Savage method is shown. Additionally, it was found that a realignment of E_{int}

under a thermal annealing occurs in polarization-reversed $LiNbO_3$:Mg faster than in undoped $LiNbO_3$ [720]. Chen et al. [721] used the method proposed by Bermúdez et al. [715] and demonstrated a smooth decrease of E_c with increasing Mg concentration up to 6.5%. Later a similar strong effect was observed for another optical-damage-resistant impurity, i.e., Zn [722]. The switching field, denoted here as the threshold field E_{th} sharply falls down at Zn concentrations exceeding 5 mol.%, thus when approaching the Zn concentration threshold, simultaneously E_{int} is lowered to about 5 kV cm^{-1}, see Fig. 7.16. However, at a further increase of the Zn concentration, E_{int} in contrast to the switching field starts to grow, which is interpreted as a result of mechanical stresses in highly doped $LiNbO_3$:Zn.

The highest effect on E_c and E_{int} in $LiNbO_3$ is provided by a combination of increasing Li content and a low Mg doping. In near-stoichiometric $LiNbO_3$-doped with 2% Mg (the Li content was increased by the VTE method) the switching field was as low as (18 ± 2) kV cm^{-1} [723]. Averages of available literature data for E_c and E_{int} are shown in Table 7.1. Here E_c denotes the field

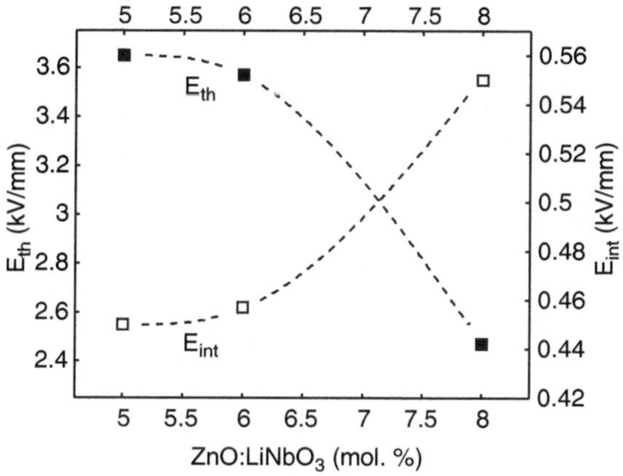

Fig. 7.16. Doping effects on the threshold E_{th} and internal E_{int} field in the polarization switching of $LiNbO_3$:Zn, data taken from [722]. The *dashed lines* are guides for the eye

Table 7.1. Coercive and bias fields in $LiNbO_3$ and $LiTaO_3$ vs. crystal composition

Crystal	E_c (kV cm^{-1})	E_{int} (kV cm^{-1})
CLN	210	2–33
SLN	60–65	$\to 0$
CLN:5%MgO	60–68	5
CLN:8%ZnO	25–35	5
SLN:1%MgO	25	$\to 0$
CLT	200–210	40–50
SLT	50	17

corresponding to the polarization reversal from the virgin single-domain state. Interestingly, that this concentration dependence qualitatively correlates with that of the photorefraction (Sect. 4), because both E_c and E_{int} values, from the one hand, and the photorefraction, from another hand, decrease regularly following the sequence *congruent* → *near stoichiometric* → *Mg-doped near stoichiometric* crystals.

Pulsed Switching in $LiNbO_3$ and $LiTaO_3$ Crystals

In congruent [625, 652, 710] and near stoichiometric $LiTaO_3$ [717] as well as in congruent $LiNbO_3$ [710] the studies of the ferroelectric switching were performed using a combination of the Merz method with the EOIM and NSOM methods. The ferroelectric switching in congruent and stoichiometric $LiTaO_3$ [724] and congruent and stoichiometric $LiNbO_3$ [725] was also studied by means of a combination of the Merz method and polarizing-optical in situ observations of the domain dynamics. The results of both research groups are similar. As the optical measurements could not clearly separate the stage of the domain nucleation itself, so Gopalan et al. [625, 710] suggested that on the polar face precursory surface microdomains with a density of 10^3–$10^5\,cm^{-2}$ exist serving as nucleation centers. This assumption was based on indirect evidences of Nozawa and Miyazawa [726]. On applying a sufficient field amplitude, multiple microdomains appear on the surface with a rather high density, for example up to $10^3\,mm^{-2}$ in congruent $LiTaO_3$ [724]. In congruent $LiTaO_3$ these are equilateral triangles with domain walls parallel to the crystallographic xz planes. In congruent $LiNbO_3$ these are almost equilateral hexagons with domain walls parallel to the yz (glide) planes. The detailed description of the domain shape in $LiNbO_3$ and $LiTaO_3$, the models of their formation and the appropriate bibliography one may find in [610,613,727]. The rate of emerging microdomains identified as the nucleation rate [652], increases with the field amplitude. According to the first publications [625,652,710,711] the switching times t_s (Fig. 7.17) obtained in the usual way from the switching currents obey an activation law. The plot $\log t_s$ vs E^{-1} in Fig. 7.17 consists of two linear parts ('low'- and 'high'-field ranges), and thus is characterized by two activation fields δ which are higher than for model ferroelectrics, for example, about 16 and $30\,kV\,cm^{-1}$ for the low- and high-field ranges, respectively, in congruent $LiTaO_3$ [652]. The forward switching from the virgin state occurs under fields sufficiently higher than the reverse one, because E_{int} impedes the former process. To reduce $t_s(1/E)$ to a general form, Gopalan et al. [625] proposed a presentation, which takes into account E_{int}, thus providing a unique activation field δ for the forward and reverse switching. It has the form

$$t_s \propto \exp\left(\frac{\delta}{E \pm E_{int}}\right), \tag{7.32}$$

where either $E = E + E_{int}$ or $E = E - E_{int}$ are substituted to the denominator in (7.32) for the reverse or forward process, respectively.

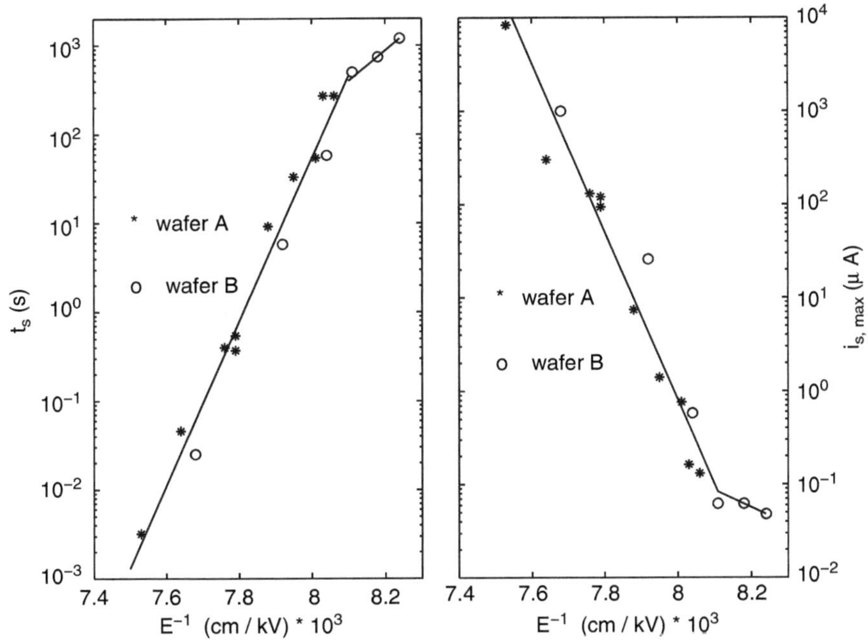

Fig. 7.17. The switching time, t_s (*left*) and maximum transient current, $i_{s,max}$ (*right*) as a function of the inverse electric field, E , during forward poling of wafers A and B of z-cut $LiTaO_3$ crystal, redrawn with data taken from [625]

Kuroda [718] studied the pulse switching and domain dynamics in $LiNbO_3$ doped with 5% MgO. The switching times t_s under $E_c \approx 44.5\,\mathrm{kV\,cm^{-1}}$ were of about 10^2 s and decreased with E following a typical exponential law. Long t_s permitted the authors to trace the domain kinetics. The observations of the etched +z- and +y surfaces revealed an almost classical picture, namely a nucleation of needle-like domains close to the +c surface, their dominant sideway broadening with a weakly pronounced frontal growth and a final coalescence occurring after more than an hour of exposure to a field.

The velocity of the sideways domain wall movement in $LiNbO_3$ and $LiTaO_3$ observed optically [625,652,710,711], as well as the switching velocities obeyed an activation law, however, with a single activation field. The difference in the plots $\log(t_s)$ vs. (E^{-1}) and $v_{sw}(E)$ was initially explained with traditional terms [625], attributing the switching times in the low-field range to the dominant role of the nucleation, and in the high-field range to the sideways domain wall movement.

Subsequent studies of the domain dynamics refined the kinetics of the sideways domain wall motion in $LiTaO_3$ [652, 711, 717, 724] and $LiNbO_3$ [609, 725]. In both crystals under relatively low fields (of the order of E_c) the domain motion occurs in two stages, namely a relatively slow initial motion of isolated domains and their consequent merging resulting in an accelerated movement of a common domain front. So, according to Shur et al. [724] the switching current may be modeled as a superposition of these two parts and is described by the equation

$$j(t) = j_1(t) + j_2(t) = At^{-\gamma} - 2\,P_\mathrm{s}\frac{\mathrm{d}q}{\mathrm{d}t}. \qquad (7.33)$$

Here the first term corresponds to the arising and merging of initial isolated domains. The second input characterizes the area involved into the switching (q is a dimensionless parameter) and consists of two fractions formed by isolated and merged domains; to fit it to the KAI model [630], Shur and coworkers [724] present this term as a transformation from the α to the β process, see Sect. 7.1.2.

The growth kinetics of an individual domain was investigated in $LiNbO_3$ [609, 725]. Under an applied external field a domain starts to move via the propagation along the wall of microdomain steps (bunches of the elementary nanoscale steps), so that the domain wall motion looks as a layer-by-layer growth. This process obviously corresponds to the on-wall domain nucleation in the Miller–Weinreich model. The consequent stage of the domain merging was investigated in details in congruent and stoichiometric $LiTaO_3$ [652, 711, 717, 724].

According to Shur et al. [724] in congruent $LiTaO_3$ the merged domain front moves preferentially along the +Y-direction. The velocity of this front significantly exceeds the velocity of widening of individual microdomains, being for example, 2,300 and 200 $\mu m\,s^{-1}$, respectively [652]. Gopalan et al. [711] propose a simple model, where they try to explain a speeding-up of the formed domain front in the framework of the Miller–Weinreich mechanism of the on-wall domain nucleation. Shur et al. [609] (and references therein) propose an

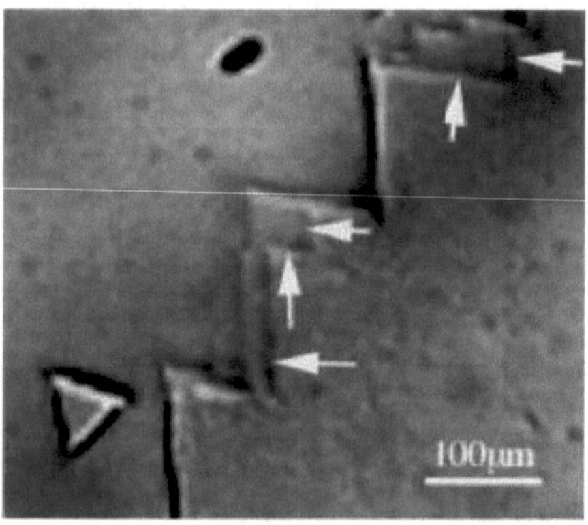

Fig. 7.18. Transient traces left behind a moving domain wall in congruent $LiTaO_3$ are indicated by *arrow* heads. These traces disappear in a second or less [711]. Reused with permission from Gopalan and Mitchell [711], Copyright 1999, American Institute of Physics

explanation of merging in terms of screening effects. The domain nucleation itself may be competitive with the initial slow stage and is negligible as compared to the fast one. The domain merging is conceptually identical to the domain coalescence in the Miller–Weinreich model.

The next step in understanding the polarization specific in $LiNbO_3$ and $LiTaO_3$ was done as a result of finding the so-called 'pinning' of the domain walls during sideways movement. The optical observations in $LiNbO_3$ [710] and $LiTaO_3$ [724] found that the sideways domain wall movement is 'jerky'. The peak of $i_s(t)$ accompanying this movement contains resolved spikes which by origin are the Barkhausen currents. This jerky motion was attributed to a 'pinning' of the domain walls by defects and subsequent 'depinning' [625, 710, 724].

It should be emphasized that the domain-wall pinning effects which are obviously common for ferroelectrics (as well for magnetics) on the whole, for the first time were directly observed and studied in details just in $LiNbO_3$ and $LiTaO_3$; the found regularities perhaps form a basis to a refined scenario of the ferroelectric switching.

According to the observations of Gopalan et al. [625, 710], the domain motion starts under external fields significantly lower than E_c, however, owing to a capture of a domain wall by a 'pinning' center, the motion is stopped. To provide the depinning at a given field E, the length of the field pulse should exceed certain $t*$, otherwise a domain would come back to its initial boundaries. This pinning effect could account for the existence of a *threshold* switching field in $LiNbO_3$ and $LiTaO_3$, which means that whenever a field $E < E_c$ is applied (where E_c is just the threshold), no switching occurs. Recall that in model ferroelectrics like $BaTiO_3$ or TGS the switching already starts at $E < E_c$ although via slow kinetics of the relaxation type.

Additional details of the pinning process were extracted from a comparison of the domain dynamics in $LiTaO_3$ under pulsed and steady-state fields of the same amplitude [652, 711]. The domain motion was observed under a pulse duration $t_p < t_s$ measured from the switching currents. As a priori expected, the pinning effect is characteristic for the pulsed rather than for steady-state fields. On applying a single pulse, a domain wall stays fixed during almost 90% of t_p, whereupon is rapidly shifted ('depinned'). Therefore, certain time is required to release a domain wall from a pinning center. No pinning occurs under steady-state fields, since a shift of the domain wall linearly grows with time. As a result, an average domain wall velocity under pulsed fields is substantially lower than under the steady-state ones. For example, in congruent $LiTaO_3$, at $212\,\mathrm{kV\,cm^{-1}}$ the motion velocities of an individual small domain are 34 and $200\,\mathrm{\mu m\,s^{-1}}$ under pulsed and steady-state fields, respectively. In the same way, the merged domain front moves with the velocities of 340 and $2{,}300\,\mathrm{\mu m\,s^{-1}}$ under pulsed and steady-state fields, respectively [652].

In this relation the domain backswitching is again worthy of mentioning, because as found in early studies [705], this effect is one of the decisive factors in formation of PPLN and PPLT structures [609]. The domain backswitching was reported repeatedly and studied in details in CLN [625, 728], CLT [710], SLT [717], and $LiNbO_3$:5%MgO [719]. In summary, the backswitching characteristics are as follows. Critical times $t*$, denoted by Gopalan and Mitchell [625] as *stabilization times* in CLT are noticeably larger than in CLN; for example, for the forward switching under $E \approx E_c \approx 210\,\mathrm{kV\,cm^{-1}}$ they are 1.5–2 s and 10–30 ms, respectively, and decrease in stoichiometric crystals to 300 ms in SLT at $E \approx 20\,\mathrm{kV\,cm^{-1}} \approx \mathrm{E_c}$. This regularity correlates with the relationship between E_{int} in these crystals, see Table 7.1. As expected, in any crystal $t*$ for forward (hampered) switching is longer than for the reverse one; for example, 1.5–2 s and 0.1–0.3 s, respectively, in CLT. An effect related to the backswitching and characteristic for all ferroelectrics, is a dependence of the switching parameters on the time interval between applying fields of opposite polarities, which was reported for CLT [729], SLT [717], and CLN [730]. If a time interval t_{g} between applying fields, denoted in [717] as a *time gap* and in [729] as a *recovery time*, is shorter than a certain value, or, in other words, if the field alternation is too fast, then with decreasing t_{g} the values of P_{s}, E_{c}, and t_{s} are lowered. In SLT [717] and CLN [730], this impact of the duty-off factor appears at t_{g} in the range of tens of seconds. Some authors [717, 729] mistakenly concluded that a decrease of E_{c} under a rapid switching promises a facilitated P_{s} reversal; however, they ignored a simultaneous decrease of the switched charge P_{s} [717], which evidenced that a part of the crystal is not involved in switching and remains clamped.

As clear from the preceding, the backswitching process may find an explanation in the framework of the pinning–depinning mechanism. Actually, the observed domain motion under a field pulse [652, 711] evidences that certain critical time is required to release a pinned domain wall; this time is involved in the parameter $t*$ of the backswitching. Another qualitative explanation of the backswitching is given by Shur [609]. As the domain motion is accompanied by a local nonequilibrium depolarizing field, so accordingly to Shur [609] the backswitching is caused by a noncomplete compensation of this field by external screening charges, which leads to a formation of a residual local field E_{rd} preventing from the domain motion. The faster is the domain motion, the lesser is the degree of screening and higher is E_{rd}, thus the backswitching is more pronounced. These two explanations of the backswitching process are complementary rather than conflicting.

A large amount of experiments was aimed to analyze the pinning–depinning mechanism. Observation by means of the NSOM method [712, 713] found that a birefringence region adjacent to a domain wall, becomes distorted under an external field as low as $20\,\mathrm{kV\,cm^{-1}}$, corresponding to $0.1\,E_{\mathrm{c}}$. This distortion was regarded as a proof of a pinning effect and interpreted as a bending and bowing of a domain wall around the pinning centers. Actually, the classic concept of the domain dynamics is based on the assumption of flat domain walls, so their distortion is a necessary condition of pinning.

A scheme depicted in Fig. 7.19, was proposed to describe this process. On the basis of the domain wall curvature, which was calculated with a rough

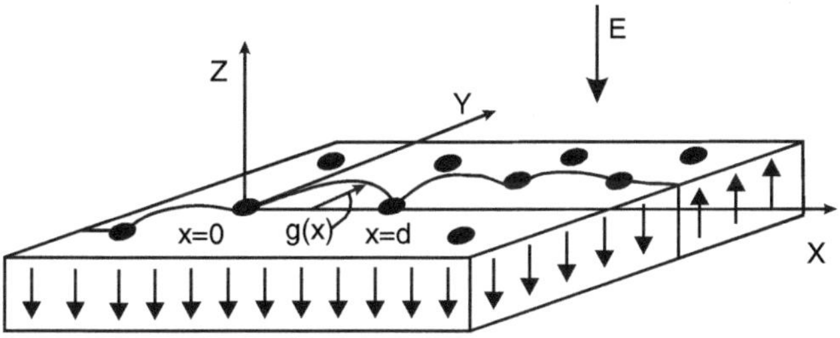

Fig. 7.19. Schematic of bowing of a pinned domain wall (defects are *dots*) under an applied electric field, redrawn from [713]

approximation from the distorted birefringence region, an estimation of the energy of the unit domain wall area $\sigma = 0.2\text{–}0.4\,\mathrm{J\,m^{-2}}$ was calculated [713]. This value is at least by an order of magnitude higher than that evaluated for 180°-domain walls in $BaTiO_3$, namely $\sigma = 0.002\text{–}0.02\,\mathrm{erg\,cm^{-2}}$ ([614] and references therein).

The data of the X-ray topography in $LiNbO_3$ seem to support the pinning mechanism [646]. In these experiments performed by means of measuring the rocking curves and topographs, the first step was the investigation of static domains created by a preliminary applying of an external field. The observed X-ray contrast is due to the existence of mechanical strains around the domain walls. Then an evolution of domains was observed in situ under $E < E_c$ applied in the course of structure studies. On applying external fields in the range from ± 30 to $\pm 90\,\mathrm{kV\,cm^{-1}}$, i.e., substantially lower than the forward and reverse switching fields, a domain evolution was registered looking as an affine widening or shrinking of domains with no movement.

An indirect evidence of a domain wall pinning was concluded from a bending of PPLN pattern after a weak temperature annealing, found by an PFM method [731]. Additionally, indirect evidences of a pinning mechanism were concluded from recording the domains in CLN and SLN by means of applying high fields to the AFM tip [731]. In SLN, the diameter of the recorded domain linearly grows with the duration of the field exposure up to about 20 s and then starts to rise much more strongly, whereas in CLN this dependence is weak, for details see Fig. 7.20. The latter may be assigned to a wall pinning preventing the domain growing.

Microscopic Models of the Polarization Process in $LiNbO_3$ and $LiTaO_3$ Crystals

The main specific features of the polarization process in these crystals are enormous coercive and internal fields drastically depending on the crystal stoichiometry and the existence of the threshold switching field. Serious efforts were mounted to analyze the microscopic origin of this specific and to squeeze all observed effects into the framework of a unified microscopic model.

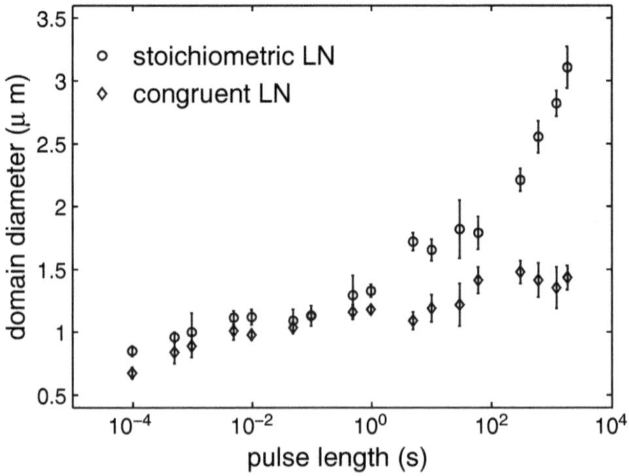

Fig. 7.20. Single domain diameter as a function of the AFM high voltage pulse length for congruent and stoichiometric single LN crystals. The error bars represent the standard deviation of results obtained over four domains tailored at different places on the sample, data were taken from [731]

As found in the pioneering works [476, 625, 707–709] the sign of E_{int} is changed, if the polarization-reversed crystal is quickly heated to certain threshold temperatures. In terms of the Arlt–Neumann model of the bias field [622] this means that the dipole moments of defects controlling E_{int} are reoriented at elevated temperatures in compliance with the reversed direction of P_s. As in $LiNbO_3$ at $T > 100$–$120°C$ a thermal activation of hydrogen takes place, see Sect. 5.2 so, it was reasonable to associate the defects governing E_{int} in $LiNbO_3$ and $LiTaO_3$ with H-ions. To prove this, the effect of external fields and thermal annealing on the IR absorption band characteristic for the OH stretch mode was analyzed [707–709], see left part of Fig. 7.21. Actually, after the polarization reversal the structure of the IR band in a virgin $LiTaO_3$ or $LiNbO_3$ was modified, namely, the intensities of three main lines composing the IR band were redistributed, see right part of Fig. 7.21. These results were recently validated [732], where the peak position of the IR band in the $LiNbO_3$ crystal was shifted by $11\,cm^{-1}$ after the polarization reversal; this shift has a maximum after the first (forward) switching, and at further field cycling the peak oscillates back and forth within $4\,cm^{-1}$. Annealing of the polarization-reversed crystals led to a slow relaxation of the IR band to its initial shape [707–709], which was assigned to a realignment of E_{int}. As the OH-bond is almost normal to the P_s direction [100] (see Fig. 2.10), OH-dipoles themselves cannot be responsible for the E_{int}. Gopalan et al. [707–709] suggested that the observed relaxation of the OH-bond is in some way caused by an evolution of Li vacancies which in its turn depends on the [Li]/[Nb] ratio. Similar slow relaxation processes in polarization-reversed $LiNbO_3$ crystals

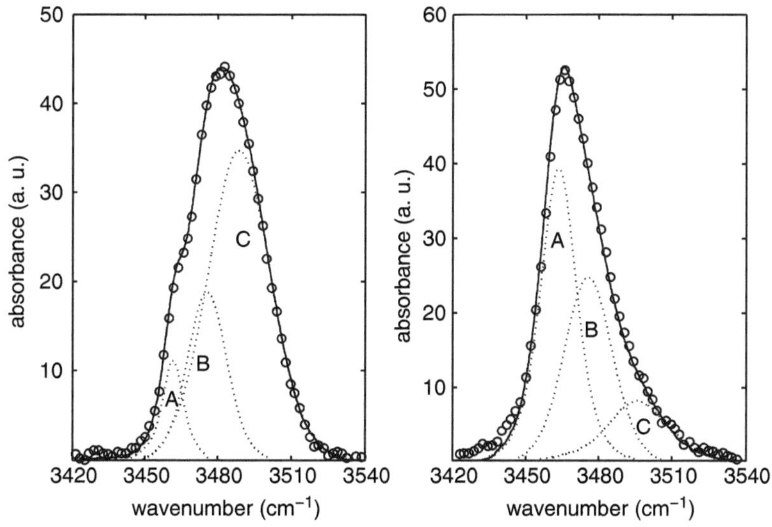

Fig. 7.21. Absorbance of the OH stretch mode in z-cut $LiTaO_3$ at rom temperature with decompositions into three components and the experimental data points for a virgin sample (*left*) and a domain-reversed crystal (*right*); redrawn from [708]

were observed by means of Raman spectroscopy [733]. After the polarization reversal the peak positions of six representative Raman lines were somewhat shifted, whereupon they were slowly recovering to the initial state with characteristic times in the range of several hours. This was attributed to the relaxation of the internal field.

However, the initial attractive conclusion that E_{int} is controlled by protons, was later ruled out, because its value in $LiNbO_3$ was shown not to depend on the H-content in the crystal [476], Fig. 7.22. At the same time, subsequent studies reaffirmed an impact of switching on the shape of OH-stretch band.

The response of the OH-band to the ferroelectric switching was found as well in the stoichiometric $LiNbO_3$ obtained by the VTE technique [734]. The intensities of three lines composing the IR band were redistributed after the polarization reversal and almost completely restored after a temperature annealing.

The most drastic effect is observed in Mg-doped $LiNbO_3$ with [Mg] above the threshold [732]. As known, in these crystals the OH-band is shifted to a shorter wavelength [100]. After the polarization reversal of a virgin crystal the band is decreased by more than 80%. After turning-off the field, it relaxes very slowly, not returning to the initial optical absorption value during tens of hours. This extremely slow relaxation of the IR band in the polarization-reversal $LiNbO_3$:Mg resembles qualitatively the slow relaxation of the Raman-lines in the polarization-reversed undoped $LiNbO_3$ [733] and is most probably related to a slow spontaneous depoling of the reversed state.

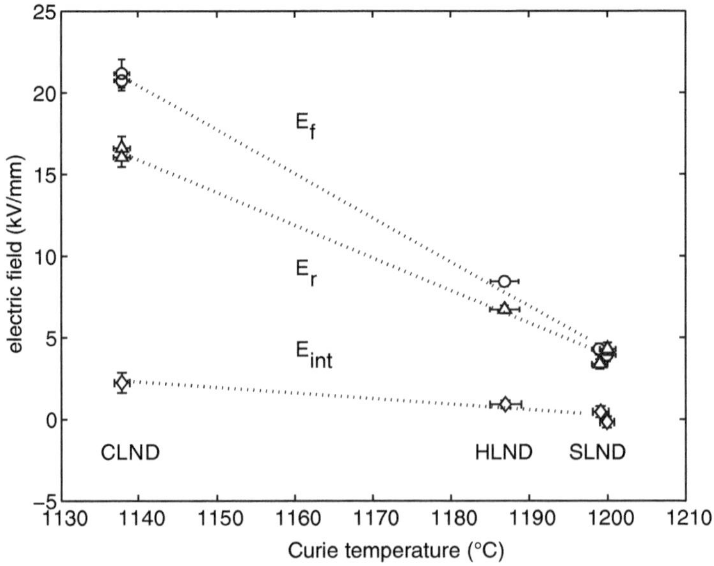

Fig. 7.22. The switching fields, E_f for forward poling and E_r for reverse poling, and internal field E_{int} as a function of the measured Curie temperatures, T_c , of $LiNbO_3$ crystals with low hydrogen content (samples CLND, SLND, and HLND). Data are taken from [476]

At room temperature the initial shape of the band in $LiNbO_3$:Mg is completely restored by the reverse poling [732]. It may be restored as well by annealing a polarization-reversed crystal at $T = 250°C$ during 2 h.

The scope of the listed results permits to conclude unambiguously that the polarization reversal somehow affects the OH-stretching bond in $LiNbO_3$ evidently because of a variation of the ion environment. At the same time, the reverse effect, that is a direct contribution from protons or hydroxyls into the switching process suggested in the early investigations, is open to question.

It is worthy to mention here that the ferroelectric switching affects not only the OH-band, but spectroscopic characteristics of other impurity ions, too. Measurements by the site-selective excitation-emission spectroscopy found a drastic reconstruction of the Er^{3+}-emission spectra after the polarization reversal in the congruent and near-stoichiometric $LiNbO_3$ crystals [735]. The scanning of the crystal finds this spectrum reconstruction within approximately 22 μm around the domain wall. In $LiNbO_3$:RuO_2 the polarization reversal produces a strong photochromic effect [736]. Both effects provide new methods for the domain observation in $LiNbO_3$ [610, 735–737].

After the discovery the stoichiometry as a guiding theme, the next step for an insight into the microscopic origin of the unipolarity and the high coercive fields $LiNbO_3$ and $LiTaO_3$, was invoking the intrinsic defect model. Actually, the [Li]/[Nb] ratio as an impact on various properties of $LiNbO_3$, is interpreted in terms of Nb_{Li}, V_{Li}, etc., so it was suggestive to extend this language to the polarization process as well. According to Kim et al. [717] the polarization reversal in $LiNbO_3$ containing structure defects cannot be presented simply as a symmetry inversion of the single-domain matrix. This reasoning is supported by the results of the X-ray topography in congruent $LiNbO_3$ crystals containing antiparallel domains [646]. These experiments revealed a difference in the intensities of the (0012) and (001$\bar{2}$) Bragg reflections (8.5 keV) which is assigned to different structure factors between antiparallel domains.

A qualitative microscopic model of E_{int} on the basis of the intrinsic defect structure was proposed for $LiTaO_3$ [717] in the framework of the Arlt–Neumann model [622]. Subsequently, this model was expanded to $LiNbO_3$ [610]. The internal field was attributed to the dipole moments of defect clusters $[Nb_{Li} - 4V_{Li}]$. The polarization reversal requires particularly moving the regular Li ions along the polar axis through the plane of the oxygen-triangles. The model proposed Kim et al. [717] postulates that at room temperature the point defects Nb_{Li} are fixed by Li vacancies, thus the cluster dipoles cannot be reoriented in consistence with P_s and conserve the initial direction of E_{int}. These dipoles may be realigned only at $T > 100 - 150°C$. So, according to this model the reason of the observed reorientation of E_{int} at elevated temperatures is a reorientation of the cluster $[Nb_{Li} - 4V_{Li}]$ dipoles, which in its turn is attributed to a thermo-activated mobility of the Li vacancies and their hopping over equivalent Li sites. Using this reasoning, the authors follow the interpretation of the $[Mn^{2+} - V_O]$-dipole reorientation in $BaTiO_3$ [623]. The validity of this model, particularly concerning the role of the Li vacancy hopping is very difficult to verify. As discussed in Sect. 3.2.1, the conductivity in this temperature range is actually dominantly ionic and attributed to the hydrogen transport. The activation energies of the diffusion of H^+ and Li^+ are close to about 1 eV and rather scattered [738]. So far no spectroscopic proofs exists for a realignment of a $[Nb_{Li} - 4V_{Li}]$ dipole at elevated temperatures, as it was established in $BaTiO_3$:Mn [623].

As repeatedly mentioned, a specific of the ferroelectric switching in $LiNbO_3$ and $LiTaO_3$ is in the freshly switched domains the appearance of a large birefringence in the vicinity of the domain wall. In congruent crystals it achieves $\Delta n \approx 10^{-4}$ and may be reduced by an order of magnitude by a thermal annealing; in near-stoichiometric crystals Δn is significantly lesser [610]. This birefringence provides a possibility to observe the domains (strictly speaking, the domain walls) in polarizing microscopes and gives such a beautiful effect as emerging star-like images (Fig. 7.23) when a laser beam propagates along the polar axis through a poly-domain $LiNbO_3$ [739]. The switching-induced birefringence is $\Delta n = 10^{-5} - 10^{-4}$ and caused by mechanical strains and space-charge fields normal to the domain wall. According to optical measurements [712] the strains are $|\epsilon_x - \epsilon_y| \approx 10^{-4}$, $\epsilon_{yz} \approx 10^{-4}$, and $\epsilon_z \approx 10^{-6}$; according to structure data $\epsilon_{zx} \approx |6 \pm 3| \times 10^{-5}$ [646]. The tangential field is $E_y \approx 10\,\mathrm{kV\,cm^{-1}}$ [712]. The birefringence region is within 2 µm [712, 713], whereas the approximate shear strain width observed

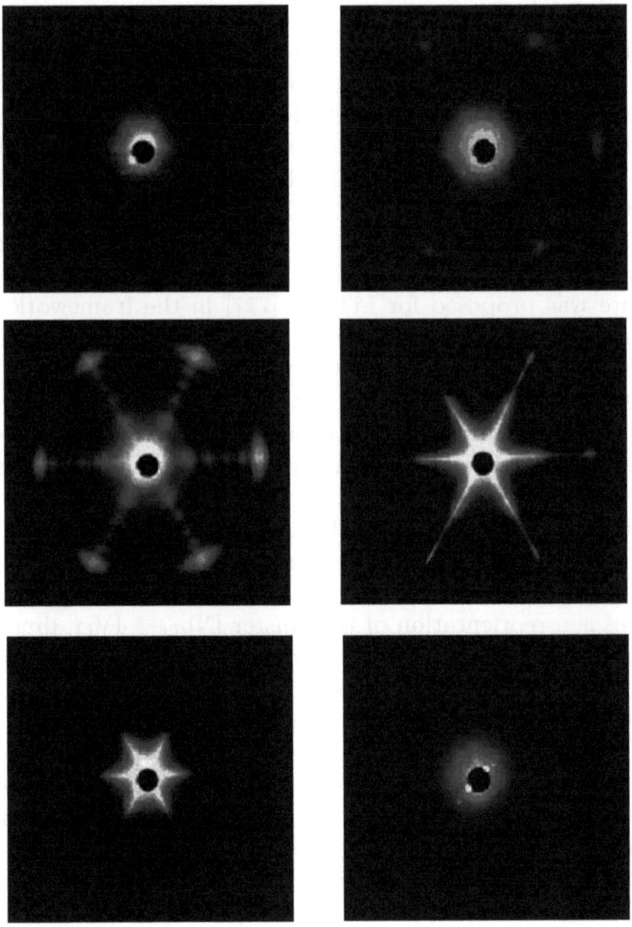

Fig. 7.23. Far-field light patterns observed on the screen for illumination with light wavelength 351.1 nm for various fields applied to the crystal. The applied fields in [kV/mm] are: (**a**) 19.08, (**b**) 19.30, (**c**) 19.53, (**d**) 19.67, (**e**) 19.69, (**f**) 19.72. In all these pictures the main beam is blocked, taken from [739]

by X-ray topography is up to 10 μm [645, 646]. At the same time, the domain wall thickness itself, is estimated (PFM method) to be in the range 65–150 nm [617–619] and even not exceeding 0.28 nm in $LiTaO_3$ according TEM measurements [740]. A long discussion on the origin of a large disturbed area around the domain wall is still in progress. Jach et al. [646] attribute the large region of shear strains adjacent to the domain walls in congruent crystals to interaction of the domain walls with the defect clusters, in other words, to pinning the walls by the intrinsic defects. This relation is supported by the shape of the spatial distribution of the piezoresponse when scanning the PFM-tip across single domain walls [617]. The piezoresponse having a maximum in the center of gravity of the domain wall is smoothly falling down on each side within a distance of 100–120 nm. In a stoichiometric crystal, the distribution of the piezoresponse with respect to the wall is symmetric, thus is identical in the regions with the initial and reversed P_s. In contrast, in a congruent crystal this distribution is noticeably asymmetric tailing to the reversed area, which may be indirectly related to the existence of E_{int}.

However, on the whole the mechanism associating the strong extensive shear strains to the intrinsic defects is qualitative, although very attractive. From the measurements of the X-ray and diffuse neutron scattering in congruent $LiNbO_3$ crystals Zotov et al. [741] estimated the size of the disorder/defect clusters formed by the structure defects (pairs of V_{Li} or V_{Li} and Nb_{Li} randomly distributed on the lattice sites with resulting displacements of the neighboring Li ions). These estimates give cluster sizes of 114–760 pm. At the same time, a simplified estimate [645] taking into account the known concentration of the intrinsic defects leads to the value of possible cluster sizes by several orders exceeding those found in [741].

The concept which relates directly the high coercive and internal fields in $LiNbO_3$ and $LiTaO_3$ to pinning by the intrinsic defects, does not answer several fundamental questions. For example, in this concept the intrinsic defect structure in $LiNbO_3$ and $LiTaO_3$ is assumed to be identical, so one may expect an identity in the polarization characteristics of both crystals. At the same time, E_{int} in $LiTaO_3$ is more than twice as much higher than in $LiNbO_3$, see Table 7.1. This concept may scarcely explain a sharp decrease of E_c and E_{int} in highly Mg- or Zn-doped congruent $LiNbO_3$ and Li-enriched $LiNbO_3$ weakly doped with Mg. Indeed, in the framework of the intrinsic defect model these crystals do not contain the hypothetical pinning centers Nb_{Li} and Li vacancies. At the same time, these crystals are spatially much less uniform than undoped ones and, as seen from the concentration dependencies of the lattice parameters, see Fig. 2.21, the lattice in highly doped crystals is stressed. At [Mg] or [Zn] above the thresholds, new intrinsic defects Mg_{Nb} or Zn_{Nb} are formed, which themselves could serve as pinning centers. So, in spite of missing Nb_{Li}, one would more likely expect a hampered switching process at high Mg or Zn concentrations. We cited these arguments to illustrate that the qualitative explanation of the switching parameters of $LiNbO_3$ and $LiTaO_3$ in terms of the intrinsic defect structure is very far from a comprehension.

7.2.3 Photoinduced Effects in Ferroelectric Switching of $LiNbO_3$ and $LiTaO_3$ (Photodomain Effects)

Besides the crystal stoichiometry, another factor affecting the value of E_c and switching is the illumination by a photoactive light, either near UV [729, 742–746], or blue-green [747, 748]. The most sensitive to the irradiation are Mg-doped $LiNbO_3$ crystals with [Mg] close to the threshold. The first observation [742] found a lowering of E_c in a congruent $LiNbO_3$:5%Mg to about 60 kV cm^{-1}, i.e., by about 10% under illumination by a nonfiltered UV light

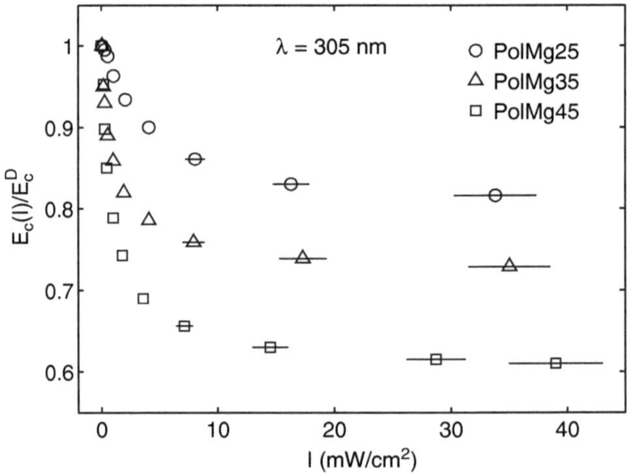

Fig. 7.24. Relative coercive field reduction $E_c(I)/E_c^D$ vs. light intensity I at a wavelength $\lambda = 305$ nm, redrawn from [745]. PolMg25, PolMg35, and PolMg45 mean congruent $LiNbO_3$ crystals doped with 2.5, 3.5, and 4.5 mol.% MgO (in the melt), respectively; the crystal doped with 5.5%MgO behaves identically to the one doped with 4.5%

with an intensity of $0.2\,W\,cm^{-2}$. Wengler et al. investigated the spectral characteristics and the compositional dependence of this effect [744,745]. Most efficient was a radiation of $\lambda = 305$ nm, resulting in a lowering of E_c in congruent $LiNbO_3$:5% Mg by more than 30% at intensities as low as $10–12\,mW\,cm^{-2}$; a comparable effect was observed in near-stoichiometric $LiNbO_3$:1%Mg. At higher intensities of about $40\,W\,cm^{-2}$ the value of E_c at 5%Mg doping fell down by 50%; a noticeable decrease of E_c was observed as well at lower Mg concentrations, see Fig. 7.24. A longer wavelength radiation $\lambda = 334$ nm up to intensities of $100\,W\,cm^{-2}$ affected E_c only in $LiNbO_3$:Mg with [Mg] above the threshold.

It is qualitatively clear, that the photoinduced effects are due to the screening by the free charges of noncompensated depolarizing fields accompanying a domain motion [606, 609]. These effects are most pronounced in the above-threshold $LiNbO_3$:Mg crystals due to a relatively high photoconductivity σ_{ph} discussed earlier in Sect. 4.2.1.

Moreover, a UV irradiation of these crystals is accompanied by increasing the optical absorption in the range of 330–350 nm [325], which additionally enhances σ_{ph} in the near UV. Increasing Li content leads as well to an enhancement of σ_{ph}. As a general result, the dielectric relaxation times τ_M in highly doped $LiNbO_3$:Mg and perhaps in near-stoichiometric $LiNbO_3$:Mg at moderate light intensities are in the range of seconds or even shorter, which is comparable to the switching times. Screening effects may contribute to both involved processes, either to facilitate the domain nucleation or to accelerate the domain wall sideways movement. According to recent observations [745], the more probable is a decrease of the nucleation field. This assumption is corroborated by the intriguing fact that under the illumination with $\lambda = 330\,\text{nm}$ the nucleation of antiparallel domains occurs always at the −z-surface independently of the illuminated z-face [742, 745], whereas in the usual case (without illumination) the antiparallel domains, as mentioned earlier, always appear nearby the +z-surface. Therefore, the screening controls the nucleation conditions at the surface. At the same time, one may expect a bulk screening effects. For example, a photoactive illumination may affect the space-charge field E_{sc} which exists in the vicinity of a domain wall [712]. In terms of the pinning mechanism, it would mean a change of the potential barrier formed by a pinning center.

The above consideration concerned a weakly absorbed UV light, so the photoinduced effects may be related both to the surface and bulk reasons. The interpretation of the events proceeding under illumination by a strongly absorbed light $\lambda = 305\,\text{nm}$ ($\alpha \geqslant 10\,\text{cm}^{-1}$) is more complicated. The specific of this case is a strong nonuniformity of the applied field in the crystals bulk owing to a nonuniform distribution of σ_{ph} along the light beam propagation. According to calculations [745], an incident energy is efficiently absorbed within the depths of 0.2–0.3 mm, so in the samples 0.5 mm in thickness, the field increases from the illuminated crystal surface to the opposite one by more than an order of magnitude. In accordance with these calculations, the domain nucleation under strongly absorbed illumination starts from the surface opposite to the illuminated face (independently of its polarity), that is in the region of a field maximum.

The irradiation by a visible blue-green light also leads to a decrease of E_c [747, 748], or, to be more precise, of the field corresponding to the appearance of antiparallel domains registered optically, denoted by the authors as the nucleation field E_n. The value of E_n noticeably decreases in the crystal spot exposed to a focused Ar-ion laser beam. This effect is again mostly pronounced in highly doped $LiNbO_3$:Mg crystals, where E_n falls down by an order of magnitude [748]. In spite of an apparent analogy to the results of a UV impact, the mechanism seems to be different. The used blue–green intensities are very high, for example of about $100\,\text{MW}\,\text{cm}^{-2}$ ($\lambda = 488\,\text{nm}$) [747] and of about $3\,\text{kW}\,\text{cm}^{-2}$ ($\lambda = 457, 488, 514\,\text{nm}$) [748], thus incomparably higher than the UV intensities mentioned earlier. As known, a visible radiation excites a photorefractive field which under these intensities may achieve very large values even in undoped $LiNbO_3$ crystals [35]. Additionally, a focused intensive irradiation produces strong local pyroelectric fields. Therefore, under these conditions several processes are involved and there is no way to separate a contribution from either field and their interaction with the applied external field. These results permit only to conclude qualitatively that very high intensities of a visible light decrease the nucleation field. Such an illumination-assisted domain switching could be very attractive for practice, because in principle it permits to record very fine domain patterns (in the range of microns) controlled by the diameter of a focused light beam [747, 748]. However, physical mechanisms of the observed effects are very far from a comprehension.

As a separate photoinduced effect one may regard the rise of surface microdomains in congruent $LiNbO_3$ crystals under intensive UV irradiation without external fields [746, 749–754]. These microdomain structures result from pulsed irradiation with a wavelength below ($\lambda = 248\,\text{nm}$) and of those from the UV bandedge ($\lambda = 298 - 329\,\text{nm}$). The used pulse energies and pulse lengths are in the range $50 - 150\,\text{mJ}\,\text{cm}^{-2}$ and $7 - 20\,\text{ns}$, respectively, thus, the power densities are as high as $\text{MW}\,\text{cm}^{-2}$, which is just below the ablation threshold of $500\,\text{mJ}\,\text{cm}^{-2}$ under these conditions [749]. The depth of the domain relief visualized either by chemical etching or by piezoelectric force microscopy depends on the optical absorption coefficient of the applied UV light and is the shallowest for $\lambda = 248\,\text{nm}$. Using phase masks and a recording wavelength $\lambda = 248\,\text{nm}$, regular domain patterns with a grating period of 726 nm were obtained [752]. With increasing light exposure (number of UV pulses), the microdomain structure transformed from rows of individually separated dots (individual microdomains) to continuous nonresolved lines. Even under a nonpatterned light-beam the emerging microdomains tend to a self-organizing forming regular patterns corresponding to the crystal point symmetry [751]. These effects are observed on either an irradiated z-surface independently of its polarity. An interpretation of these self-organization processes leading to coalescence of UV-induced nano- and micro-domains into regular *beams* is given by Shur [754] in terms of the screening effects in a nonequilibrium state. The intensities required to produce these surface micro-structures are too high to relate the observed effects to the photoinduced charge transport or pyroelectric effect, they have obviously another origin, perhaps a slight Li outdiffusion. Interestingly, that the visualized domain micro-structures observed in these works qualitatively remind of the patterns observed on the polar cuts of $LiNbO_3$:Fe crystals when irradiation by photoactive light (Ar-ion laser, $0.1–1\,\text{W}\,\text{cm}^{-2}$) and in situ chemical etching, see [755] and references therein. This effect caused by a suppression of the etching velocity in the illuminated part of the −z-surface, was attributed to an impact of the photovoltaic field on the etching process. At the same time, the surface patterns visualized in this case by means of the PFM and SEM methods resemble the domain patterns induced by UV light in the above listed experiments, so, it is not inconceivable that in [755] a near-surface polarization reversal occurred as well.

7.2.4 Recording of Ferroelectric Domains by Electron Beams and AFM Voltages: Microstructured Domain Patterns

From the aforesaid, one sees that the regularities of the switching process in $LiNbO_3$ and $LiTaO_3$ are studied sufficiently to apply them in practice. In spite of this, a practical elaboration of PPLN and PPLT devices meets serious difficulties. The first obstacle is the effect of domain merging which makes hard to control their final width, thus the duty cycle. This problem was repeatedly discussed in publications, starting with the basic review [756]. Another obstacle is the backswitching, which under a given strip electrode is especially pronounced at the electrode edges [728]. Therefore, the field regime requires a field exposure to be relatively long to avoid the backswitching, and at the same time relatively short to suppress the domain merging and coalescence. Certain technological difficulties are caused by the necessity of a preparation of regular fine electrode patterns by the lithographic techniques. Beyond, certain future applications such as photonic crystal structures require submicron domain periods, whereas the minimal achieved domain sizes under field poling are within 2 μm. All these reasons initiate the development of alternative methods, the eldest of which is the recording of the domain pattern by an electron beam promising in principle a resolution up to 5 nm. As mentioned in the beginning of this section, the first results on PPLN and PPLT recording by an electron beam were obtained in 1990–1992 [696, 697, 699, 700]. PPLN and PPLT structures were created in samples placed into a scanning electron microscope (SEM) modified for these aims [697,699,700], or recorded by electron beam lithography method [698]. Two modes of scanning, either continuous line scanning with a constant speed, or a dotted-line scanning (hopping) were used [757]. For $LiNbO_3$:Mg the electron-beam recording of a domain pattern was reported in [758]. The experimental conditions used in these and subsequent works are on the whole very similar: the beam voltage is within 30 kV, as at higher voltages the samples are cracked, the +z-cut is covered by a thin metal electrode and grounded; the −z-cut is scanned by an recording electron beam 0.5–1 μm in diameter with a velocity of 0.01–1 $mm\,s^{-1}$; a designated spatial period is program-controlled. A necessary condition is a high purity of the −z-surface (scanned). To produce the domain reversal, the fluence should exceed $8 \times 10^{14}\,e\,cm^{-2}$ [759]. The characteristics of the arising domain pattern are controlled by a compromise between the scanning velocity and fluence (i.e., the current). Even in early works, the required periods of PPLN in the range of 6–8 μm were reported, for example, in [697] the recorded PPLN structure with a period of 7.9 μm demonstrated a normalized conversion efficiency of 40 $W\,cm^{-2}$ for SHG from a Ti-sapphire laser (861.6 nm).

A further development of the domain recording by an electron beam bombardment permitted to obtain extremely fine domain patterns. One-dimensional (1D) and two-dimensional (2D) domain arrays were created with domain diameters of few hundreds of nanometers [759], which are considered as prototype of photonic-like structures. Glickman et al. [760] proposed an advanced indirect polarization reversal by an electron beam, in which the −z-cut is covered by a thin dielectric layer with a high trap density. This provides a high localization of the electron beam, because of a high local capturing of the electrons. As a result, the formed domain *dots* 18 μm in diameter and a penetration depth into the crystal of about 350 μm gave a fine structure consisting of numerous nano-domains. The produced pattern is also regarded as a prototype of a 2D photonic-like structure. A similar method of coating the −z-surface with a resistive layer before recording domain arrays with an electron beam was recently used [761, 762].

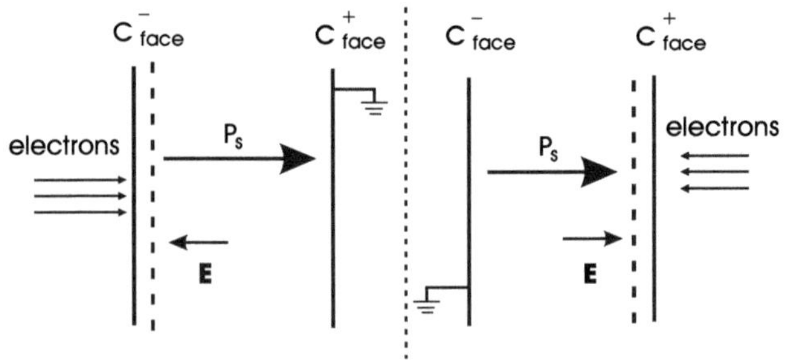

Fig. 7.25. Scheme of the domain inversion process according to [700]

The attempts were made to record PPLN and PPLT with the aid of ion beams, for example, by an Si^{2+} ion beam [763]. The thickness of the recorded PPLT and PPLN pattern was of 0.3 mm with a duty cycle close to 0.5. Recently a micro- and nano-domain engineering in 800 nm thick $LiNbO_3$ crystals with the use of a Ga^{2+}-ion beam was reported [764].

The reason of the polarization reversal under charged particles is not clarified yet. Figure 7.25 shows a qualitative illustration of electric fields arising under electron irradiation either of polar faces. Ito et al. [697] following Haycock and Townsend [686] suggested that a local oxygen outdiffusion under the electron beam produces a pathway for a directed Li ion transport over the bulk with a consequent formation of an antiparallel domain. Fujimura et al. [757] assume that a local action of the electron beam produces a seed of an antiparallel domain. Attempts were made to calculate the electron-beam induced local fields. Nutt et al. [699] estimated the local field developed under an electron beam with a diameter of 0.5 μm. Scanning with a velocity of $300\,\mu\mathrm{m\,s^{-1}}$ gives $E = 10^6\,\mathrm{V\,cm^{-1}}$, which satisfies the required condition for a domain nucleation and further frontal growth. An estimate of the surface charge induced by an electron beam when recording a line with a width of 1 μm gives $100\,\mu\mathrm{C\,cm^{-2}}$ which is comparable to P_s [757]. According to the consideration of Nutt et al. [699], the field is decreasing over the crystal thickness. Actually, in consistence with this conclusion a tilt of domain walls with respect to the polar axis was repeatedly reported in electron-beam produced PPLN [698, 699, 765]. Particularly, the structure recorded by means of a continuous line scanning of an electron beam and visualized by the chemical etching, looked as parallel hill rows on the −z-surface (scanned), and as a regular matrix of isolated pits on the opposite, +z-face [765].

A domain wall tilting, thus a decrease of the domain cross-section along the polar axis, is obviously a fundamental drawback of the volumetric electron-beam recorded PPLNs and PPLTs, which makes this method seemingly more suitable for PP structures in waveguides or very thin crystals. For example, a PP structure on a nanoscale level was formed in a congruent $LiTaO_3$ single-crystal as thin as 100 nm by means of the electron-beam lithography technique. The recorded domain linewidth was 100 nm with a period of 200 nm [766].

As a method being a relative of the action of charged beams, one may regard producing PP structures with the use of a corona discharge, which was employed in $LiNbO_3$:Mg [767]. In this case regular metal electrodes were deposited on the +z-cut and grounded, whereas the −z-cut was exposed to the corona discharge. Under exposure with the voltage of 5.5 kV at 100°C a regular domain pattern was formed spreading through the crystal thickness of 0.4 mm. This technique was used as well for a creation of OPOs on the base of congruent and near stoichiometric $LiNbO_3$:Mg [768]. The drawback of these structures is again an insufficient regularity because of a domain wall tilting with respect to the polar plane.

We now shortly dwell on the domain pattern recording using the atomic force microscopy in which the polarization reversal occurs under a dc-voltage applied to the AFM tip itself. The appearing domains are subsequently scanned with the use of a standard SPM method, either in the piezoresponse mode probing for a surface charge distribution, or, after a chemical etching, in the topograph mode. Domain recording using the AFM tip promises patterns at a nanoscale level. At the same time, the linear sizes of these nano-domain patterns are extremely small, in the range of μm, thus being far behind ones recorded by electron or ion beams, which are in the range of millimeters, e.g. $6 \times 6\,\mathrm{mm}^2$ [766].

We should mention that as a qualitative prototype of the polarization reversal of $LiNbO_3$ under an AFM tip, one may regard a formation of microdomains under voltages applied to a spike with a thickness of about 1 μm of a needle-like electrode contacted to a $LiNbO_3$ surface ([769] and references therein). Under voltages in the range from 500 to 2,500 V, the antiparallel domains appeared when contacting the electrode either to polar (z) or non-polar (y)-faces. In the first case the penetrating depth of the domains into the crystal thickness is about 50 μm, in the second case they appear within a surface layer of 0.15 μm thickness on the (010) face, the domain length being in the range of 15 μm. The calligraphic poling method proposed by Mohageg et al. [770] is as a matter of fact a development of Fregatov and Sherman [769]. In this method, a thin pen-electrode of 1 μm radius is moving along a polar face of a $LiNbO_3$ crystal; which results in recording stable regular domain patterns in 100 μm thick $LiNbO_3$ specimens under voltages of 2–3 kV for some seconds.

The polarization reversal under a dc-voltage applied between a conductive AFM tip and the bottom-electroded crystal surface was formerly observed in several ferroelectric crystals having usual E_c values of about several kV cm^{-1}, e.g. [771–773] and ferroelectric films, e.g. [774] and references therein. In these cases, a tip-voltage of a standard AF microscope (of the order of dozens of volts) is sufficient to switch the ferroelectric films and thin crystal samples. However, in $LiNbO_3$ or $LiTaO_3$ crystals, even near-stoichiometric ones, the standard AFM voltages are not comparable to E_c, therefore, to realize the ferroelectric switching in these materials one should use two alternatives, either to deal with extremely thin samples, or to increase the tip voltages.

The first possibility was realized by Terabe et al. [775], who were the first to switch $LiNbO_3$ crystals under an AFM tip. It was done in near-stoichiometric $LiNbO_3$ and Li-enriched $LiNbO_3$:1%Mg in samples as thin as 5μm with the use of a maximum tip voltage of 60 V. The authors observed very regular hexagonal domains whose diameter smoothly increased from 0.5 to 3 μm with increasing the time exposure from 1 s to 80 min or by increasing the voltage at a constant pulse duration. The authors succeeded in obtaining regular domain dot arrays and even in recording ring-shaped domain patterns following scanning trajectories of the recording tip. To provide a temporal stability of the recorded domains, the time exposure required not less than 1 s. Similar experiments were performed in extremely thin $LiTaO_3$ samples, for both congruent and stoichiometric compositions of about 50–150 nm thick under short pulses with lengths varying from 4 ns to 150 μs and amplitudes in the range from 7 to 15 V [776–781].

Even under field pulses as short as several ns the antiparallel domains did appear; note that such an ultrafast switching has never been reported before in volumetric ferroelectric crystals. The lowest diameter of recorded domain dots achieved in these experiments was less than 10 nm, which could be estimated with the aid of a precise method of scanning nonlinear dielectric microscopy (SNDM) developed by this research group [776]. The achieved ultra-fine domain patterns permitted the authors to claim a realization of a high density (up to 10 Tbit $inch^{-2}$) storage system, see Fig. 7.26. The precision of the SNDM method permitted Daimon and Cho [782] estimated the domain wall thickness H of AFM-recorded domains in CLT and NSLT ultrathin crystals. An intriguing result is a decrease of H with decreasing crystal thickness and domain radius. As expected, H is sufficiently lower in NSLT than in CLT for the same domain radii; for illustration, $H = 2.9$ and 10.7 nm, respectively, for the same domain radius of 500 nm.

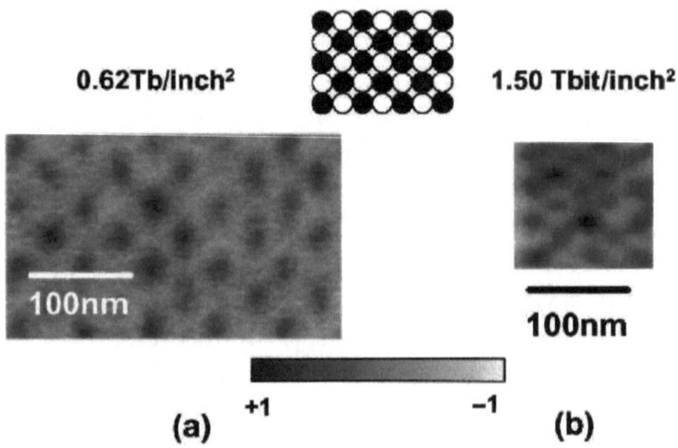

Fig. 7.26. SNDM images of close-packed array of domain dots (normalized amplitude) at a data density of (**a**) 0.62 Tbit $in.^{-2}$ (11 V, 10 μs) and (**b**) 1.50 Tbit $in.^{-2}$ (12 V, 80 ns) using 70-nm-thick single-crystal CLT according to [776]. Reused with permission from Cho et al. [776], Copyright 2002, American Institute of Physics

Interestingly, that in these experiments a dependence of the ferroelectric properties on the sample thickness was detected [777]. The P–E hysteresis loop in a congruent $LiTaO_3$ 6.5 µm thick sample was very similar to one observed in volumetric crystals with approximately the same E_c and E_{int}, see Sect. 7.2.2. In contrast, in the 0.5 µm thick sample the parameters of the P–E loop noticeably depended on the field frequency. We should note that the observation of ferroelectricity in single-crystals as thin as 50 nm, is to our knowledge reported for the first time. As follows from the results of this research group, the temporal domain stability is provided by a large enough duration of the voltage pulse. Another crucial parameter appeared to be the domain radius r_d [783]. For AFM-domain recording in NSLN crystals with a thickness from 25 to 150 μm, Kan et al. [783] found that the decay time of domains is controlled by certain critical radius r_c inherent to each crystal thickness, namely, the domains having $r_d < r_c$ decay very fast after turning-off the voltage, whereas for $r_d > r_c$ the domain lifetime dramatically increased. As the value of r_c rows with the crystal thickness, so r_c was interpreted as the domain radius corresponding to the domain penetration through the whole crystal thickness.

Some authors elaborated a modified high-voltage AF microscope [HVAFM] permitting to enhance a tip dc-voltage up to 5 kV and deal with thicker samples [731, 784–787]. Using this original equipment, this group succeeded in tailoring submicron domain structures in $LiNbO_3$ plates of 150–250 µm thickness. Under a single rectangular voltage pulse of 40 ms duration and 3.2 kV amplitude they recorded 2D domain arrays consisting of triangular domains with linear sizes less than 1.5 µm. When moving the HVAFM tip along the +z-surface the domain strips were recorded with lateral sizes in between 0.5–1.5 µm and periods up to about 1 µm. In both cases the induced domains were stable and penetrated through the whole crystal volume. The possibility of varying the amplitudes and pulse duration in wide limits, permitted to deduce the characteristics of the polarization reversal under these conditions, particularly to compare the switching behavior in relatively thick and very thin crystals. These dependencies are fundamentally different and on a whole may be formulated as follows. In thick crystals (0.5–1 mm) the domain radius r_d for a given pulse duration t_p grows with the tip voltage U_t by a power law, e.g. $\propto U^{2/3}$ [786], whereupon tending to a saturation [731, 786]. In very thin $LiNbO_3$ crystals (0.85 µm thick) as well as in ultrathin $LiTaO_3$ crystals in the reports sited above, r_d for a given t_p grows linearly with U_t without any saturation [619, 787], see Fig. 7.27.

The dependencies of r_d on t_p are different as well, since in ultrathin samples r_d grows linearly with t_p [619, 787] in a wide pulse duration range, whereas in thick samples this dependence is seemingly logarithmic [786, 787]. The curves for r_d on the pulse duration t_p were compared in thick congruent and stoichiometric $LiNbO_3$ plates [731]. As mentioned earlier (see Fig. 7.20), in the former this dependence is rather weak, whereas in the latter r_d pronouncedly grows with t_p by a power law in a wide t_p range from ms to 10^4 s without saturation.

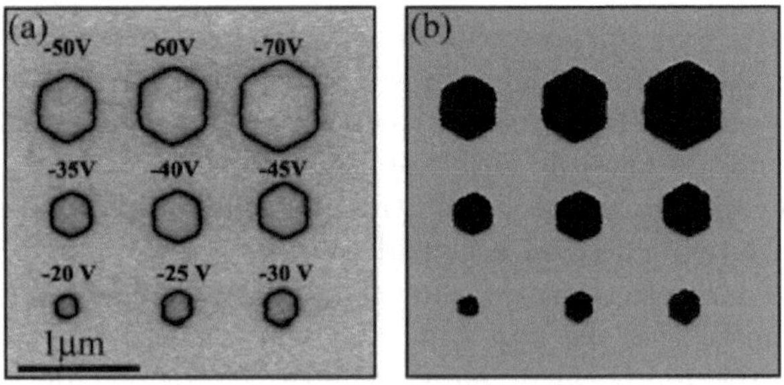

Fig. 7.27. PFM (**a**) amplitude and (**b**) phase images of ferroelectric domains fabricated by 10 ms voltage pulses of various amplitudes [619]. Reused with permission from Rodriguez [619], Copyright 2004, American Institute of Physics

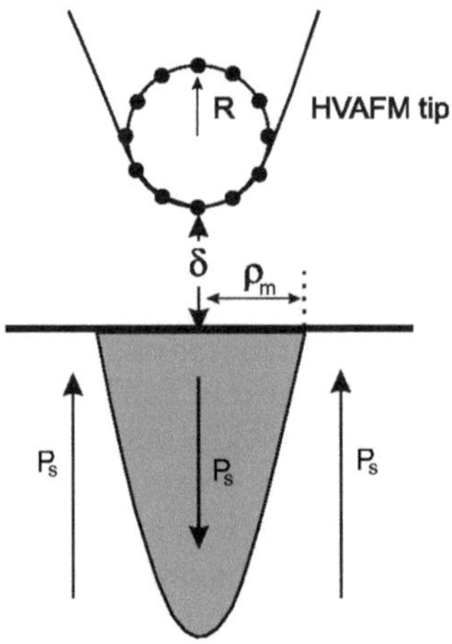

Fig. 7.28. Schematic diagram to illustrate the calculation of the electric field generated by a HVAFM tip, redrawn from [784]

A specific of the switching process and domain dynamics under a dc-voltage of an AFM tip is due to a dramatic nonuniformity of the developed field inside the crystal bulk. The tip-induced field distribution is calculated as described below, for details see [617, 642, 784] and Fig. 7.28. The tip is assumed to be a charged sphere with a radius R located at a distance δ from the sample surface. The capacitance C_t between the tip and a dielectric surface is calculated following [788, 789]; in the case of a ferroelectric the anisotropy of the dielectric permittivity should be taken into account. Then the electric field of the charge $q = C_t U_t$ is calculated following [790], a radial distribution of the field E_z according to (7.34) is obtained

$$E(r) = \frac{C_{\mathrm{t}} U_{\mathrm{t}}}{2\pi\epsilon_0\sqrt{\epsilon_{\mathrm{c}}\epsilon_{\mathrm{a}}}+1}\sqrt{\frac{\epsilon_{\mathrm{a}}}{\epsilon_{\mathrm{c}}}}\frac{R+\delta}{\left[(R+\delta)^2+r^2\right]^{3/2}}, \tag{7.34}$$

where r is the radial coordinate normal to the z-axis, ϵ_a and ϵ_c are the dielectric permittivity along the polar and nonpolar axes, respectively; the meaning of other notations was mentioned earlier. Even for standard AFM voltages in the range of several volts and $R \approx 40$–$50\,\mathrm{nm}$, the fields developed immediately under the tip are rather high; for example, for $U_t = 5\,V$ and $\delta = 1\,\mathrm{nm}$ the field E_z is about $10^5\,\mathrm{V\,cm^{-1}}$, but falls down very rapidly, approximately by orders of magnitude at a distance of $r \approx 20\,\mathrm{nm}$ [617]. Under a HVAFM tip the fields are enormous, for example, at $R = 50\,\mathrm{nm}$, $\delta = 1\,\mathrm{nm}$ and with account to ϵ_a and ϵ_c in $LiNbO_3$, the calculations [784, 785] give $E_z = 5 \times 10^7\,\mathrm{V\,cm^{-1}}$ for $U_t = 3.2\,\mathrm{kV}$ and $r = 0$, that is directly under the tip. This field exceeds by more than two orders of magnitude E_c. At the same time, E_z very steeply falls down with z and becomes about $100\,\mathrm{V\,cm^{-1}}$ and several $\mathrm{V\,cm^{-1}}$ at distances, of 10 and $100\,\mu\mathrm{m}$ below the surface [784, 785], respectively. Nevertheless, the domain reversal occurs over the whole crystals thickness of about 150–250 $\mu\mathrm{m}$, thus extending to the zero-field region [784, 785]. Moreover, the tailored domain arrays are sufficiently uniform along the polar axis [784, 785], so do not respond to a drastic nonuniformity of the field. Certain parallel may be drawn with the domain recording by an electron beam, because in this case an achieved length of a grown domain along the polar axis is of about 0.5 mm, thus by orders of magnitude exceeding the electron penetration depth being of the order of microns.

So, both the experimental switching characteristics and the specific of the field distribution require a refinement of the model as applied to the polarization reversal under AFM voltages. In the literature this model is widely discussed, e.g., [642, 774, 785, 787, 791–794]. Particularly, the group of Molotskii [785, 787, 792, 794] developed a switching model for $LiNbO_3$ and $LiTaO_3$ crystals under super-high voltages of HVAFM. The main distinctions in switching scenarios under uniform and AFM fields may be formulated as follows. As mentioned in Sect. 7.1.2, in the classical model the activation energy required for the domain nucleation is by an order of magnitude higher than that provided by relatively low coercive fields, so a nucleation occurs due to *triggering* by an interface, point defects, seed microdomains on the surface, etc. Among three competitive processes (nucleation, sideways and frontal domain-wall motion), the nucleation is the slowest, thus limiting the switching velocity. By contrast, the calculations performed for HVAFM conditions have shown that applying such high fields lowers the activation energy of the nucleation down to $10^{-3}\,\mathrm{eV}$, which leads to nucleation times as short as $10^{-13}\,\mathrm{s}$. Therefore, the domain nucleation being the fastest stage, in no way limits the switching velocity. Unlike the activation law (7.20) controlling the sideways domain-wall motion under uniform fields, the process directly under a HVAFM-tip is nonactivated, because $E_t \gg E_c$. For example, Molotskii et al. [787, 791, 794] suggest a viscous friction model of the domain velocity (7.22), thus $v_{DW} \sim E_z$. At the same time, according to (7.34) E_z attenuates with r very steeply and the sideways wall kinetics changes as a wall is moving apart from a tip, so the mechanism of $v_{DW}(E_z)$ may change. Actually, a plot of v_{sw} vs. the domain radius r_d in thin $LiNbO_3$ crystals [619], sharply falls down at distinct r_d, which is for example about 0.5 μm for a tip voltage of 100 V. According to the used model [791] this is due to the transformation of the viscous drag (7.22) to the usual activation law (7.20). However, Rodriguez et al. [619] interpret this curve as a result of the dependence of the activation field δ in (7.20) on E.

The most debatable is the mechanism of the domain frontal growth under the AFM tip. As mentioned earlier, in spite of a practical vanishing of E_z at the depth of about 100 μm below the surface, the domains grow through the whole crystals thickness 150–250 μm. Note that in the classical scenario the driving force of the polarization reversal is determined by a compromise between the depolarizing energy W_E and the domain wall energy W_w, see (7.7)–(7.11), so at $E = 0$ no domain motion occurs. To account for the frontal domain growth within the zero-field region observed under a HVAFM-tip, Molotskii and coauthors ([794] and references therein) advanced a concept named by them *the ferroelectric domain breakdown* by a visual analogy with the electric breakdown, since the formed string-shaped domains remind the electric breakdown tunnels. The main idea of the model is that the driving force of the domain motion in this case is not the tip voltage itself. The attained domain length is controlled now by the equilibrium domain configuration corresponding to the given conditions, which in its turn is dictated by an extremely high potential of the tip serving as an almost point charge. In other words, the domain elongation towards a zero-field region means its tendency to an equilibrium shape. The highly charged AFM-tip in this model is not simply a passive electrode, but plays the role of an active nucleation center similarly to a charged nucleation center in a crystallization process. The domain energy now additionally to the depolarizing energy W_E and the domain wall energy W_W (7.7)–(7.11) contains the energy of the interaction between the domain and the field of the AFM tip, which in general is expressed as

$$W_t = -2 \int P_s E_z d^3 r. \qquad (7.35)$$

Minimizing the whole domain energy and using very rough approximations, the author deduce the domain equilibrium dimensions assuming the energetically favorable domain shape as a strongly elongated semi-ellipsoid, following the classic Landauer model [624]. The obtained equilibrium domain lengths and radii are 170–200 μm and 0.5–0.8 μm, respectively. So, the calculated equilibrium domain length actually extends to the zero-field region which argues in favor of this exotic model.

As mentioned in Sect. 7.2.2 the measurements of switching under AFM tip are very informative with regard to the domain pinning effects. Rosenman et al. [784] estimated in terms of the pinning model a characteristic size of the domain captured by a pinning center (so to say, a seed size); this estimate gave 0.3 μm. With account to the domain wall mobility in $LiNbO_3$ $\mu_{sw} = 1.6 \times 10^{-5}\,\mathrm{cm^2\,V^{-1}s^{-1}}$, a time required for a domain wall passing this distance is $t_m \sim 10^{-5}$ s. This time is incomparably lower than the experimentally observed time of about 40 ms needed to provide a stable microdomain, not disappearing after turning-off the voltage. Therefore, it was concluded that in the case of a domain evolution under an AFM tip in an $LiNbO_3$ crystal, the decisive kinetic factor is a certain time required to stabilize a domain wall, in other words, a pinning time. In the opinion of the authors their results on the HVAFM tip domain recording in $LiNbO_3$ support the pinning mechanism of the domain wall motion [731, 784, 786]. As already mentioned, a more steep dependence of r_d and t_p in stoichiometric $LiNbO_3$ compared to congruent $LiNbO_3$ [731], was attributed to pinning effects in the latter. As mentioned above, the data of the AFM-domain recording in ultrathin samples NSLN and NSLN:1%Mg [775] also evidence of pinning effects, because the temporal stability of the domain was provided by field pulses not shorter than a critical time of one second.

7.2.5 Conclusion

Concluding this section, we should note that the problems of polarization reversal and domain dynamics in ferroelectrics are still far from a comprehension, particularly with regard to the nucleation process. This is true for LN and LT as well, although the body of experimental investigations is enormous, perhaps comparable to that accumulated during the preceding three decades for the model ferroelectrics as a whole. An unresolved problem in LN and LT are in our opinion the properties of E_c and E_{int}, particularly their dependence on the crystal composition and temperature. Usually E_c decreases when approaching the phase transition, whereas in $LiNbO_3$ and $LiTaO_3$ E_c and E_{int} start to decrease far below T_c. This is seemingly related to the pinning effects, which so far have been studied only at room temperature. Another unclear property is an extremely high stability of PP structures against external actions, especially heating. No information is available of the thermal stability of switched single or random (nonregular) antiparallel domains formed after applying a high field. So, the question stands, whether PP structures are stable due to an inherent stability of switched domains (owing, say, to pinning effects), or it is caused by some kind of a coupling. Anyhow, the PPLN thermal stability seems to be in certain contradiction with the disappearance or reorientation of E_{int} at relatively low temperatures in the range of 150–250°C.

In spite of vagueness in physical mechanisms, a significant progress in elaboration of some practical devices seems to be achieved. As already mentioned, we do not intend to review these researches and simply illustrate the domain engineering by several recent examples. The bibliography on the PPLN, PPLT and their formation one may find, for instance, in the reviews [477,674,756,795]. The special review of researches in the third-harmonic generation (THG) in quasi-periodical domain structure (for example, of a Fibonacci type) is given in [796]. Useful references on the PPLN and PPLT modulators and deflectors are presented in [797, 798].

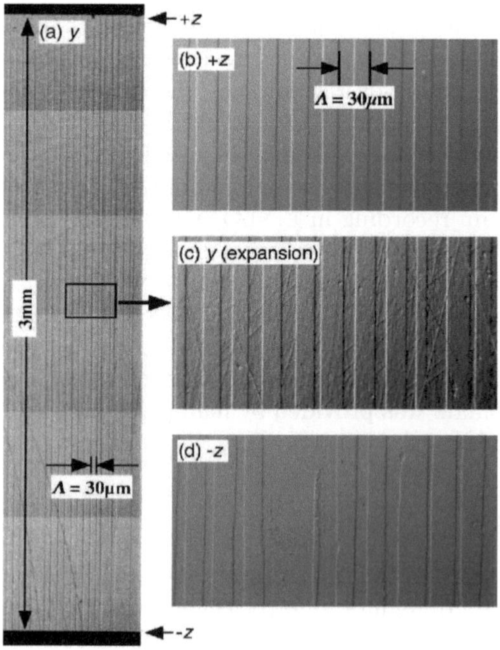

Fig. 7.29. Photographs of etched surfaces of periodically poled 3 mm thick MgO-doped $LiNbO_3$ crystal: (**a**) y face from +z to −z surface, (**b**) +z surface, (**c**) y face (expansion), and (**d**) −z surface [799]. Reused with permission from Ishizuki et al. [799], Copyright 2003, American Institute of Physics

A progress in PPLN and PPLT sizes is worthy of mentioning. As seen from Sect. 7.2.1, former samples of PPLN and PPLT did not exceed 0.5 mm in thickness, because the use of thicker crystal plates required too high fields leading to an electric breakdown. Finding a decrease of E_c in Mg- or Zn-doped and stoichiometric crystals facilitated the elaboration of PP structures with a thickness of a few mm, see for instance the references [715, 799]. Figure 7.29 demonstrates a rather regular PP structure in a 3 mm thick $LiNbO_3$:5%Mg crystal created at about 250°C; E_c was as low as $12\,kV\,cm^{-1}$. S. Kurimura [800] has reported on a large-aperture OPO formed in an SLT:1%Mg plate of 2 mm thickness; $E_c = 17\,kV\,cm^{-1}$. When pumping this element with $\lambda = 1{,}064$ nm, a conversion to the near-infrared range 2,500–3,500 nm was obtained, tuning the wavelength by a variation of the domain period and the temperature.

As the technology of producing the controlled domain arrays was developed, their practical potentials were enhancing. For example, one of the exciting tasks of the laser technique is the simultaneous generation of three primary colors – red, green, blue (RGD) from a single laser source. This was achieved with the aid of a QPM on appropriately organized domain arrays, particularly on aperiodic or quasi-periodic structures, for references see [477, 796].

A multiwave conversion of the fundamental radiation in this case is based on an optical parametric process followed by a subsequent sum-frequency doubling or mixing. An example of an aperiodic structure created in a congruent $LiTaO_3$ is shown in Fig. 7.30. An efficient RGB light was obtained with this structure using the fundamental wavelengths 1,342 and 1,063 nm from a diode-pumped Q-switched Nd–$GdVO_4$ laser.

Recently, the concept of the nonlinear photonic crystals [679] generated great interest. As mentioned in Sect. 7.1.4, a 2D nonlinear photonic crystal in $LiNbO_3$ was for the first time realized by Broderick et al. [680]. Following the idea of Berger [679], an $LiNbO_3$ crystal was coated by a honeycomb metallic mask with hexagonal holes, whereupon a switching voltage was applied. The area under the whole electrode was reversed, whereas the nonswitched spots under uncoated holes formed a 2D regular pattern of hexagonal domains called by the authors a *HeXLN crystal.* Its optical properties were consistent with the predictions of Berger [679], particularly a multiwave noncollinear frequency conversion was observed. These results stimulated searching for other routes of creation of 2D domain arrays, particularly by means of the electron beam

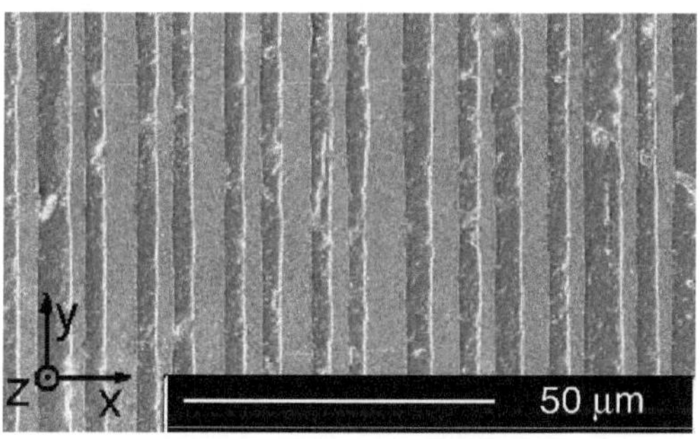

Fig. 7.30. Micrograph of an $LiTaO_3$ sample revealed by etching (the side of +c). The direction of the beam inside the sample was parallel to the x-axis and the polarization of the beam was parallel to the z-axis. This sample was used to generate three colors, according to [801]. Reused with permission from Ren et al. [801], Copyright 2004, IOP Publishing Limited

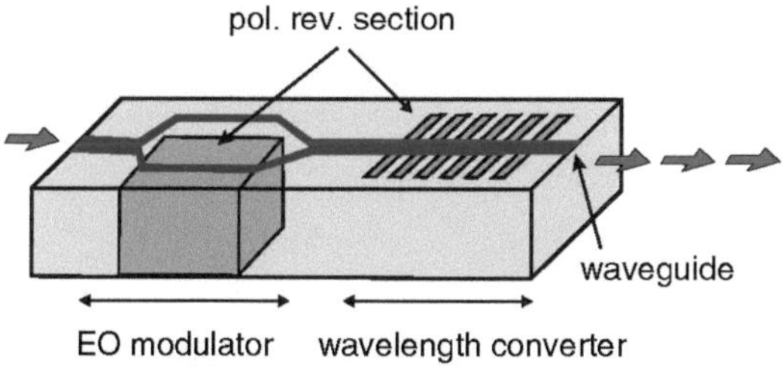

Fig. 7.31. Concept of an integrated polarization reversed optical device (PRO) according to Kurimura [800]

or AFM recording, as was exemplified in Sect. 7.2.4. Further theoretical and experimental developments of the concept of nonlinear photonic crystals was performed, for example, in the following references [802–806].

Concluding this section, we may following Kurimuras dream of possible future applications of periodic domain arrays in integrated devices (Fig. 7.31), which would use their potentials for both a spatial modulation and a frequency conversion of an optical wave [800].

References

1. A.A. Ballman, J. Am. Ceramic Soc. **48**(2), 112 (1965)
2. E. Fatuzzo, W.J. Merz, *Ferroelectricity, Series of Monographs on Selected Topics in Solid State Physics*, vol. VII (North-Holland, Amsterdam, 1967)
3. T. Mitsui, I. Tatsuzaki, E. Nakamura, *An Introduction to the Physics of Ferroelectrics* (Gordon and Breach, N.Y., London, Paris, 1976)
4. M.E. Lines, A.M. Glass, *Principles and Applications of Ferroelectrics and Related Materials* (Clarendon, Oxford, 1977)
5. Y. Xu, *Ferroelectric Materials and Their Applications* (Elsevier, Amsterdam, 1991)
6. Y.S. Kuzminov, *Lithium Niobate Crystals* (Cambridge International Science, Cambridge, 1999)
7. A. Räuber, in *Current Topics in Materials Science*, ed. by E. Kaldis (North-Holland, Amsterdam, 1978), p. 481
8. R.S. Weis, T.K. Gaylord, Appl. Phys. A **37**(4), 191 (1985)
9. B.I. Sturman, V.M. Fridkin, *The photovoltaic and photorefractive effects in noncentrosymmetric materials, Ferroelectricity and related phenomena*, vol. 8 (Gordon and Breach, Philadelphia, 1992)
10. I. of Electrical Engineers, *Properties of lithium niobate.* EMIS datareviews series; 5 (INSPEC, Institution of Electrical Engineers, London: INSPEC, 1989)
11. K.K.W. (Ed.), *Properties of lithium niobate.* EMIS datareviews series; 28 (INSPEC, Institution of Electrical Engineers, London: INSPEC, 2002)
12. P. Günter, J.P.H. (Eds.), *Photorefractive Materials and Their Applications I, Fundamental Phenomena, Topics in Applied Physics*, vol. 61 (Springer, Berlin, 1988)
13. P. Günter, J.P.H. (Eds.), *Photorefractive Materials and Their Applications II, Survey of applications, Topics in Applied Physics*, vol. 62 (Springer, Berlin, 1989)
14. P. Günter, J.P.H. (Eds.), *Photorefractive Materials and Their Applications 1, Basic Effects, Springer Series in Optical Sciences*, vol. 113 (Springer, Berlin, 2006)
15. P. Günter, J.P.H. (Eds.), *Photorefractive Materials and Their Applications 2, Materials, Springer Series in Optical Sciences*, vol. 114 (Springer, Berlin, 2007)
16. P. Günter, J.P.H. (Eds.), *Photorefractive Materials and Their Applications 3, Applications, Springer Series in Optical Sciences*, vol. 115 (Springer, Berlin, 2007)

17. D.N. Nikogosjan, *Nonlinear Optical Crystals: A Complete Survey* (Springer, New York, 2005)
18. Y. Furukawa, K. Kitamura, K. Niwa, H. Hatano, P. Bernasconi, G. Montemezzani, P. Günter, Jpn. J. Appl. Phys. Part 1 **38**, 1816 (1999)
19. I. Takanaga, J. Hirohashi, J. Kushibi, Jpn. J. Appl. Phys. Part 1 **38**, 3201 (1999)
20. P.K. Gallagher, H.M. Obryan, E.M. Gyorgy, J.T. Krause, Ferroelectrics **75** (1–2), 71 (1987)
21. M. Nakamura, M. Sekita, S. Takekawa, K. Kitamura, J. Cryst. Growth **290**(1), 144 (2006)
22. Y.L. Chen, J.J. Xu, X.J. Chen, Y.F. Kong, G.Y. Zhang, Opt. Commun. **188**, 359 (2001)
23. I. Tomeno, S. Matsumura, Phys. Rev. B **38**(1), 606 (1988)
24. K. Kitamura, Y. Furukawa, K. Niwa, V. Gopalan, T.E. Mitchell, Appl. Phys. Lett. **73**, 3073 (1998)
25. Y.S. Kuzminov, V.V. Osiko, Kristallografiya **39**(3), 530 (1994)
26. A. Bruner, D. Eger, M.B. Oron, P. Blau, M. Katz, S. Ruschin, Opt. Lett. **28**(3), 194 (2003)
27. Y. Kondo, K. Fukuda, Y. Yamashita, K. Yokoyama, K. Arita, M. Watanabe, Y. Furukawa, K. Kitamura, H. Nakajima, Jpn. J. Appl. Phys. Part 1 **39**, 1477 (2000)
28. M. Nakamura, S. Higuchi, S. Takekawa, K. Terabe, Y. Furukawa, K. Kitamura, Jpn. J. Appl. Phys. Part 2 **41**, L465 (2002)
29. H. Hatano, Y. Liu, K. Kitamura, in *Photorefractive Materials and Their Applications 2, Materials*, ed. by P. Günter, J.P. Huignard, no. 114 in Springer Series in Optical Sciences (Springer, 2007), pp. 127–164
30. P.F. Bordui, R.G. Norwood, D.H. Jundt, M.M. Fejer, J. Appl. Phys. **71**, 875 (1992)
31. P.F. Bordui, R.G. Norwood, C.D. Bird, J.T. Carella, J. Appl. Phys. **78**, 4647 (1995)
32. X. Liang, X.W. Xu, T.C. Chong, S.N. Yuan, F.L. Yu, Y.S. Tay, J. Cryst. Growth **260**(1–2), 143 (2004)
33. K. Kitamura, Y.K. Yamamoto, N. Iyi, S. Kimura, T. Hayashi, J. Cryst. Growth **116**, 327 (1992)
34. G.I. Malovichko, V.G. Grachev, L.P. Yurchenko, V.Y. Proshko, E.P. Kokanyan, V.T. Gabrielyan, Phys. Stat. Solidi (a) **133**, K29 (1992)
35. G.I. Malovichko, V.G. Grachev, E.P. Kokanyan, O.F. Schirmer, K. Betzler, B. Gather, F. Jermann, S. Klauer, U. Schlarb, M. Wöhlecke, Appl. Phys. A **56**, 103 (1993)
36. K. Polgár, Á. Péter, L. Kovács, G. Corradi, Z. Szaller, J. Cryst. Growth **177**, 211 (1997)
37. M. Wöhlecke, G. Corradi, K. Betzler, Appl. Phys. B **63**, 323 (1996)
38. C. Bäumer, C. David, A. Tunyagi, K. Betzler, H. Hesse, E. Krätzig, M. Wöhlecke, J. Appl. Phys. **93**(5), 3102 (2003)
39. C. Bäumer, C. David, K. Betzler, H. Hesse, K. Lengyel, L. Kovács, M. Wöhlecke, Phys. Stat. Solidi (a) **201**(4), R13 (2004)
40. R.D. Shannon, C.T. Prewitt, Acta Crystallogr., Sect. B: Struct. Sci. **25**, 925 (1969)
41. S.C. Abrahams, P. Marsh, Acta Crystallogr., Sect. B: Struct. Sci. **42**, 61 (1986)

42. N. Iyi, K. Kitamura, F. Izumi, J.K. Yamamoto, T. Hayashi, H. Asano, S. Kimura, J. Solid State Chem. **101**, 340 (1992)
43. A.P. Wilkinson, A.K. Cheetham, R.H. Jarman, J. Appl. Phys. **74**, 3080 (1993)
44. N. Zotov, H. Boysen, F. Frey, T. Metzger, E. Born, J. Phys. Chem. Solids **55**, 145 (1994)
45. G.G. DeLeo, J.L. Dobson, M.F. Masters, L.H. Bonjack, Phys. Rev. B **37**, 8394 (1988)
46. P. Lerner, C. Legras, J.P. Dumas, J. Cryst. Growth **3**, 231 (1968)
47. G.E. Peterson, A. Carnevale, J. Chem. Phys. **56**, 4848 (1972)
48. H. Donnerberg, S.M. Tomlinson, C.R.A. Catlow, O.F. Schirmer, Phys. Rev. B **40**, 11909 (1989)
49. F.A. Kröger, *The Chemistry of Imperfect Crystals Vol. 2* (North-Holland, Amsterdam, 1974)
50. D.M. Smyth, Prog. Solid State Chem. **15**, 145 (1984)
51. J. Blumel, E. Born, T. Metzger, J. Phys. Chem. Solids **55**, 589 (1994)
52. H. Boysen, F. Altorfer, Acta Crystallogr., Sect. B: Struct. Sci. **50**, 405 (1994)
53. H. Lehnert, H. Boysen, F. Frey, A. Hewat, P. Radaelli, Z. Kristallogr. **212**, 712 (1997)
54. B. Etschmann, N. Ishizawa, V. Streltsov, S. Oishi, Z. Kristallogr. **216**, 455 (2001)
55. O.F. Schirmer, O. Thiemann, M. Wöhlecke, J. Phys. Chem. Solids **52**, 185 (1991)
56. T. Volk, N. Rubinina, M. Wöhlecke, J. Opt. Soc. Am. B **11**, 1681 (1994)
57. G.K. Kitaeva, K.A. Kuznetsov, A.N. Penin, A.V. Shepelev, Phys. Rev. B **65**, 054304 (2002)
58. P. Herth, D. Schaniel, T. Woike, T. Granzow, M. Imlau, E. Krätzig, Phys. Rev. B **71**(12), 125128 (2005)
59. I.G. Austin, N.F. Mott, Adv. Phys. **18**(71), 41 (1969)
60. W. Mühlstrom, H.G. Reik, Phys. Rev. **162**(3), 703 (1967)
61. H.G. Reik, D. Heese, J. Phys. Chem. Solids **28**, 581 (1967)
62. D. Emin, Phys. Rev. B **48**(18), 13691 (1993)
63. O.F. Schirmer, D. von der Linde, Appl. Phys. Lett. **33**, 35 (1978)
64. D. von der Linde, O.F. Schirmer, H. Kurz, Appl. Phys. A **15**, 153 (1978)
65. L.E. Halliburton, K.L. Sweeney, C.Y. Chen, Nucl. Instr. and Meth. B **1**, 344 (1984)
66. O.F. Schirmer, J. Phys.: Condens. Matter **18**(43), R667 (2006)
67. L. Arizmendi, J.M. Cabrera, F. Agulló-López, J. Phys. C: Solid State Phys. **17**, 515 (1984)
68. A. García-Cabañes, J.A. Sanz-García, J.M. Cabrera, F. Agulló-López, C. Zaldo, R. Pareja, K. Polgár, K. Raksanyi, I. Földváry, Phys. Rev. B **37**, 6085 (1988)
69. S. Bredikhin, S. Scharner, M. Klingler, V. Kveder, B. Red'kin, W. Weppner, J. Appl. Phys. **88**, 5687 (2000)
70. D.A. Dutt, F.J. Feigl, G.G. Deleo, J. Phys. Chem. Solids **51**, 407 (1990)
71. D.M. Smyth, Ferroelectrics **50**(1–4), 419 (1983)
72. O.F. Schirmer, S. Juppe, J. Koppitz, Crystal Lattice Defects and Amorphous Materials **16**, 353 (1987)
73. J. Koppitz, O.F. Schirmer, A.I. Kuznetsov, Europhys. Lett. **4**, 1055 (1987)
74. J. Ketchum, K. Sweeney, L. Halliburton, A. Armington, Phys. Letts. A **94**, 450 (1983)

75. K. Sweeney, L. Halliburton, Appl. Phys. Lett. **43**, 336 (1983)
76. K.L. Sweeney, L.E. Halliburton, D.A. Bryan, R. Rice, R. Gerson, H.E. Tomaschke, J. Appl. Phys. **57**, 1036 (1985)
77. X. Feng, J. Ying, J. Liu, Z. Yin, Science in China (series A) **33**, 113 (1990)
78. B. Faust, H. Müller, O. Schirmer, Ferroelectrics **153**, 297 (1994)
79. A. García Cabañes, E. Dieguez, J.M. Cabrera, F. Agulló-López, J. Phys.: Condens. Matter **1**(36), 6453 (1989)
80. M. Lu, B.N. Makarenko, Y.Z. Hu, J.W. Rabalais, J. Chem. Phys. **118**(6), 2873 (2003)
81. X.J. Chen, D.S. Zhu, B. Li, T. Ling, Z.K. Wu, Opt. Lett. **26**, 998 (2001)
82. D.S. Zhu, J.J. Xu, H.J. Qiao, W. Li, Y.L. Shi, F. Gao, Z.H. Wang, B. Fu, G.Q. Zhang, Opt. Commun. **272**(2), 391 (2007)
83. C. Merschjann, B. Schoke, M. Imlau, Phys. Rev. B **76**(8), 085114 (2007)
84. H. Guenther, R. Macfarlane, Y. Furukawa, K. Kitamura, R. Neurgaonkar, Appl. Opt. **37**, 7611 (1998)
85. R. Macfarlane, H. Guenther, Y. Furukawa, K. Kitamura, OSA TOPS **27**, 484 (1999)
86. D. Berben, K. Buse, S. Wevering, P. Herth, M. Imlau, T. Woike, J. Appl. Phys. **87**, 1034 (2000)
87. L. Hesselink, S.S. Orlov, A. Liu, A. Akella, D. Lande, R.R. Neurgaonkar, Science **282**, 1089 (1998)
88. A. Winnacker, R.M. Macfarlane, Y. Furukawa, K. Kitamura, Appl. Opt. **41**, 4891 (2002)
89. Y.S. Bai, R.R. Neurgaonkar, R. Kachru, Opt. Lett. **22**(5), 334 (1997)
90. H. Guenther, G. Wittmann, R.M. Macfarlane, R.R. Neurgaonkar, Opt. Lett. **22**(17), 1305 (1997)
91. D. Lande, S.S. Orlov, A. Akella, L. Hesselink, R.R. Neurgaonkar, Opt. Lett. **22**, 1722 (1997)
92. M. Lee, S. Takekawa, Y. Furukawa, K. Kitamura, H. Hatano, S. Tanaka, Appl. Phys. Lett. **76**, 1653 (2000)
93. M. Lee, S. Takekawa, Y. Furukawa, K. Kitamura, H. Hatano, Optical Materials **18**, 53 (2001)
94. D. von der Linde, A.M. Glass, K.F. Rodgers, Appl. Phys. Lett. **25**(3), 155 (1974)
95. D. von der Linde, A.M. Glass, K.F. Rodgers, J. Appl. Phys. **47**(1), 217 (1976)
96. P. Herth, T. Granzow, D. Schaniel, T. Woike, M. Imlau, E. Krätzig, Phys. Rev. Lett. **95**(6), 067404 (2005)
97. O. Beyer, D. Maxein, T. Woike, K. Buse, Appl. Phys. B **83**(4), 527 (2006)
98. F. Jermann, J. Otten, J. Opt. Soc. Am. B **10**, 2085 (1993)
99. M. Wöhlecke, L. Kovács, Crit. Rev. Solid State Mater. Sci. **26**, 1 (2001)
100. J.M. Cabrera, J. Olivares, M. Carracosca, J.Rams, R. Müller, E. Diéguez, Advances in Physics **45**, 349 (1996)
101. R.G. Smith, D.B. Fraser, R.T. Denton, T.C. Rich, J. Appl. Phys. **39**, 4600 (1968)
102. J.R. Herrington, B. Dischler, A. Räuber, J. Schneider, Solid State Commun. **12**, 351 (1973)
103. C.E. Rice, J. Solid State Chem. **64**, 188 (1986)
104. S. Klauer, M. Wöhlecke, Phys. Rev. B **49**, 158 (1994)
105. L. Kovács, V. Szalay, R. Capelletti, Solid State Commun. **52**, 1029 (1984)

106. M. Wierschem, C. Bäumer, A. Jovanović, M. Wöhlecke, S. Kapphan, L. Kovács, Phys. Stat. Solidi (b) **215**, R7 (1999)
107. L. Kovács, M. Wöhlecke, A. Jovanović, K. Polgár, S. Kapphan, J. Phys. Chem. Solids **52**, 797 (1991)
108. K. Lengyel, L. Kovács, G. Mandula, R. Rupp, Ferroelectrics **257**, 255 (2001)
109. S. Klauer, M. Wöhlecke, S. Kapphan, Phys. Rev. B **45**(6), 2786 (1992)
110. G. Dravecz, L. Kovács, Appl. Phys. B **88**(2), 305 (2007)
111. R. Richter, T. Bremer, P. Hertel, E. Krätzig, Phys. Stat. Solidi (a) **114**, 765 (1989)
112. S. Kapphan, A. Breitkopf, Phys. Stat. Solidi (a) **133**, 159 (1992)
113. W. Bollmann, K. Schlothauer, J. Zogal, Kristall und Technik **11**, 1327 (1976)
114. R. Müller, L. Arizmendi, M. Carrascosa, J.M. Cabrera, Appl. Phys. Lett. **60**, 3212 (1992)
115. E. Krätzig, O.F. Schirmer, in *Photorefractive Materials and Their Applications I*, ed. by P. Günter, J.P. Huignard, no. 61 in Topics in Appl. Phys. (Springer, Berlin, 1988), pp. 131–166
116. H. Kurz, E. Krätzig, W. Keune, H. Engelmann, U. Gonser, B. Dischler, A. Räuber, Appl. Phys. **12**, 355 (1977)
117. L. Rebouta, P.J.M. Smulders, D.O. Boerma, F. Agulló-López, M.F. da Silva, J.C. Soares, Phys. Rev. B **48**, 1993 (3600)
118. G. Corradi, A.V. Chadwick, A.R. West, K. Cruickshank, M. Paul, Radiat. Eff. Defects Solids **134**, 219 (1995)
119. A. Kling, J.C. Soares, M.F. da Silva, in *Insulating Materials for Optoelectronics. New Developments*, ed. by F. Agulló-López (World Scientific Publishing, 1995), pp. 175–200
120. J. García-Solé, L. Bausá, D. Jaque, E. Montoya, H. Murrieta, F. Jaque, Spectrochimica Acta Part A-Molecular Spectroscopy **54**, 1571 (1998)
121. G. Malovichko, V. Grachev, E. Kokanyan, O. Schirmer, Phys. Rev. B **59**, 9113 (1999)
122. G. Malovichko, V. Grachev, O. Schirmer, Appl. Phys. B **68**, 785 (1999)
123. G. Malovichko, V. Grachev, E. Kokanyan, O. Schirmer, Ferroelectrics **239**, 1227 (2000)
124. M.G. Clark, F.J. DiSalvo, A.M. Glass, G.E. Peterson, J. Chem. Phys. **59**, 6209 (1973)
125. M. Falk, K. Buse, Appl. Phys. B **81**(6), 853 (2005)
126. O.F. Schirmer, H.J. Reyher, M. Wöhlecke, in *Insulating Materials for Optoelectronics. New Developments*, ed. by F. Agulló-López (World Scientific Publishing, 1995), p. 93
127. F. Agulló-López, K.A. Müller, Crystal Lattice Defects and Amorphous Materials **15**(1–4), 89 (1987)
128. T.H. Yeom, Y.M. Chang, S.H. Choh, C. Rudowicz, Phys. Stat. Solidi (b) **185**(2), 409 (1994)
129. M.G. Zhao, M. Chiu, Phys. Rev. B **49**(18), 12556 (1994)
130. C.R.A. Catlow, A.V. Chadwick, M. Cole, S.M. Tomlinson, Radiat. Eff. Defects Solids **119**, 565 (1991)
131. T.S. Bush, C.R.A. Catlow, A.V. Chadwick, M. Cole, R.M. Geatches, G.N. Greaves, S.M. Tomlinson, J. Mater. Chem. **2**(3), 309 (1992)
132. H. Söthe, J.M. Spaeth, J. Phys.: Condens. Matter **4**(49), 9901 (1992)
133. G.I. Malovichko, V.G. Grachev, O.F. Schirmer, B. Faust, J. Phys.: Condens. Matter **5**, 3971 (1993)

134. H. Donnerberg, S.M. Tomlinson, C.R.A. Catlow, O.F. Schirmer, Phys. Rev. B **44**, 4877 (1991)
135. G. Malovichko, R. Petersen, C. Bauman, V. Grachev, J. Appl. Phys. **100**(2), 023911 (2006)
136. D.J. Keeble, M. Loyo-Menoyo, Y. Furukawa, K. Kitamura, Phys. Rev. B **71**(22), 224111 (2005)
137. K. Peithmann, J. Hukriede, K. Buse, E. Krätzig, Phys. Rev. B **61**, 4615 (2000)
138. Y.P. Yang, K. Buse, D. Psaltis, Opt. Lett. **27**(3), 158 (2002)
139. Y.P. Yang, D. Psaltis, M. Luennemann, D. Berben, U. Hartwig, K. Buse, J. Opt. Soc. Am. B **20**(7), 1491 (2003)
140. G. Corradi, H. Söthe, J.M. Spaeth, K. Polgár, J. Phys.: Condens. Matter **2**(31), 6603 (1990)
141. L.E. Bausa, J. García-Solé, in *Insulating Materials for Optoelectronics. New Developments*, ed. by F. Agulló-López (World Scientific Publishing, 1995), pp. 147–173
142. V. Grachev, G. Malovichko, Phys. Rev. B **62**, 7779 (2000)
143. G. Malovichko, V. Grachev, A. Hofstaetter, E. Kokanyan, A. Scharmann, O. Schirmer, Phys. Rev. B **65**(22) (2002)
144. V. Grachev, G. Malovichko, O. Schirmer, Ferroelectrics **185**, 5 (1996)
145. A.M. Glass, J. Chem. Phys. **50**(4), 1501 (1969)
146. C. Fischer, S. Kapphan, X.Q. Feng, N. Cheng, Radiat. Eff. Defects Solids **135**, 697 (1995)
147. P. Bourson, G. Malovichko, A. Ridah, E. Kokanyan, Ferroelectrics **185**, 273 (1996)
148. F. Lhommé, P. Bourson, M.D. Fontana, G. Malovichko, M. Aillerie, E. Kokanyan, J. Phys.: Condens. Matter **10**, 1137 (1998)
149. F. Lhommé, P. Bourson, G. Boulon, Y. Guyot, R. Burlot-Loison, M.D. Fontana, M. Aillerie, G. Malovichko, J. Luminescence **83–84**, 441 (1999)
150. G.M. Salley, S.A. Basun, A.A. Kaplyanskii, R.S. Meltzer, K. Polgár, U. Happek, J. Luminescence **87–89**, 1133 (2000)
151. V. Dierolf, A.A. Kaplyanskii, S. Kapphan, A.B. Kutsenko, Radiat. Eff. Defects Solids **155**, 241 (2001)
152. F. Lhommé, P. Bourson, G. Boulon, Y. Guyot, M.D. Fontana, Eur. Phys. J. Appl. Phys. **20**, 29 (2002)
153. M. Loyo-Menoyo, D.J. Keeble, Y. Furukawa, K. Kitamura, **16**(49), 9047 (2004)
154. M. Loyo-Menoyo, D.J. Keeble, Y. Furukawa, K. Kitamura, J. Appl. Phys. **97**(12), 123905 (2005)
155. I. Kaminow, L. Stulz, IEEE J. Quantum Electronics **11**, 306 (1975)
156. T.Y. Fan, A. Cordova-Plaza, M.J.F. Digonnet, R.L. Byer, H.J. Shaw, J. Opt. Soc. Am. B **3**, 140 (1986)
157. A. Cordova-Plaza, M.J.F. Digonnet, H.J. Shaw, IEEE J. Quantum Electronics **QE-23**, 262 (1987)
158. J.K. Yamamoto, A. Sugimoto, K. Yamagishi, Opt. Lett. **19**, 1311 (1994)
159. J.O. Tocho, E. Camarillo, F. Cussó, F. Jaque, J. García-Solé, Solid State Commun. **80**, 575 (1991)
160. A. Lorenzo, H. Loro, J.E. Muñoz Santiuste, M.C. Terrile, G. Boulon, L.E. Bausá, J. García-Solé, Optical Materials **8**, 55 (1997)
161. H. Loro, J. Hernandez A., E. Camarillo, H. Murrieta S., M. Voda, F. Jaque, J. García-Solé, Ferroelectrics **185**, 33 (1996)

162. J.O. Tocho, J.A. Sanz-García, F. Jaque, J. García–Solé, J. Appl. Phys. **70**, 5582 (1991)
163. C. Prieto, Optical Materials **12**, 135 (1999)
164. M. Vila, A. de Bernabe, C. Prieto, J. Alloys Compd. **323–324**, 331 (2001)
165. G. Malovichko, V. Grachev, S. Okulov, E. Kokanyan, F. Henecker, A. Hofstaetter, O. Schirmer, Phys. Stat. Solidi (b) **243**(2), 409 (2006)
166. D.M. Gill, L. McCaughan, J.C. Wright, Phys. Rev. B **53**, 2334 (1996)
167. V. Dierolf, M. Koerdt, Radiat. Eff. Defects Solids **149**, 323 (1999)
168. V. Dierolf, M. Koerdt, Phys. Rev. B **61**, 8043 (2000)
169. V. Dierolf, C. Sandmann, A.B. Kutsenko, T. Troster, Radiat. Eff. Defects Solids **155**, 253 (2001)
170. J.E.M. Santiuste, B. Macalik, J.G. Sole, Phys. Rev. B **47**(1), 88 (1993)
171. G.G. Zhong, J. Jin, Z.K. Wu, J. Opt. Soc. Am.p. 631 (1980)
172. D.A. Bryan, R. Gerson, H.E. Tomaschke, Appl. Phys. Lett. **44**, 847 (1984)
173. D.A. Bryan, R.R. Rice, R. Gerson, H.E. Tomaschke, K.L. Sweeney, L.E. Halliburton, Opt. Eng. **24**, 138 (1985)
174. H.X. Feng, J.K. Wen, H.F. Wang, S. Y.Hand, Y.X. Xu, J. Phys. Chem. Solids **51**, 397 (1990)
175. T.R. Volk, N.M. Rubinina, V.I. Pryalkin, V.V. Krasnikov, V.V. Volkov, Ferroelectrics **109**, 345 (1990)
176. T.R. Volk, V.I. Pryalkin, N.M. Rubinina, Opt. Lett. **15**, 996 (1990)
177. Y. Zhang, Y. Xu, M. Li, Y. Zhao, J. Cryst. Growth **233**, 537 (2001)
178. T.R. Volk, N.M. Rubinina, Ferroelectric Letters **14**, 37 (1992)
179. T.R. Volk, M. Wöhlecke, N. Rubinina, N. V.Razumovskiĭ, F. Jermann, C. Fischer, R. Böwer, Appl. Phys. **A60**, 217 (1995)
180. Y. Kong, J. Wen, H. Wang, Appl. Phys. Lett. **66**, 280 (1995)
181. J.K. Yamamoto, K. Kitamura, N. Iyi, S. Kimura, Y. Furukawa, M. Sato, Appl. Phys. Lett. **61**, 2156 (1992)
182. J.K. Yamamoto, T. Yamazaki, K. Yamagishi, Appl. Phys. Lett. **64**, 3228 (1994)
183. E.P. Kokanyan, L. Razzari, I. Cristiani, V. Degiorgio, J.B. Gruber, Appl. Phys. Lett. **84**(11), 1880 (2004)
184. L. Razzari, P. Minzioni, I. Cristiani, V. Degiorgio, E.P. Kokanyan, Appl. Phys. Lett. **86**(13), 131914 (2005)
185. P. Galinetto, F. Rossella, P. Minzioni, L. Razzari, I. Cristiani, V. Degiorgio, E.P. Kokanyan, J. Nonlinear Opt. Phys. & Materials **15**(1), 9 (2006)
186. P. Minzioni, I. Cristiani, V. Degiorgio, E.P. Kokanyan, J. Appl. Phys. **101**(11), 116105 (2007)
187. P. Minzioni, I. Cristiani, J. Yu, J. Parravicini, E.P. Kokanyan, V. Degiorgio, Optics Express **15**, 14171 (2007)
188. Y.F. Kong, S.G. Liu, Y.J. Zhao, H.D. Liu, S.L. Chen, J.J. Xu, Appl. Phys. Lett. **91**(8), 081908 (2007)
189. F. Nitanda, Y. Furukawa, S. Makio, M. Sato, K. Ito, Jpn. J. Appl. Phys. Part 1 **34**, 1546 (1995)
190. Y.F. Kong, J.K. Wen, Z.F. Yang, Chin. Phys. Lett. **14**, 629 (1997)
191. X. Zhen, R. Wang, W. Xu, Y. Xu, L. Zhao, Optical Materials **19**, 427 (2002)
192. J.L. Jackel, D.H. Olson, A.M. Glass, J. Appl. Phys. **52**, 4855 (1981)
193. I. Savatinova, S. Tonchev, P. Kircheva, Ferroelectrics **249**, 257 (2001)
194. T.R. Volk, M. Wöhlecke, Ferroelectrics Review **1**, 195 (1998)

195. T. Volk, M. Wöhlecke, N. Rubinina, in *Photorefractive Materials and Their Applications 2, Materials*, ed. by P. Günter, J.P. Huignard, no. 114 in Springer Series in Optical Sciences (Springer, 2007), pp. 165–203
196. T. Volk, M. Ivanov, V. Pryalkin, N. Rubinina, Ferroelectrics **126**, 57 (1992)
197. I. Földváry, K. Polgár, A. Mecseki, Acta Phys. Hungarica **55**, 321 (1984)
198. K. Polgár, L. Kovács, I. Földváry, I. Cravero, Solid State Commun. **59**, 375 (1986)
199. T. Volk, M. Wöhlecke, N. Rubinina, A. Reichert, N. Razumovski, Ferroelectrics **183**, 291 (1996)
200. B.C. Grabmaier, F. Otto, J. Cryst. Growth **79**, 682 (1986)
201. B.C. Grabmaier, W. Wersing, W. Koestler, J. Cryst. Growth **110**, 339 (1991)
202. H. Donnerberg, J. Solid State Chem. **123**, 208 (1996)
203. L. Zhao, X. Wang, B. Wang, W. Wen, T.Y. Zhang, Appl. Phys. B **78**(6), 769 (2004)
204. D. Redfield, W.J. Burke, J. Appl. Phys. **45**, 4566 (1974)
205. W. Rossner, B. Grabmaier, W. Wersing, Ferroelectrics **93**, 57 (1989)
206. N. Iyi, K. Kitamura, Y. Yajima, S. Kimura, Y. Furukawa, M. Sato, J. Solid State Chem. **118**, 148 (1995)
207. S. Shimamura, Y. Watanabe, T. Sota, K. Suzuki, N. Iyi, K. Kitamura, T. Yamazaki, A. Sugimoto, K. Yamagishi, J. Phys.: Condens. Matter **8**, 6825 (1996)
208. C. Prieto, C. Zaldo, P. Fessler, H. Dexpert, J.A. Sanzgarcia, E. Diéguez, Phys. Rev. B **43**(4), 2594 (1991)
209. J.G. Marques, A. Kling, J.C. Soares, L. Rebouta, M.F. da Silva, E. Diéguez, F. Agulló-López, Nucl. Instr. and Meth. B **136–138**, 431 (1998)
210. L.J. Hu, Y.H. Chang, F.S. Yen, S.P. Lin, I.N. Lin, W.Y. Lin, J. Appl. Phys. **69**, 7635 (1992)
211. J.J. Liu, W.L. Zhang, G.Y. Zhang, Phys. Stat. Solidi (a) **156**, 285 (1996)
212. J. Wen, L. Wang, Y. Tang, H. Wang, Appl. Phys. Lett. **53**(4), 260 (1988)
213. Y. Kong, B. Li, Y. Chen, Z. Huang, S. Chen, L. Zhang, S. Liu, J. Xu, W. Yan, H. Liu, Y. Wang, X. Xie, W. Zhang, G. Zhang, J. Infrared Millim. Waves **22**, 40 (2003)
214. X. Feng, D. Wang, J. Zhang, Phys. Stat. Solidi (b) **157**(2), K127 (1990)
215. T. Chernaya, B. Maximov, T. Volk, N. Rubinina, V. Simonov, JETP Letters **73**, 103 (2001)
216. T. Volk, B. Maximov, T. Chernaya, N. Rubinina, M. Wöhlecke, V. Simonov, Appl. Phys. B **72**, 647 (2001)
217. S. Sulyanov, B. Maximov, T. Volk, H. Boysen, J. Schneider, N. Rubinina, T. Hansen, Appl. Phys. **A 74 [Suppl.]**, S1031 (2002)
218. T. Volk, B. Maximov, S. Sulyanov, N. Rubinina, M. Wöhlecke, Optical Materials **23**, 229 (2003)
219. J.G. Marques, A. Kling, L. Rebouta, M.F. da Silva, A.A. Melo, J.C. Soares, M.D. Serrano, E. Diéguez, F. Agulló-López, Materials Science Forum **248**, 395 (1997)
220. U. Schlarb, M. Wöhlecke, B. Gather, A. Reichert, K. Betzler, T. Volk, N. Rubinina, Optical Materials **4**, 791 (1995)
221. F. Abdi, M. Aillerie, M. Fontana, P. Bourson, T. Volk, B. Maximov, S. Sulyanov, N. Rubinina, M. Wöhlecke, Appl. Phys. B **68**, 795 (1999)
222. U. Schlarb, B. Matzas, A. Reichert, K. Betzler, M. Wöhlecke, B. Gather, T. Volk, Ferroelectrics **185**, 269 (1996)

223. X.K. He, D.F. Xue, Opt. Commun. **265**(2), 537 (2006)
224. N.E. Brese, M. O'Keeffe, Acta Crystallogr., Sect. B: Struct. Sci. **47**, 192 (1991)
225. Y. Furukawa, K. Kitamura, S. Takekawa, A. Miyamoto, M. Terao, N. Suda, Appl. Phys. Lett. **77**, 2494 (2000)
226. Y. Furukawa, K. Kitamura, S. Takekawa, K. Niwa, Y. Yajima, N. Iyi, I. Mnushkina, P. Guggenheim, J.M. Martin, J. Cryst. Growth **211**, 230 (2000)
227. Y. Furukawa, K. Kitamura, S. Takekawa, K. Niwa, H. Hatano, Opt. Lett. **23**, 1892 (1998)
228. L. Palfalvi, J. Hebling, G. Almasi, Á. Péter, K. Polgár, K. Lengyel, R. Szipocs, J. Appl. Phys. **95**(3), 902 (2004)
229. Á. Péter, K. Polgár, L. Kovács, K. Lengyel, J. Cryst. Growth **284**(1–2), 149 (2005)
230. R. Sommerfeldt, L. Holtmann, E. Krätzig, B. Grabmaier, Phys. Stat. Solidi (a) **106**, 89 (1988)
231. A. Böker, H. Donnerberg, O.F. Schirmer, X.Q. Feng, J. Phys.: Condens. Matter **2**, 6865 (1990)
232. G.I. Malovichko, V.G. Grachev, E.P. Kokonyan, Ferroelectrics **125**, 289 (1992)
233. G. Corradi, H. Söthe, J.M. Spaeth, K. Polgár, J. Phys.: Condens. Matter **3**, 1901 (1991)
234. A. Martin, F.J. Lopez, F. Agulló-López, J. Phys.: Condens. Matter **4**(3), 847 (1992)
235. J. DiazCaro, J. García-Solé, D. Bravo, J.A. SanzGarcia, F.J. Lopez, F. Jaque, Phys. Rev. B **54**(18), 13042 (1996)
236. J. DiazCaro, J. García-Solé, D. Bravo, T.P.J. Han, F. Jaque, B. Henderson, Ferroelectric Letters **23**(1–2), 27 (1997)
237. G.A. Torchia, J.A. Sanz-García, F.J. López, D. Bravo, J. García-Solé, F. Jaque, H.G. Gallagher, T.P.J. Han, J. Phys.: Condens. Matter **10**, L341 (1998)
238. G. Corradi, H. Söthe, J.M. Spaeth, K. Polgár, Ferroelectrics **125**(1–4), 295 (1992)
239. E. Camarillo, J. Tocho, I. Vergara, E. Diéguez, J.G. Sole, F. Jaque, Phys. Rev. B **45**(9), 4600 (1992)
240. G.M. Salley, S.A. Basun, G.F. Imbusch, A.A. Kaplyanskii, S. Kapphan, R.S. Meltzer, U. Happek, J. Luminescence **83–84**, 423 (1999)
241. T.P.J. Han, F. Jaque, L. Arizmendi, V. Bermúdez, A. Suchocki, A. Kaminska, S. Kobyakov, Phys. Rev. B **68**(13), 132103 (2003)
242. D. Jaque, J.J. Romero, J.A. Sanz-García, I.A.D. Carcer, E. Camarillo, F. Jaque, Radiat. Eff. Defects Solids **155**, 235 (2001)
243. S. Juppe, O.F. Schirmer, Phys. Lett. A **117**(3), 150 (1986)
244. H.J. Reyher, R. Schulz, O. Thiemann, Phys. Rev. B **50**(6), 3609 (1994)
245. O. Thiemann, H. Donnerberg, M. Wöhlecke, O.F. Schirmer, Phys. Rev. B **49**(9), 5845 (1994)
246. G. Corradi, I.M. Zaritskii, A. Hofstaetter, K. Polgár, L.G. Rakitina, Phys. Rev. B **58**(13), 8329 (1998)
247. G. Corradi, M. Meyer, L. Kovács, K. Polgár, Appl. Phys. B **78**(5), 607 (2004)
248. K. Kasemir, K. Betzler, B. Matzas, B. Tiegel, T. Wahlbrink, M. Wöhlecke, B. Gather, N. Rubinina, T. Volk, J. Appl. Phys. **84**, 5191 (1998)
249. Y. Kong, J. Deng, W. Zhang, J.W.G. Zhang, H. Wang, Phys. Letters A **196**, 128 (1994)
250. G. Lifante, F. Cussó, F. Jaque, J.A. Sanz–García, A. Monteil, B. Varrel, G. Boulon, J. García–Solé, Chem. Phys. Letters **176**, 482 (1991)

251. J. García-Solé, T. Petit, H. Jaffrezic, G. Boulon, Europhys. Lett. **24**, 719 (1993)
252. D. Bravo, A. Martin, F.J. Lopez, Solid State Commun. **112**, 541 (1999)
253. C. Bonardi, C.J. Magon, E.A. Vidoto, M.C. Terrile, L.E. Bausá, E. Montoya, D. Bravo, A. Martin, F.J. Lopez, J. Alloys Compd. **323**, 340 (2001)
254. Y. Watanabe, T. Sota, K. Suzuki, N. Iyi, K. Kitamura, S. Kimura, J. Phys.: Condens. Matter **7**, 3627 (1995)
255. J.R. Herrington, B. Dischler, A. Räuber, J. Schneider, Solid State Commun. **12**, 351 (1973)
256. A. Gröne, S. Kapphan, J. Phys. Chem. Solids **56**, 687 (1995)
257. M.J. de Rosendo, L. Arizmendi, J.M. Cabrera, F. Agulló-López, Solid State Commun. **59**, 499 (1986)
258. L. Kovács, K. Polgár, R. Capelletti, Crystal Lattice Defects and Amorphous Materials **15**, 115 (1987)
259. L. Kovács, I. Földváry, K. Polgár, Acta Phys. Hungarica **61**, 223 (1987)
260. T.R. Volk, N.M. Rubinina, Sov. Phys. Solid State **33**, 674 (1991)
261. J.K. Yamamoto, K. Kitamura, N. Iyi, S. Kimura, J. Cryst. Growth **128**, 920 (1993)
262. S.Q. Li, S.G. Liu, Y.F. Kong, J.J. Xu, G.Y. Zhang, Appl. Phys. Lett. **89**(10), 101126 (2006)
263. L. Kovács, Z. Szaller, I. Cravero, I. Földvári, C. Zaldo, J. Phys. Chem. Solids **51**(5), 417 (1990)
264. L. Kovács, L. Rebouta, J.C. Soares, M.F. Dasilva, M. Hageali, J.P. Stoquert, P. Siffert, J.A. Sanzgarcia, G. Corradi, Z. Szaller, K. Polgár, J. Phys.: Condens. Matter **5**(7), 781 (1993)
265. Y. Furukawa, A. Yokotani, T. Sasaki, H. Yoshida, K. Yoshida, F. Nitanda, M. Sato, J. Appl. Phys. **69**, 3372 (1991)
266. M.P. Petrov, S.I. Stepanov, A.V. Khomenko, *Photorefractive Crystals in Coherent Optical Systems, Springer Series in Optical Sciences*, vol. 59 (Springer, Berlin, 1991)
267. L. Solymar, D.J. Webb, A. Grunnet-Jepsen, *The Physics and Applications of Photorefractive Materials, Oxford Series in Optical and Imaging Sciences*, vol. 11 (Clarendon, Oxford, 1996)
268. G. Montemezzani, C. Medrano, M. Zgonik, P. Günter, in *Nonlinear optical effects and materials*, ed. by P. Günter, no. 72 in Springer Series in Optical Sciences (Springer, 2000), pp. 301–373
269. K. Buse, J. Imbrock, E. Krätzig, K. Peithmann, in *Photorefractive Materials and Their Applications 2, Materials*, ed. by P. Günter, J.P. Huignard, no. 114 in Springer Series in Optical Sciences (Springer, 2007), pp. 83–126
270. P. Bernasconi, G. Montemezzani, P. Günter, Y. Furukawa, K. Kitamura, Ferroelectrics **223**, 373 (1999)
271. P. Dittrich, L. Koziarska-Glinka, G. Montemezzani, P. Günter, S. Takekawa, K. Kitamura, Y. Furukawa, J. Opt. Soc. Am. B **21**(3), 632 (2004)
272. P. Dittrich, G. Montemezzani, M. Habu, M. Matsukura, S. Takekawa, K. Kitamura, P. Günter, Opt. Commun. **234**(1–6), 131 (2004)
273. P.G.t. G. Montemezzani, P. Dittrich, in *Photorefractive Materials and Their Applications 1, Basic Effects*, ed. by P. Günter, J.P. Huignard, no. 113 in Springer Series in Optical Sciences (Springer, 2006), pp. 203–230
274. T.R. Volk, S.A. Shramchenko, L.A. Shuvalov, Ferroelectric Letters **2**(2), 55 (1984)

275. T.R. Volk, S.A. Shramchenko, L.A. Shuvalov, Sov. Phys. Solid State **26**(12), 2135 (1984)
276. T.R. Volk, S.A. Shramchenko, Ferroelectrics **64**(1–3), 479 (1985)
277. L.B. Schein, P.J. Cressman, L.E. Cross, J. Appl. Phys. **49**(2), 798 (1978)
278. V.E. Vood, P.J. Cressman, R.L. Holman, C.M. Verber, in *Photorefractive Materials and Their Applications II, Survey of applications*, ed. by P. Günter, J.P. Huignard, no. 62 in Topics in Applied Physics (Springer, 1989), p. 45
279. S.G. Odoulov, Sov. Phys. JETP Lett. **35**, 10 (1982)
280. N.V. Kukhtarev, Tech. Phys. Lett. **2**, 438 (1976)
281. N.V. Kukhtarev, V.B. Markov, S.G. Odulov, M.S. Soskin, V.L. Vinetskii, Ferroelectrics **22**, 949 (1979)
282. N.V. Kukhtarev, V.B. Markov, S.G. Odulov, M.S. Soskin, V.L. Vinetskii, Ferroelectrics **22**, 961 (1979)
283. H. Kogelnik, Bell System Technical J. **48**(9), 2909 (1969)
284. P. Günter, J.P. Huignard, in *Photorefractive Materials and Their Applications II, Survey of applications*, ed. by P. Günter, J.P. Huignard, no. 62 in Topics in Applied Physics (Springer, 1989), pp. 205–271
285. F.H. Mok, G.W. Burr, D. Psaltis, Opt. Lett. **21**(12), 896 (1996)
286. E.S. Maniloff, K.M. Johnson, J. Appl. Phys. **70**(9), 4702 (1991)
287. D. Psaltis, D. Brady, K. Wagner, Appl. Opt. **27**(9), 1752 (1988)
288. S.G. Odulov, M.S. Soskin, in *Photorefractive Materials and Their Applications II, Survey of applications*, ed. by P. Günter, J.P. Huignard, no. 62 in Topics in Applied Physics (Springer, Berlin, 1989), pp. 5–43
289. P. Yeh, *Introduction to Photorefractive Nonlinear Optics* (Wiley, New York, 1993)
290. L.B. Au, L. Solymar, Appl. Phys. B **45**(3), 125 (1988)
291. J. Feinberg, D. Heiman, A.R. Tanguay, R.W. Hellwarth, J. Appl. Phys. **51**(3), 1297 (1980)
292. K. Peithmann, A. Wiebrock, K. Buse, Appl. Phys. B **68**(5), 777 (1999)
293. Y. Furukawa, M. Sato, M.C. Bashaw, M.M. Fejer, N. Iyi, K. Kitamura, Jpn. J. Appl. Phys. Part 1 **35**, 2740 (1996)
294. T.R. Volk, N.V. Razumovskiĭ, A.V. Mamaev, N.M. Rubinina, J. Opt. Soc. Am. B **13**, 1457 (1996)
295. K. Buse, F. Jermann, E. Krätzig, Appl. Phys. A **58**, 191 (1994)
296. Y. Ohmori, M. Yamaguchi, K. Yoshino, Y. Inuishi, Jpn. J. Appl. Phys. **15**(11), 2263 (1976)
297. J.C. Burfoot, *Ferroelectrics* (D. Van Nostrand, London, Princeton, Toronto, 1967)
298. W. Jorgensen, R.W. Bartlett, J. Phys. Chem. Solids **30**(12), 2639 (1969)
299. D.L. Staebler, J.J. Amodei, Ferroelectrics **3**(2–3), 107 (1972)
300. Y.P. Yang, I. Nee, K. Buse, D. Psaltis, Appl. Phys. Lett. **78**(26), 4076 (2001)
301. A.P.R. I. Lapshin, Izv. Akad. Nauk SSSR, ser. Neorg. Materials **12**, 2199 (1976)
302. W. Bollmann, H.J. Stöhr, Phys. Stat. Solidi (a) **39**(2), 477 (1977)
303. N. Schmidt, K. Betzler, M. Grabs, S. Kapphan, F. Klose, J. Appl. Phys. **65**(3), 1253 (1989)
304. I. Nee, M. Müller, K. Buse, E. Krätzig, J. Appl. Phys. **88**(7), 4282 (2000)
305. I. Nee, M. Müller, K. Buse, Appl. Phys. B **72**(2), 195 (2001)
306. I.B. Barkan, M.V. Entin, S.I. Marennikov, Phys. Stat. Solidi (a) **44**(1), K91 (1977)

307. E. Krätzig, R. Orlowski, AP **15**(2), 133 (1978)
308. I.B. Barkan, A.V. Vorob'ev, S.I. Marennikov, Soviet Journal of Quantum Electronics **9**, 492 (1979)
309. D. Kip, J. Hukriede, E. Krätzig, Phys. Stat. Solidi (a) **168**(1), R3 (1998)
310. M. Falk, J. Japs, T. Woike, K. Buse, Appl. Phys. B **87**(1), 119 (2007)
311. K. Peithmann, K. Buse, E. Krätzig, Appl. Phys. B **74**(6), 549 (2002)
312. L. Kovács, K. Polgár, R. Capelletti, C. Mora, Phys. Stat. Solidi (a) **120**(1), 97 (1990)
313. C. Canali, A. Carnera, G. Dellamea, P. Mazzoldi, S.M. Alshukri, A.C.G. Nutt, R.M. Delarue, J. Appl. Phys. **59**(8), 2643 (1986)
314. E.M. de Miguel-Sanz, M. Carrascosa, L. Arizmendi, Phys. Rev. B **65**(16) (2002)
315. M.A. Ellabban, G. Mandula, R.A. Rupp, M. Fally, E. Hartmann, L. Kovács, K. Polgár, in *Selected Papers from Fifth International Conference on Correlation Optics, Proceedings of SPIE*, vol. 4607, ed. by O.V. Angelsky (2002), vol. 4607, pp. 313–321
316. M.A. Ellabban, G. Mandula, M. Fally, R.A. Rupp, L. Kovács, Appl. Phys. Lett. **78**(6), 844 (2001)
317. R. Orlowski, E. Krätzig, Solid State Commun. **27**, 1351 (1978)
318. R. Orlowski, E. Krätzig, Ferroelectrics **26**, 831 (1980)
319. M.A. Ellabban, T. Woike, M. Fally, R.A. Rupp, Europhys. Lett. **70**(4), 471 (2005)
320. E. Krätzig, K. Kurz, Optica Acta **24**, 475 (1977)
321. M. Simon, S. Wevering, K. Buse, E. Krätzig, J. Phys. D: Appl. Phys. **30**(1), 144 (1997)
322. S.A. Basun, D.R. Evans, T.J. Bunning, S. Guha, J.O. Barnes, G. Cook, R.S. Meltzer, J. Appl. Phys. **92**(12), 7051 (2002)
323. E. Krätzig, R. Orlowski, Ferroelectrics **27**(1–4), 241 (1980)
324. K. Buse, S. Breer, K. Peithmann, S. Kapphan, M. Gao, E. Krätzig, Phys. Rev. B **56**(3), 1225 (1997)
325. T.R. Volk, N.M. Rubinina, Phys. Stat. Solidi (a) **108**, 437 (1988)
326. T.R. Volk, M.A. Ivanov, F.Y. Shchapov, N.M. Rubinina, Ferroelectrics **126**, 185 (1992)
327. H. Wang, J. Wen, J. Li, H. Wang, J. Jing, Appl. Phys. Lett. **57**, 344 (1990)
328. J.C. Deng, J.K. Wen, Z.K. Wu, H.F. Wang, Appl. Phys. Lett. **64**, 2622 (1994)
329. E. Krätzig, Ferroelectrics **21**, 635 (1978)
330. B.I. Sturman, Zhurnal Eksperimentalnoi I Teoreticheskoi Fiziki **83**(5), 1930 (1982)
331. B.I. Sturman, Uspekhi Fizicheskikh Nauk **144**(3), 497 (1984)
332. J.F. Heanue, M.C. Bashaw, L. Hesselink, Science **265**(5173), 749 (1994)
333. M. Carrascosa, J.M. Cabrera, F. Agulló-López, Opt. Commun. **69**(1), 83 (1988)
334. G.C. Valley, IEEE J. Quantum Electronics **19**(11), 1637 (1983)
335. H. Okamura, J. Opt. Soc. Am. B **18**(7), 960 (2001)
336. B. Sturman, O. Beyer, D. Maxein, K. Buse, J. Opt. Soc. Am. B **24**(3), 419 (2007)
337. F. Jermann, E. Krätzig, Appl. Phys. A **55**, 114 (1992)
338. R. Göring, Y.L. Zhan, S. Steinberg, Appl. Phys. A **55**(1), 97 (1992)
339. M.B. Klein, G.C. Valley, J. Appl. Phys. **57**(11), 4901 (1985)
340. S. Ducharme, J. Feinberg, J. Appl. Phys. **56**(3), 839 (1984)
341. D. Mahgerefteh, J. Feinberg, Opt. Lett. **13**(12), 1111 (1988)

342. A. Motes, J.J. Kim, J. Opt. Soc. Am. B **4**(9), 1379 (1987)
343. G.A. Brost, R.A. Motes, J.R. Rotge, J. Opt. Soc. Am. B **5**(9), 1879 (1988)
344. G.C. Valley, Appl. Opt. **22**, 3160 (1983)
345. T.R. Volk, S.B. Astaf'ev, N.V. Razumovskiĭ, Sov. Phys.: Solid State **37**, 583 (1995)
346. F.P. Strohkendl, J. Appl. Phys. **65**(10), 3773 (1989)
347. L. Holtmann, Phys. Stat. Solidi (a) **113**(1), K89 (1989)
348. D. Mahgerefteh, J. Feinberg, Phys. Rev. Lett. **64**(18), 2195 (1990)
349. K. Buse, T. Bierwirth, J. Opt. Soc. Am. B **12**(4), 629 (1995)
350. D.M. Kim, J.G. Gallagher, T.A. Rabson, F.K. Tittel, Appl. Phys. **17**(4), 413 (1978)
351. C.T. Chen, D.M. Kim, D. von der Linde, Appl. Phys. Lett. **34**(5), 321 (1979)
352. O. Althoff, A. Erdmann, L. Wiskott, P. Hertel, Phys. Stat. Solidi (a) **128**, K41 (1991)
353. V.E. Wood, N.F. Hartman, C.M. Verber, Ferroelectrics **27**, 237 (1980)
354. R. Goering, A.S. Rasch, W. Karthe, Proc. SPIE **1274**, 18 (1990)
355. I. Kanaev, V. Malinovskii, A. Pugachev, Sov. Phys. Solid State **27**(6), 1063 (1985)
356. Y. Ming, E. Krätzig, R. Orlowski, Phys. Stat. Solidi (a) **92**(1), 221 (1985)
357. F. Jermann, M. Simon, E. Krätzig, J. Opt. Soc. Am. B **12**, 2066 (1995)
358. M. Simon, F. Jermann, E. Krätzig, Optical Materials **3**, 243 (1994)
359. J. Carnicero, O. Caballero, M. Carrascosa, J.M. Cabrera, Appl. Phys. B **79**(3), 351 (2004)
360. I.B. Barkan, S.I. Marennikov, M.V. Entin, Phys. Stat. Solidi (a) **45**, K17 (1978)
361. F. Laeri, R. Jungen, G. Angelow, U. Vietze, T. Engel, M. Wurtz, D. Hilgenberg, Appl. Phys. B **61**(4), 351 (1995)
362. N.V. Anghert, V.A. Pashkov, N.M. Solov'yeva, Zh. Exp. Teor. Fiz. **26**, 1666 (1972)
363. Y. Furukawa, M. Sato, K. Kitamura, Y. Yajima, M. Minakata, J. Appl. Phys. **72**, 3250 (1992)
364. Y. Furukawa, M. Sato, M.C. Bashaw, M.M. Fejer, N. Iyi, K. Kitamura, Jpn. J. Appl. Phys. **35**, 2740 (1996)
365. Y. Furukawa, M. Sato, K. Kitamura, Y. Yajima, M. Minakata, J. Appl. Phys. **72**, 3250 (1992)
366. Y. Furukawa, K. Kitamura, Y. Ji, G. Montemezzani, M. Zgonik, C. Medrano, P. Günter, Opt. Lett. **22**, 501 (1997)
367. M.H. Garrett, I. Mnushkina, Y. Furukawa, K. Kitamura, L.E. Halliburton, N.C. Giles, C.D. Setzler, in *in Proceedings of 1997 Topical Meeting on Photorefractive Materials, Effects and Devices, 1997 (Chiba, Japan)* (1997), pp. 295–298
368. X. Chen, B. Li, J. Xu, D. Zhu, S. Pan, Z. Wu, J. Appl. Phys. **90**, 1516 (2001)
369. A. García-Cabañes, J.M. Cabrera, J. Phys.: Condens. Matter **5**, 2267 (1993)
370. Y. Furukawa, M. Sato, F. Nitanda, K. Ito, J. Cryst. Growth **99**, 832 (1990)
371. F. Holtmann, J. Imbrock, C. Bäumer, H. Hesse, E. Krätzig, D. Kip, J. Appl. Phys. **96**(12), 7455 (2004)
372. C.Q. Xu, H. Okayama, Y. Ogawa, J. Appl. Phys. **87**(7), 3203 (2000)
373. J.K. Yamamoto, K. Kitamura, N. Iyi, S. Kimura, J. Cryst. Growth **128**, 920 (1993)
374. M. Nakamura, S. Takekawa, Y.W. Liu, K. Kitamura, Jpn. J. Appl. Phys. Part 2 **43**(7B), L947 (2004)

375. L.D. Sshearer, M. Leduc, Z. Zachorowski, IEEE J. Quantum Electronics **QE-23**, 1996 (1987)
376. M. Simon, F. Jermann, T. Volk, E. Krätzig, Phys. Stat. Solidi (a) **149**, 723 (1995)
377. G.A. Torchia, J.O. Tocho, F. Jaque, J. Phys. Chem. Solids **63**, 555 (2002)
378. W. Zheng, N.D. Zhang, L.C. Zhao, Y.H. Xu, Opt. Commun. **227**(4–6), 259 (2003)
379. I. Zhang, B. Wang, S.Q. Fang, D.C. Ma, Y.Q. Zhao, Y.H. Xu, Mater. Lett. **58**(24), 3074 (2004)
380. X.H. Zhen, H.T. Li, Z.J. Sun, S.J. Ye, L.C. Zhao, Y.H. Xu, J. Phys. D: Appl. Phys. **37**(4), 634 (2004)
381. L. Arizmendi, C.D.L. Heras, F. Jaque, A. Suchocki, S. Kobyakov, T.P.J. Han, Appl. Phys. B **87**(1), 123 (2007)
382. S. Fang, Y. Qiao, X. Zhang, F. Ling, Phys. Stat. Solidi (a) **204**(3), 833 (2007)
383. K. Niwa, Y. Furukawa, S. Takekawa, K. Kitamura, J. Cryst. Growth **208**, 493 (2000)
384. Y. Furukawa, M. Sato, K. Kitamura, F. Nitanda, J. Cryst. Growth **128**, 909 (1993)
385. S.L. Chen, H.D. Liu, Y.F. Kong, Z.H. Huang, J.J. Xu, Cryst. Res. Technol. **41**(8), 790 (2006)
386. S.S. Chen, H.D. Liu, Y.F. Kong, Z.H. Huang, J.J. Xu, G.Y. Zhang, Optical Materials **29**(7), 885 (2007)
387. J. Koppitz, O.F. Schirmer, M. Wöhlecke, A.I. Kusnetsov, B.C. Grabmaier, Ferroelectrics **92**, 233 (1989)
388. S.I. Bae, J. Ichikawa, K. Shimamura, H. Onodera, T. Fukuda, J. Cryst. Growth **180**, 94 (1997)
389. X.H. Zhen, Q. Li, L.C. Zhao, Y.H. Xu, J. Mater. Sci. **42**(10), 3670 (2007)
390. L. Arizmendi, F. Agulló-López, Mater. Res. Bull. **19**(3), 32 (1994)
391. A. Adibi, K. Buse, D. Psaltis, Appl. Phys. B **72**, 653 (2001)
392. R. Gerson, J.F. Kirchhoff, L.E. Halliburton, D.A. Bryan, J. Appl. Phys. **60**, 3553 (1986)
393. E. Krätzig, Ferroelectrics **21**, 635 (1978)
394. T.R. Volk, S.A. Shramchenko, M.A. Ivanov, M.L. Meilman, N.M. Rubinina, Ferroelectrics **68**, 325 (1986)
395. G.Q. Zhang, Y. Tomita, J. Appl. Phys. **91**(7), 4177 (2002)
396. G.Q. Zhang, Y. Tomita, J. Appl. Phys. **93**(12), 9456 (2003)
397. G.I. Malovichko, V.G. Grachev, E.P. Kokanyan, O.F. Schirmer, K. Betzler, B. Gather, F. Jermann, S. Klauer, U. Schlarb, M. Wöhlecke, Appl. Phys. A **56**, 103 (1993)
398. H. Donnerberg, S.M. Tomlinson, C.R.A. Catlow, O.F. Schirmer, Phys. Rev. B **40**, 11909 (1989)
399. G. Zhang, J. Xu, S. Liu, Q. Sun, G. Zhang, Q. Fang, C. Ma, in *Proceedings of SPIE*, vol. 14 (1995), vol. 14, p. 2529
400. Y.B. Guo, Y. Liao, L.C. Cao, G.D. Liu, Q.S. He, G.F. Jin, Optics Express **12**(22), 5556 (2004)
401. A.M. Glass, Opt. Eng. **17**(5), 470 (1978)
402. A.L. Aleksandrovski, I.I. Naumova, Proc. SPIE **1841**, 188 (1991)
403. J. Hirohashi, V. Pasiskevicius, S. Wang, F. Laurell, J. Appl. Phys. **101**(3), 033105 (2007)

404. J.C. Deng, W.L. Zhang, J.K. Wen, G.Y. Zhang, H.F. Wang, Opt. Lett. **19**(13), 933 (1994)
405. J.L. Nightingale, W.J. Silva, G.E. Reade, A. Rybicki, W.J. Kozlovsky, R.L. Byer, SPIE, Laser and Nonlinear Optical Materials **681**, 20 (1986)
406. W.J. Kozlovsky, C.D. Nabors, R.C. Eckhardt, R.L. Byer, Opt. Lett. **14**, 66 (1989)
407. J.Q. Yao, W.Q. Shi, J.E. Millerd, G.F. Xu, E. Garmire, M. Birnbaum, Opt. Lett. **15**, 1339 (1990)
408. J. Li, X.J. Chen, B. Wu, B. Li, S.H. Pan, Appl. Phys. Lett. **67**(23), 3384 (1995)
409. J.C. Deng, Y.F. Kong, J. Li, J.K. Wen, B. Li, J. Appl. Phys. **79**, 9334 (1996)
410. T.R. Volk, V.V. Krasnikov, V.I. Pryalkin, N.M. Rubinina, Kvantovaya Elektronika (Quantum Elektronics) **17**, 262 (1990)
411. Y. Furukawa, K. Kitamura, A. Alexandrovski, R.K. Route, M.M. Fejer, G. Foulon, Appl. Phys. Lett. **78**(14), 1970 (2001)
412. S. Campbell, P. Yeh. Photorefractive and thermal fixing properties of 4 mol.% MgO doped $LiNbO_3$. CLEO'94 report CTuJ3 (1994)
413. G.Q. Zhang, G.Y. Zhang, S.M. Liu, J.J. Xu, G.Y. Tian, Q. Sun, Opt. Lett. **22**(22), 1666 (1997)
414. N.Y. Kamber, J.J. Xu, C.Q. Zhang, X.Z. Zhang, Q.A. Sun, S.M. Liu, G.Y. Zhang, Opt. Commun. **161**(4–6), 271 (1999)
415. N.Y. Kamber, J.J. Xu, S.M. Mikha, G.Q. Zhang, X.Z. Zhang, S.M. Liu, G.Y. Zhang, J. Appl. Phys. **87**(6), 2684 (2000)
416. G.Q. Zhang, G.Y. Zhang, S.M. Liu, J.J. Xu, Q. Sun, X.Z. Zhang, J. Appl. Phys. **83**(8), 4392 (1998)
417. H.J. Qiao, J.J. Xu, Q. Wu, X.Y. Yu, Q. Sun, X.Z. Zhang, G.Y. Zhang, T.R. Volk, Optical Materials **23**(1–2), 269 (2003)
418. R. Jungen, G. Angelow, F. Laeri, C. Grabmaier, Appl. Phys. A **55**(1), 101 (1992)
419. J.J. Xu, G.Y. Zhang, F.F. Li, X.Z. Zhang, Q.A. Sun, S.M. Liu, F. Song, Y.F. Kong, X.J. Chen, H.J. Qiao, J.H. Yao, L.J. Zhao, Opt. Lett. **25**(2), 129 (2000)
420. H.J. Qiao, J.J. Xu, G.Q. Zhang, X.Z. Zhang, Q. Sun, G.Y. Zhang, Phys. Rev. B **70**(9), 094101 (2004)
421. E.K. Gulanyan, I.R. Dorosh, V.D. Iskin, A.L. Mikaelyan, M.A. Maiorchik, Sov. J. Quantum Electron **9**, 647 (1979)
422. M.P. Petrov, S.I. Stepanov, A.A. Kamshilin, Opt. Laser Technology **11**, 149 (1979)
423. H.C. Külich, Opt. Commun. **64**, 407 (1987)
424. S. Fries, Appl. Phys. A **55**, 104 (1992)
425. A. Adibi, K. Buse, D. Psaltis, Opt. Lett. **25**, 539 (2000)
426. D. von der Linde, A.M. Glass, K.F. Rodgers, Appl. Phys. Lett. **26**, 22 (1975)
427. A.M. Glass, D. von der Linde. U. S. Patent No 3922061 (1975)
428. H. Vormann, E. Krätzig, Solid State Commun. **49**, 843 (1984)
429. K. Buse, F. Jermann, E. Krätzig, Ferroelectrics **141**, 197 (1993)
430. K. Buse, F. Jermann, E. Krätzig, Optical Materials **4**(2–3), 237 (1995)
431. J. Imbrock, S. Wevering, K. Buse, E. Krätzig, J. Opt. Soc. Am. B **16**(9), 1392 (1999)
432. Y.S. Bai, R.R. Neurgaonkar, R. Kachru, Opt. Lett. **21**(8), 567 (1996)
433. Y.S. Bai, R. Kachru, Phys. Rev. Lett. **78**(15), 2944 (1997)

434. M. Lee, S. Takekawa, Y. Furukawa, K. Kitamura, H. Hatano, J. Appl. Phys. **87**, 1291 (2000)
435. M. Lee, S. Takekawa, Y. Furukawa, Y. Uchida, K. Kitamura, H. Hatano, S. Tanaka, J. Appl. Phys. **88**, 4476 (2000)
436. M. Lee, S. Takekawa, Y. Furukawa, K. Kitamura, H. Hatano, Phys. Rev. Lett. **84**, 875 (2000)
437. M. Lee, S. Takekawa, Y. Furukawa, K. Kitamura, H. Hatano, S. Tao, Opt. Lett. **25**, 1337 (2000)
438. Y. Tomita, M. Hoshi, S. Sunarno, Jpn. J. Appl. Phys. Part 2 **40**(10A), L1035 (2001)
439. V.T. Pham, S.K. Lee, M.T. Trinh, K.S. Lim, D.S. Hamilton, K. Polgár, Opt. Commun. **248**(1–3), 89 (2005)
440. M. Lee, I.G. Kim, S. Takekawa, Y. Furukawa, Y. Uchida, K. Kitamura, H. Hatano, J. Appl. Phys. **89**(10), 5311 (2001)
441. Y.W. Liu, K. Kitamura, S. Takekawa, G. Ravi, M. Nakamura, H. Hatano, T. Yamaji, Appl. Phys. Lett. **81**, 2686 (2002)
442. Y.W. Liu, K. Kitamura, G. Ravi, S. Takekawa, M. Nakamura, H. Hatano, J. Appl. Phys. **96**(11), 5996 (2004)
443. Y.W. Liu, K. Kitamura, S. Takekawa, G. Ravi, M. Nakamura, H. Hatano, J. Appl. Phys. **97**(7) (2005)
444. Y. Tomita, S. Sunarno, G.Q. Zhang, J. Opt. Soc. Am. B **21**(4), 753 (2004)
445. H.J. Qiao, Y. Tomita, Optics Express **14**(12), 5773 (2006)
446. Y.W. Liu, K. Kitamura, S. Takekawa, M. Nakamura, Y. Furukawa, H. Hatano, Appl. Phys. Lett. **82**(24), 4218 (2003)
447. Y.W. Liu, K. Kitamura, S. Takekawa, M. Nakamura, Y. Furukawa, H. Hatano, J. Appl. Phys. **95**(12), 7637 (2004)
448. Y.W. Liu, K. Kitamura, S. Takekawa, G. Ravi, M. Nakamura, Y. Furukawa, H. Hatano, Appl. Opt. **43**(31), 5778 (2004)
449. G.Q. Zhang, Y. Tomita, W.S. Xu, C.H. Yang, Appl. Phys. Lett. **77**, 3508 (2000)
450. G.Q. Zhang, S. Sunarno, M. Hoshi, Y. Tomita, C.H. Yang, W.S. Xu, Appl. Opt. **40**(29), 5248 (2001)
451. S. Sunarno, Y. Tomita, G. Zhang, Appl. Phys. Lett. **81**, 4505 (2002)
452. G.Q. Zhang, Y. Tomita, X.Z. Zhang, J.J. Xu, Appl. Phys. Lett. **81**(8), 1393 (2002)
453. K. Buse, A. Adibi, D. Psaltis, Nature **393**(6686), 665 (1998)
454. A. Adibi, K. Buse, D. Psaltis, Appl. Phys. Lett. **74**(25), 3767 (1999)
455. A. Adibi, K. Buse, D. Psaltis, Opt. Lett. **24**(10), 652 (1999)
456. A. Adibi, K. Buse, D. Psaltis, J. Opt. Soc. Am. B **18**(5), 584 (2001)
457. D.L. Staebler, W. Phillips, Appl. Phys. Lett. **24**(6), 268 (1974)
458. O. Momtahan, A. Adibi, J. Opt. Soc. Am. B **20**(3), 449 (2003)
459. Y.W. Liu, L.R. Liu, J. Opt. Soc. Am. B **19**(10), 2413 (2002)
460. Y.W. Liu, L.R. Liu, C.H. Zhou, L.Y. Xu, Opt. Lett. **25**(12), 908 (2000)
461. Y. Liu, L. Liu, L. Xu, C. Zhou, Opt. Commun. **181**(1–3), 47 (2000)
462. Y. Liu, L. Liu, D. Liu, L. Xu, C. Zhou, Opt. Commun. **190**, 339 (2001)
463. I.G. Kim, M. Lee, S. Takekawa, Y. Furukawa, K. Kitamura, L. Galambos, L. Hesselink, Jpn. J. Appl. Phys. Part 2 **39**, L1094 (2000)
464. X.F. Yue, A. Adibi, T. Hudson, K. Buse, D. Psaltis, J. Appl. Phys. **87**(9), 4051 (2000)
465. J.P. Herriau, J.P. Huignard, Appl. Phys. Lett. **49**(18), 1140 (1986)

466. A. Delboulbe, C. Fromont, J.P. Herriau, S. Mallick, J.P. Huignard, Appl. Phys. Lett. **55**(8), 713 (1989)
467. F. Micheron, G. Bismuth, Appl. Phys. Lett. **20**(2), 79 (1972)
468. F. Micheron, G. Bismuth, Appl. Phys. Lett. **23**(2), 71 (1973)
469. F. Micheron, C. Mayeux, J.C. Trotier, Appl. Opt. **13**(4), 784 (1974)
470. B. Fischer, M. Cronin-Golomb, J.O. White, A. Yariv, R. Neurgaonkar, Appl. Phys. Lett. **40**(10), 863 (1982)
471. R. Matull, R.A. Rupp, J. Phys. D: Appl. Phys. **21**(11), 1556 (1988)
472. Y. Qiao, S. Orlov, D. Psaltis, R.R. Neurgaonkar, Opt. Lett. **18**(12), 1004 (1993)
473. R.S. Cudney, J. Fousek, M. Zgonik, P. Gunter, M.H. Garrett, D. Rytz, Appl. Phys. Lett. **63**(25), 3399 (1993)
474. H.A. Eggert, B. Hecking, K. Buse, Opt. Lett. **29**(21), 2476 (2004)
475. H.A. Eggert, F. Kalkum, B. Hecking, K. Buse, J. Opt. Soc. Am. B **22**(12), 2553 (2005)
476. V. Gopalan, T.E. Mitchell, Y. Furukawa, K. Kitamura, Appl. Phys. Lett. **72**, 1981 (1998)
477. L. Arizmendi, Phys. Stat. Solidi (a) **201**, 253 (2004)
478. T. Volk, M. Wöhlecke, Crit. Rev. Solid State Mater. Sci. **30**(3), 125 (2005)
479. J.M.C.M. Carrascosa, L. Arizmendi, in *Photorefractive Materials and Their Applications 1, Basic Effects*, ed. by P. Günter, J.P. Huignard, no. 113 in Springer Series in Optical Sciences (Springer, 2006), pp. 369–396
480. J.J. Amodei, D.L. Staebler, Appl. Phys. Lett. **18**(12), 540 (1971)
481. J.J. Amodei, D.L. Staebler, W. Phillips, Appl. Opt. **11**(2), 390 (1972)
482. D.L. Staebler, W.J. Burke, W. Phillips, J.J. Amodei, Appl. Phys. Lett. **26**(4), 182 (1975)
483. B. Liu, L. Liu, L. Xu, J. Ma, S.H. Lee, Appl. Opt. **37**, 1342 (1998)
484. J.J. Amodei, D.L. Staebler, RCA Review **33**(1), 71 (1972)
485. W. Phillips, D.L. Staebler, J.J. Amodei, RCA Review **33**(1), 94 (1972)
486. P. Hertel, K.H. Ringhofer, R. Sommerfeldt, Phys. Stat. Solidi (a) **104**(2), 855 (1987)
487. M. Carrascosa, F. Agulló-López, J. Opt. Soc. Am. B **7**(12), 2317 (1990)
488. B.I. Sturman, M. Carrascosa, F. Agulló-López, J. Limeres, Phys. Rev. B **57**(20), 12792 (1998)
489. A. Yariv, S. Orlov, G. Rakuljic, V. Leyva, Opt. Lett. **20**(11), 1334 (1995)
490. A. Yariv, S.S. Orlov, G.A. Rakuljic, J. Opt. Soc. Am. B **13**(11), 2513 (1996)
491. G.A. Rakuljic, Opt. Lett. **22**(11), 825 (1997)
492. W. Meyer, P. Würfel, R. Munser, G. Müller-Vogt, Phys. Stat. Solidi (a) **53**(1), 171 (1979)
493. V.V. Kulikov, S.I. Stepanov, Sov. Phys. Solid State **21**(11), 1849 (1979)
494. G. Montemezzani, M. Zgonik, P. Günter, J. Opt. Soc. Am. B **10**(2), 171 (1993)
495. M.C. Bashaw, J.F. Heanue, J. Opt. Soc. Am. B **14**(8), 2024 (1997)
496. F. Micheron, J. Trotier, Ferroelectrics **8**, 441 (1974)
497. D. Kirillov, J. Feinberg, Opt. Lett. **16**(19), 1520 (1991)
498. D.F. Zhang, Y.H. Zhang, C.X. Li, Y.S. Chen, Y. Zhu, Appl. Opt. **34**(23), 5241 (1995)
499. G. Montemezzani, P. Günter, J. Opt. Soc. Am. B **7**(12), 2323 (1990)
500. T. Nikolajsen, P.M. Johansen, Opt. Lett. **24**(20), 1419 (1999)
501. V. Leyva, A. Agranat, A. Yariv, Opt. Lett. **16**(8), 554 (1991)
502. T. Imai, S. Yagi, H. Yamazaki, J. Opt. Soc. Am. B **13**(11), 2524 (1996)

503. M. Ivker, A.J. Agranat, OSA TOPS **62**, 107 (2001)
504. L. Arizmendi, R.C. Powell, J. Appl. Phys. **62**(3), 896 (1987)
505. L. Arizmendi, J. Appl. Phys. **65**(2), 423 (1989)
506. L. Arizmendi, J.F. Lopez-Barbera, M. Carrascosa, J. Appl. Phys. **97**(7), 073505 (2005)
507. S.W. McCahon, D. Rytz, G.C. Valley, M.B. Klein, B.A. Wechsler, Appl. Opt. **28**(11), 1967 (1989)
508. A.I. Grachev, A.V. Chamrai, OSA TOPS **62**, 203 (2001)
509. M. Carrascosa, L. Arizmendi, J. Appl. Phys. **73**(6), 2709 (1993)
510. M. Jeganathan, L. Hesselink, J. Opt. Soc. Am. B **11**(9), 1791 (1994)
511. J.F. Heanue, M.C. Bashaw, A.J. Daiber, R. Snyder, L. Hesselink, Opt. Lett. **21**(19), 1615 (1996)
512. C. Gu, J. Hong, H.Y. Li, D. Psaltis, P. Yeh, J. Appl. Phys. **69**(3), 1167 (1991)
513. I.F. Kanaev, V.I. Malinovskii, Sov. Phys. Solid State **24**(7), 1223 (1982)
514. E.M. de Miguel, T. Limeres, M. Carrascosa, L. Arizmendi, J. Opt. Soc. Am. B **17**(7), 1140 (2000)
515. V.I. Kovalevich, L.A. Shuvalov, T.R. Volk, Phys. Stat. Solidi (a) **45**(1), 249 (1978)
516. L. Arizmendi, P.D. Townsend, M. Carrascosa, J. Baquedano, J.M. Cabrera, J. Phys.: Condens. Matter **3**(28), 5399 (1991)
517. R. Muller, L. Arizmendi, M. Carrascosa, J.M. Cabrera, J. Appl. Phys. **77**(1), 308 (1995)
518. L. Arizmendi, A. Mendez, J.V. Alvarez-Bravo, Appl. Phys. Lett. **70**(5), 571 (1997)
519. L. Arizmendi, E.M. de Miguel-Sanz, M. Carrascosa, Opt. Lett. **23**(12), 960 (1998)
520. S. Breer, K. Buse, F. Rickermann, Opt. Lett. **23**(1), 73 (1998)
521. A. Mendez, L. Arizmendi, Optical Materials **10**(1), 55 (1998)
522. R. Sommerfeldt, R.A. Rupp, H. Vormann, E. Kratzig, Phys. Stat. Solidi (a) **99**(1), K15 (1987)
523. B.F. Williams, W.J. Burke, D.L. Staebler, Appl. Phys. Lett. **28**(4), 224 (1976)
524. H. Vormann, G. Weber, S. Kapphan, E. Kratzig, Solid State Commun. **40**(5), 543 (1981)
525. I. Nee, K. Buse, F. Havermeyer, R.A. Rupp, M. Fally, R.P. May, Phys. Rev. B **60**(14), R9896 (1999)
526. V.V. Zhdanova, V.P. Klyuev, V.V. Lemanov, I.A. Smirnov, V.V. Tikhonov, Sov. Phys. Solid State **10**(6), 1360 (1968)
527. A.A. Blistanov, E.V. Makarevskaya, V.V. Geras'kin, O. Kamalov, M.M. Koblova, Sov. Phys. Solid State **20**(9), 1489 (1978)
528. S.N. Kaul, K. Singh, Solid State Commun. **26**(6), 365 (1978)
529. G.I. Rozenman, E.I. Boǐkova, Sov. Phys. Solid State **21**(6), 1086 (1979)
530. E. Schempp, G.E. Peterson, W. Carruthe.jr, J. Chem. Phys. **53**(1), 306 (1970)
531. B.Q. Li, Y.N. Wang, Z.R. Xu, J. Phys. C: Solid State Phys. **21**(9), L251 (1988)
532. H. Engelmann, N. Krämer, H. Yanfu, L. Rongchuan, U. Gonser, Ferroelectrics **69**, 217 (1986)
533. G. Montemezzani, M. Ingold, H. Looser, P. Günter, Ferroelectrics **92**, 281 (1989)
534. P. Günter, Phys. Reports **93**(4), 199 (1982)
535. V.M. Fridkin, A.A. Kaminskii, V.G. Lazarev, S.B. Astafev, A.V. Butashin, Appl. Phys. Lett. **55**(6), 545 (1989)

536. L. d'Auria, J.P. Huignard, C. Slezak, E. Spitz, Appl. Opt. **13**(4), 808 (1974)
537. F.H. Mok, M.C. Tackitt, H.M. Stoll, Opt. Lett. **16**(8), 605 (1991)
538. F.H. Mok, Opt. Lett. **18**(11), 915 (1993)
539. A. Kewitsch, M. Segev, A. Yariv, R.R. Neurgaonkar, Opt. Lett. **18**(7), 534 (1993)
540. L. Hesselink, M.C. Bashaw, Opt. Quantum Electronics **25**(9), S611 (1993)
541. D. Psaltis, F. Mok, H.Y.S. Li, Opt. Lett. **19**(3), 210 (1994)
542. D. Psaltis, F. Mok, Scientific Am. **273**(5), 70 (1995)
543. I. McMichael, W. Christian, D. Pletcher, T.Y. Chang, J.H. Hong, Appl. Opt. **35**(14), 2375 (1996)
544. E.S. Bjornson, M.C. Bashaw, L. Hesselink, Appl. Opt. **36**(14), 3090 (1997)
545. X. An, D. Psaltis, G.W. Burr, Appl. Opt. **38**(2), 386 (1999)
546. J. Limeres, E.M. de Miguel-Sanz, A. Suchocki, L. Arizmendi, M. Carrascosa, Optical Materials **18**(1), 115 (2001)
547. A.A. Freschi, J. Frejlich, Opt. Lett. **20**(6), 635 (1995)
548. J. Frejlich, P.M. Garcia, A.A. Freschi, Optical Materials **4**(2–3), 410 (1995)
549. S. Breer, K. Buse, K. Peithmann, H. Vogt, E. Krätzig, Rev. Sci. Instrum. **69**(4), 1591 (1998)
550. G.A. Rakuljic, V. Leyva, Opt. Lett. **18**(6), 459 (1993)
551. R. Muller, M.T. Santos, L. Arizmendi, J.M. Cabrera, J. Phys. D: Appl. Phys. **27**(2), 241 (1994)
552. R. Muller, J.V. Alvarez Bravo, L. Arizmendi, J.M. Cabrera, J. Phys. D: Appl. Phys. **27**(8), 1628 (1994)
553. S. Breer, H. Vogt, I. Nee, K. Buse, Electronics Lett. **34**(25), 2419 (1998)
554. S. Breer, K. Buse, Appl. Phys. B **66**(3), 339 (1998)
555. E.M. de Miguel-Sanz, M. Tebaldi, S. Granieri, N. Bolognini, L. Arizmendi, Appl. Phys. B **70**(3), 379 (2000)
556. J.W. An, N. Kim, K.W. Lee, Opt. Commun. **197**(4–6), 247 (2001)
557. P. Yeh, *Optical Waves in Layered Media* (Wiley, New York, 1988)
558. I. Nee, O. Beyer, M. Müller, K. Buse, J. Opt. Soc. Am. B **20**(8), 1593 (2003)
559. D.T.H. Liu, L.J. Cheng, Appl. Opt. **31**(26), 5675 (1992)
560. H. Rajbenbach, S. Bann, P. Refregier, P. Joffre, J.P. Huignard, H.S. Buchkremer, A.S. Jensen, E. Rasmussen, K.H. Brenner, G. Lohman, Appl. Opt. **31**(26), 5666 (1992)
561. J. Hukriede, D. Runde, D. Kip, J. Phys. D: Appl. Phys. **36**(3), R1 (2003)
562. J.R. Carruthers, G.E. Peterson, M. Grasso, P.M. Bridenbaugh, J. Appl. Phys. **42**, 1846 (1971)
563. J.C. Brice, J. Cryst. Growth **42**, 427 (1977)
564. I. Gyoo Kim, S. Takekawa, Y. Furukawa, M. Lee, K. Kitamura, J. Cryst. Growth **229**, 243 (2001)
565. L. Kovács, G. Ruschhaupt, K. Polgár, G. Corradi, M. Wöhlecke, Appl. Phys. Lett. **70**, 2801 (1997)
566. D.E. Zelmon, D.L. Small, D. Jundt, J. Opt. Soc. Am. B **14**, 3319 (1997)
567. D.H. Jundt, in *Properties of Lithium Niobate*, ed. by K.K. Wong, no. 28 in EMIS datareviews series; 28 (INSPEC, Institution of Electrical Engineers, 2002), pp. 115–118
568. U. Schlarb, K. Betzler, Phys. Rev. B **48**, 15613 (1993)
569. U. Schlarb, K. Betzler, Phys. Rev. B **48**, 15613 (1993)
570. U. Schlarb, K. Betzler, J. Appl. Phys. **73**, 3472 (1993)

571. U. Schlarb, K. Betzler, Ferroelectrics **156**, 99 (1994)
572. U. Schlarb, K. Betzler, Phys. Rev. B **50**, 751 (1994)
573. D.H. Jundt, Opt. Lett. **22**, 1553 (1997)
574. O. Paul, A. Quosig, T. Bauer, M. Nittmann, J. Bartschke, G. Anstett, J.A. L'Huillier, Appl. Phys. B **86**(1), 111 (2007)
575. E. Wiesendanger, G. Güntherodt, Solid State Commun. **14**, 303 (1974)
576. D. Xue, N. Iyi, K. Kitamura, J. Appl. Phys. **92**, 4638 (2002)
577. U. Schlarb, M. Wöhlecke, B. Gather, A. Reichert, K. Betzler, T. Volk, N. Rubinina, Optical Materials **4**, 791 (1995)
578. K. Kasemir, K. Betzler, B. Matzas, B. Tiegel, M. Wöhlecke, N. Rubinina, T. Volk, Phys. Stat. Solidi (a) **166/1**, R7 (1998)
579. U. Schlarb, B. Matzas, A. Reichert, K. Betzler, M. Wöhlecke, B. Gather, T. Volk, Ferroelectrics **185**, 269 (1996)
580. R. Nevado, E. Camarillo, D. Bravo, I.A. de Carcer, G. Lifante, F. Jaque, Mater. Lett. **52**(1–2), 1 (2002)
581. Y. Furukawa, M. Sato, F. Nitanda, K. Ito, J. Cryst. Growth **99**, 832 (1990)
582. M. Nakamura, S. Higuchi, S. Takekawa, K. Terabe, Y. Furukawa, K. Kitamura, Jpn. J. Appl. Phys. **41**, L49 (2002)
583. W.M. Young, R.S. Feigelson, M.M. Fejer, M.J.F. Digonnet, H.J. Shaw, Opt. Lett. **16**, 995 (1991)
584. W.M. Young, M.M. Fejer, M.J.F. Digonnet, A.F. Marshall, R.S. Feigelson, J. Lightwave Technol. **10**, 1238 (1992)
585. M.M. Fejer, M.J.F. Digonnet, R.L. Byer, Opt. Lett. **11**, 230 (1986)
586. R. Nevado, G. Lifante, Appl. Phys. A **72**, 725 (2001)
587. D.W. Yoon, O. Eknoyan, J. Lightwave Technol. **6**(6), 877 (1988)
588. W.X. Hou, T.C. Chong, Appl. Opt. **36**(21), 5083 (1997)
589. T. Suhara, T. Fujieda, M. Fujimura, H. Nishihara, Jpn. J. Appl. Phys. Part 1 **39**, L864 (2000)
590. R.E. DiPaolo, E. Cantelar, R. Nevado, J.A. Sanz-García, M. Domenech, P.L. Pernas, G. Lifante, F. Cusso, Ferroelectrics **273**, 2607 (2002)
591. T. Kawaguchi, K. Mizuuchi, T. Yoshino, M. Imaeda, K. Yamamoto, T. Fukuda, J. Cryst. Growth **203**, 173 (1999)
592. T.R. Volk, M. Wöhlecke, N. Rubinina, N. V.Razumovskiĭ, F. Jermann, C. Fischer, R. Böwer, Appl. Phys. **A60**, 217 (1995)
593. I. Shoji, T. Kondo, A. Kitamoto, M. Shirane, R. Ito, J. Opt. Soc. Am. B **14**(9), 2268 (1997)
594. E.L. Wooten, K.M. Kissa, A. Yi-Yan, E.J. Murphy, D.A. Lafaw, P.F. Hallemeier, D. Maack, D.V. Attanasio, D.J. Fritz, G.J. McBrien, D.E. Bossi, IEEE J. Sel. Top. Quantum Electron. **6**(1), 69 (2000)
595. L.N. Binh, J. Cryst. Growth **288**(1), 180 (2006)
596. F. Abdi, M. Aillerie, P. Bourson, M.D. Fontana, K. Polgár, J. Appl. Phys. **84**(4), 2251 (1998)
597. I.V. Kityk, M. Makowska-Janusik, M.D. Fontana, M. Aillerie, F. Abdi, J. Appl. Phys. **90**(11), 5542 (2001)
598. T. Fujiwara, M. Takahashi, M. Ohama, A.J. Ikushima, Y. Furukawa, K. Kitamura, Electronics Lett. **35**, 499 (1999)
599. K. Chah, M. Aillerie, M.D. Fontana, G. Malovichko, Opt. Commun. **176**(1–3), 261 (2000)
600. K. Chah, M.D. Fontana, M. Aillerie, P. Bourson, G. Malovichko, Appl. Phys. B **67**(1), 65 (1998)

601. H.L. Saadon, N. Theofanous, M. Aillerie, M. Abarkan, J.P. Salvestrini, M.D. Fontana, J. Opt. A: Pure Appl. Opt. **8**(8), 677 (2006)
602. K. Yonekura, L. Jin, K. Takizawa, Opt. Rev. **14**(4), 194 (2007)
603. H.D. Megaw, *Ferroelectricity in Crystals* (Methuen, London, 1957)
604. F. Jona, G. Shirane, *Ferroelectric Crystals* (Pergamon, NY, 1962)
605. J.C. Burfoot, G.W. Taylor, *Polar Dielectrics and Their Applications* (Macmillan, London, 1979)
606. V.M. Fridkin, *Ferroelectric Semiconductors* (Consultants Bureau, New York, 1980)
607. G.A. Smolenskii, *Ferroelectrics and Related Materials, Ferroelectricity and Related Phenomena*, vol. 3 (Gordon and Breach Science, New York, 1984)
608. V.Y. Shur, Ferroelectrics **340**, 3 (2006)
609. V.Y. Shur, J. Mater. Sci. **41**(1), 199 (2006)
610. V. Gopalan, V. Dierolf, D.A. Scrymgeour, Annu. Rev. Materials Research **37**, 449 (2007)
611. S. Kim, V. Gopalan, A. Gruverman, Appl. Phys. Lett. **80**(15), 2740 (2002)
612. J. Fousek, V. Janovec, J. Appl. Phys. **40**, 135 (1969)
613. D.A. Scrymgeour, V. Gopalan, A. Itagi, A. Saxena, P.J. Swart, Phys. Rev. B **71**(18), 184110 (2005)
614. J. Padilla, W. Zhong, D. Vanderbilt, Phys. Rev. B **53**(10), R5969 (1996)
615. B. Meyer, D. Vanderbilt, Phys. Rev. B **65**(10), 104111 (2002)
616. V.A. Zhirnov, Sov. Phys. JETP **8**, 822 (1959)
617. D.A. Scrymgeour, V. Gopalan, Phys. Rev. B **72**(2), 024103 (2005)
618. J. Wittborn, C. Canalias, K.V. Rao, R. Clemens, H. Karlsson, F. Laurell, Appl. Phys. Lett. **80**(9), 1622 (2002)
619. B.J. Rodriguez, R.J. Nemanich, A. Kingon, A. Gruverman, S.V. Kalinin, K. Terabe, X.Y. Liu, K. Kitamura, Appl. Phys. Lett. **86**(1), 012906 (2005)
620. A.K. Tagantsev, G. Gerra, J. Appl. Phys. **100**(5), 051607 (2006)
621. P.V. Lambeck, G.H. Jonker, Ferroelectrics **22**(1–2), 729 (1978)
622. G. Arlt, H. Neumann, Ferroelectrics **87**, 109 (1988)
623. W.L. Warren, G.E. Pike, K. Vanheusden, D. Dimos, B.A. Tuttle, J. Robertson, J. Appl. Phys. **79**, 9250 (1996)
624. R. Landauer, J. Appl. Phys. **28**(2), 227 (1957)
625. V. Gopalan, T.E. Mitchell, J. Appl. Phys. **83**(2), 941 (1998)
626. V. Janovec, Czechosl. J. Phys. **9**, 468 (1959)
627. R. Abe, J. Phys. Soc. Japan **14**(5), 633 (1959)
628. R.C. Miller, G. Weinreich, Phys. Rev. **117**(6), 1460 (1960)
629. W.J. Merz, Phys. Rev. **95**(3), 690 (1954)
630. W. Ishibashi, Y. Takagi, J. Phys. Soc. Japan **31**(2), 506 (1971)
631. A.N. Kolmogorov, Izvestia ANSSSR, Ser. Math **3**, 355 (1937)
632. M. Avrami, J. Chem. Phys. **7**(12), 1103 (1939)
633. M. Avrami, J. Chem. Phys. **8**, 212 (1940)
634. R.C. Miller, A. Savage, J. Appl. Phys. **32**(4), 714 (1961)
635. S.A. Semiletov, N.G. Bocharova, E.V. Rakova, Crystallogr. Rep. **32**(1), 189 (1987)
636. I. Camlibel, J. Appl. Phys. **40**, 1690 (1969)
637. T. Volk, D. Isakov, L. Ivleva, M. Wöhlecke, Appl. Phys. Lett. **83**(11), 2220 (2003)
638. E. Soergel, Appl. Phys. B **81**(6), 729 (2005)

639. K. Nassau, W. Levinstein, G.M. Loiacono, Appl. Phys. Lett. **6**(11), 228 (1965)
640. S. Grilli, P. Ferraro, S.D. Nicola, A. Finizio, G. Pierattini, P.D. Natale, M. Chiarini, Optics Express **11**(4), 392 (2003)
641. R.C. Miller, A. Savage, Phys. Rev. **115**(5), 1176 (1959)
642. A. Gruverman, S.V. Kalinin, J. Mater. Sci. **41**(1), 107 (2006)
643. V.Y. Shur, A.I. Lobov, A.G. Shur, S. Kurimura, Y. Nomura, K. Terabe, X.Y. Liu, K. Kitamura, Appl. Phys. Lett. **87**(2) (2005)
644. C.A. Wallace, J. Appl. Crystallogr. **3**(DEC1), 546 (1970)
645. S. Kim, V. Gopalan, B. Steiner, Appl. Phys. Lett. **77**(13), 2051 (2000)
646. T. Jach, S. Kim, V. Gopalan, S. Durbin, D. Bright, Phys. Rev. B **69**(6), 064113 (2004)
647. M. Gonzalez-Manas, B. Vallejo, M.A. Caballero, J. Appl. Crystallogr. **38**, 1012 (2005)
648. Z.W. Hu, P.A. Thomas, W.P. Risk, Phys. Rev. B **59**(22), 14259 (1999)
649. M. Drakopoulos, Z.W. Hu, S. Kuznetsov, A. Snigirev, I. Snigireva, P.A. Thomas, J. Phys. D: Appl. Phys. **32**(10A), A160 (1999)
650. Z.H. Hu, P.A. Thomas, A. Snigirev, I. Snigireva, A. Souvorov, P.G.R. Smith, G.W. Ross, S. Teat, Nature **392**(6677), 690 (1998)
651. Z.W. Hu, P.A. Thomas, P.Q. Huang, Phys. Rev. B **56**(14), 8559 (1997)
652. V. Gopalan, S.S.A. Gerstl, A. Itagi, T.E. Mitchell, Q.X. Jia, T.E. Schlesinger, D.D. Stancil, J. Appl. Phys. **86**(3), 1638 (1999)
653. W. DiDomenico, S.H. Wemple, J. Appl. Phys. **40**(2), 720 (1969)
654. F. Zernike, J.E. Midwinter, *Applied Nonlinear Optics* (J. Willey & Sons, New York-Sidney-Toronto-London, 1973)
655. T.R. Volk, N.R. Ivanov, D.V. Isakov, L.I. Ivleva, P.A. Lykov, Phys. Solid State **47**(2), 305 (2005)
656. O. Tikhomirov, B. Red'kin, A. Trivelli, J. Levy, J. Appl. Phys. **87**(4), 1932 (2000)
657. B. Sugg, F. Kahmann, R. Pankrath, R.A. Rupp, Appl. Opt. **33**(23), 5386 (1994)
658. S. Grilli, P. Ferraro, M. Paturzo, D. Alfieri, P.D. Natale, Optics Express **12**(9), 1832 (2004)
659. M. Paturzo, P. Ferraro, S. Grilli, D. Alfieri, P.D. Natale, M. de Angelis, A. Finizio, S.D. Nicola, G. Pierattini, F. Caccavale, D. Callejo, A. Morbiato, Optics Express **13**(14), 5416 (2005)
660. S. MacCormack, J. Feinberg, Appl. Opt. **35**(30), 5961 (1996)
661. T. Volk, T. Woike, U. Dörfler, R. Pankrath, L. Ivleva, M. Wöhlecke, Ferroelectrics **203**(1–4), 457 (1997)
662. M.Y. Goulkov, T. Granzow, U. Dörfler, T. Woike, M. Imlau, R. Pankrath, Appl. Phys. B **76**(4), 407 (2003)
663. R.C. Miller, Phys. Rev. A **134**(5A), 1313 (1964)
664. Y. Uesu, S. Kurimura, Y. Yamamoto, Appl. Phys. Lett. **66**(17), 2165 (1995)
665. S. Kurimura, Y. Uesu, J. Appl. Phys. **81**(1), 369 (1997)
666. S.I. Bozhevolnyi, J.M. Hvam, K. Pedersen, F. Laurell, H. Karlsson, T. Skettrup, M. Belmonte, Appl. Phys. Lett. **73**(13), 1814 (1998)
667. M. Florsheimer, R. Paschotta, U. Kubitscheck, C. Brillert, D. Hofmann, L. Heuer, G. Schreiber, C. Verbeek, W. Sohler, H. Fuchs, Appl. Phys. B **67**(5), 593 (1998)
668. G.K.H. Kitaeva, V.V. Tishkova, I.I. Naumova, A.N. Penin, C.H. Kang, S.H. Tang, Appl. Phys. B **81**(5), 645 (2005)

669. G.K. Kitaeva, V.V. Tishkova, A.N. Penin, J. Raman Spectrosc. **36**(2), 116 (2005)
670. M. Müller, E. Soergel, K. Buse, Appl. Phys. Lett. **83**(9), 1824 (2003)
671. T. Volk, D. Isakov, N. Ivanov, L. Ivleva, K. Betzler, A. Tunyagi, M. Wöhlecke, J. Appl. Phys. **97**(7), 074102 (2005)
672. I.I. Smolyaninov, A.V. Zayats, C.C. Davis, Opt. Lett. **22**(21), 1592 (1997)
673. T.J. Yang, U. Mohideen, M.C. Gupta, Appl. Phys. Lett. **71**(14), 1960 (1997)
674. D.S. Hum, M.M. Fejer, Comptes Rendus Physique **8**(2), 180 (2007)
675. Y.L. Lu, Y.Q. Lu, X.F. Cheng, C.C. Xue, N.B. Ming, Appl. Phys. Lett. **68**(20), 2781 (1996)
676. S. Kawai, T. Ogawa, H.S. Lee, R.C. DeMattei, R.S. Feigelson, Appl. Phys. Lett. **73**(6), 768 (1998)
677. J.J. Romero, D. Jaque, J.G. Sole, A.A. Kaminskii, Appl. Phys. Lett. **78**(14), 1961 (2001)
678. U. Völker, K. Betzler, Phys. Rev. B **74**(13) (2006)
679. V. Berger, Phys. Rev. Lett. **81**(19), 4136 (1998)
680. N.G.R. Broderick, G.W. Ross, H.L. Offerhaus, D.J. Richardson, D.C. Hanna, Phys. Rev. Lett. **84**(19), 4345 (2000)
681. H.D. Megaw, Acta Crystallogr. **7**(2), 187 (1954)
682. A.A. Ballman, H. Brown, Ferroelectrics **4**(3), 189 (1972)
683. M. Houé, P.D. Townsend, Appl. Phys. Lett. **66**, 2667 (1995)
684. C.C. Battle, S. Kim, V. Gopalan, K. Barkocy, M.C. Gupta, Q.X. Jia, T.E. Mitchell, Appl. Phys. Lett. **76**(17), 2436 (2000)
685. Y.L. Chen, J. Guo, C.B. Lou, Z.H. Huang, S.L. Chen, G.Y. Zhang, J. Rare Earths **24**, 128 (2006)
686. P.W. Haycock, P.D. Townsend, Appl. Phys. Lett. **48**(11), 698 (1986)
687. D. Feng, N.B. Ming, J.F. Hong, Y.S. Yang, J.S. Zhu, Z. Yang, Y.N. Wang, Appl. Phys. Lett. **37**(7), 607 (1980)
688. A. Feisst, P. Koidl, Appl. Phys. Lett. **47**(11), 1125 (1985)
689. S. Miyazawa, J. Appl. Phys. **50**, 4599 (1979)
690. K. Nakamura, H. Ando, H. Shimizu, Appl. Phys. Lett. **50**, 1413 (1987)
691. M. Fujimura, T. Suhara, H. Nishihara, Electronics Lett. **27**(13), 1207 (1991)
692. E.J. Lim, M.M. Fejer, R.L. Byer, W.J. Kozlovsky, Electronics Lett. **25**(11), 731 (1989)
693. Y. Ishigame, T. Suhara, H. Nishihara, Opt. Lett. **16**(6), 375 (1991)
694. X.F. Cao, B. Rose, R.V. Ramaswamy, R. Srivastava, Opt. Lett. **17**(11), 795 (1992)
695. J. Webjörn, F. Laurell, G. Arvidsson, J. Lightwave Technol. **7**, 1597 (1989)
696. R.W. Keys, A. Loni, R.M. De La Rue, J.H. Marsh, B.J. Luff, R.D. Townsend, Electronics Lett. **26**, 188 (1990)
697. H. Ito, C. Takyu, H. Inaba, Electronics Lett. **27**, 1221 (1991)
698. M. Yamada, K. Kishima, Electronics Lett. **27**, 828 (1991)
699. A. Nutt, V. Gopalan, M. Gupta, Appl. Phys. Lett. **60**, 2828 (1992)
700. W.Y. Hsu, M.C. Gupta, Appl. Phys. Lett. **60**(1), 1 (1992)
701. M. Yamada, N. Nada, M. Saitoh, K. Watanabe, Appl. Phys. Lett. **62**, 435 (1993)
702. J. Webjörn, V. Pruneri, P.S. Russell, J.R.M. Barr, D.C. Hanna, Electronics Lett. **30**(11), 894 (1994)
703. J. Webjörn, V. Pruneri, P. Russel, D. Hanna, Electronics Lett. **31**, 669 (1995)

704. V. Pruneri, R. Koch, P. Kazansky, W. Clarkson, P. Russell, D. Hanna, Opt. Lett. **20**, 2375 (1995)
705. L. Myers, R. Eckardt, M. Fejer, R.L. Byer, W. Bosenberg, J. Pierce, J. Opt. Soc. Am. B **12**, 2102 (1995)
706. K. Mizuuchi, K. Yamamoto, M. Kato, Electronics Lett. **32**, 2091 (1996)
707. V. Gopalan, M.C. Gupta, Appl. Phys. Lett. **68**(7), 888 (1996)
708. V. Gopalan, M.C. Gupta, J. Appl. Phys. **80**(11), 6099 (1996)
709. V. Gopalan, M.C. Gupta, Ferroelectrics **198**(1–4), 49 (1997)
710. V. Gopalan, T.E. Mitchell, K.E. Sicakfus, Solid State Commun. **109**(2), 111 (1999)
711. V. Gopalan, T.E. Mitchell, J. Appl. Phys. **85**(4), 2304 (1999)
712. T.J. Yang, U. Mohideen, Phys. Lett. A **250**(1–3), 205 (1998)
713. T.J. Yang, V. Gopalan, P.J. Swart, U. Mohideen, Phys. Rev. Lett. **82**(20), 4106 (1999)
714. H.M. OBryan, P.K. Gallagher, C.D. Brandle, Journal of the American Ceramic Society **68**, 493 (1985)
715. V. Bermúdez, L. Huang, D. Hui, S. Field, E. Diéguez, Appl. Phys. A **70**(5), 591 (2000)
716. Y.L. Chen, J.J. Xu, Y.F. Kong, S.L. Chen, G.Y. Zhang, J.P. Wen, Appl. Phys. Lett. **81**, 700 (2002)
717. S. Kim, V. Gopalan, K. Kitamura, Y. Furukawa, J. Appl. Phys. **90**(6), 2949 (2001)
718. A. Kuroda, S. Kurimura, Y. Uesu, Appl. Phys. Lett. **69**, 1565 (1996)
719. K. Nakamura, J. Kurz, K. Parameswaran, M.M. Fejer, J. Appl. Phys. **91**, 4528 (2002)
720. J.H. Yao, Y.H. Chen, B.X. Yan, H.L. Deng, Y.F. Kong, S.L. Chen, J.J. Xu, G.Y. Zhang, Physica B **352**(1–4), 294 (2004)
721. Y.L. Chen, W.G. Yan, J. Guo, S.L. Chen, G.Y. Zhang, Z.G. Xia, Appl. Phys. Lett. **87**(21), 212904 (2005)
722. L.H. Peng, Y.C. Zhang, Y.C. Lin, Appl. Phys. Lett. **78**(1), 4 (2001)
723. Y.L. Chen, C. Lou, J.J. Xu, S.L. Chen, Y.F. Kong, G.Y. Zhang, J.P. Wen, J. Appl. Phys. **94**(5), 3350 (2003)
724. V.Y. Shur, E.V. Nikolaeva, E.I. Shishkin, V.L. Kozhevnikov, A.P. Chernykh, K. Terabe, K. Kitamura, Appl. Phys. Lett. **79**, 3146 (2001)
725. V.Y. Shur, E.L. Rumyantsev, E.V. Nikolaeva, E.I. Shishkin, R.G. Batchko, M.M. Fejer, R.L. Byer, I. Mnushkina, Ferroelectrics **269**, 1037 (2002)
726. T. Nozawa, S. Miyazawa, Jpn. J. Appl. Phys. Part 1 **35**(1A), 107 (1996)
727. A.I. Lobov, V.Y. Shur, I.S. Baturin, E.I. Shishkin, D.K. Kuznetsov, A.G. Shur, M.A. Dolbilov, K. Gallo, Ferroelectrics **341**, 109 (2006)
728. R.G. Batchko, V.Y. Shur, M.M. Fejer, R.L. Byer, Appl. Phys. Lett. **75**(12), 1673 (1999)
729. S. Chao, W. Davis, D.D. Tuschel, R. Nichols, M. Gupta, H.C. Cheng, Appl. Phys. Lett. **67**(8), 1066 (1995)
730. M.C. Wengler, M. Müller, E. Soergel, K. Buse, Appl. Phys. B **76**(4), 393 (2003)
731. A. Agronin, Y. Rosenwaks, G. Rosenman, Appl. Phys. Lett. **88**(7), 072911 (2006)
732. U. Heinemeyer, M.C. Wengler, K. Buse, Appl. Phys. Lett. **89**(11) (2006)
733. J.G. Scott, S. Mailis, C.L. Sones, R.W. Eason, Appl. Phys. A **79**(3), 691 (2004)
734. W.B. Yan, Y.F. Kong, L.H. Shi, X.C. Li, J.J. Xu, C.B. Lou, H.D. Liu, W.L. Zhang, G.Y. Zhang, Phys. Stat. Solidi (a) **201**(9), 2013 (2004)

735. V. Dierolf, C. Sandmann, S. Kim, V. Gopalan, K.K. Polgár, J. Appl. Phys. **93**(4), 2295 (2003)
736. Q.X. Xi, D. Liu, Y.N. Zhi, Z. Luan, L.R. Liu, Appl. Phys. Lett. **87**(12), 121103 (2005)
737. V. Dierolf, C. Sandmann, J. Luminescence **125**(1–2), 67 (2007)
738. D.P. Birnie, J. Mater. Sci. **28**, 302 (1993)
739. M. Müller, E. Soergel, M.C. Wengler, K. Buse, Appl. Phys. B **78**(3–4), 367 (2004)
740. L.A. Bursill, P.J. Lin, Ferroelectrics **70**(3–4), 191 (1986)
741. N. Zotov, F. Frey, H. Boysen, H. Lehnert, A. Hornsteiner, B. Strauss, R. Sonntag, H.M. Mayer, F. Guthoff, D. Hohlwein, Acta Crystallogr., Sect. B: Struct. Sci. **51**, 961 (1995)
742. A. Fujimura, T. Sohmura, T. Suhara, Electronics Lett. **39**(9), 719 (2003)
743. M.L. Müller, E. Soergel, K. Buse, Opt. Lett. **28**(24), 2515 (2003)
744. M.C. Wengler, B. Fassbender, E. Soergel, K. Buse, J. Appl. Phys. **96**(5), 2816 (2004)
745. M.C. Wengler, U. Heinemeyer, E. Soergel, K. Buse, J. Appl. Phys. **98**(6), 064104 (2005)
746. I.T. Wellington, C.E. Valdivia, T.J. Sono, C.L. Sones, S. Mailis, R.W. Eason, Appl. Surf. Sci. **253**(9), 4215 (2007)
747. V. Dierolf, C. Sandmann, Appl. Phys. Lett. **84**(20), 3987 (2004)
748. C.L. Sones, M.C. Wengler, C.E. Valdivia, S. Mailis, R.W. Eason, K. Buse, Appl. Phys. Lett. **86**(21), 212901 (2005)
749. P.T. Brown, S. Mailis, I. Zergioti, R.W. Eason, Optical Materials **20**(2), 125 (2002)
750. S. Mailis, C. Riziotis, I.T. Wellington, P.G.R. Smith, C.B.E. Gawith, R.W. Eason, Opt. Lett. **28**(16), 1433 (2003)
751. C.E. Valdivia, C.L. Sones, J.G. Scott, S. Mailis, R.W. Eason, D.A. Scrymgeour, V. Gopalan, T. Jungk, E. Soergel, I. Clark, Appl. Phys. Lett. **86**(2), 022906 (2005)
752. C.L. Sones, C.E. Valdivia, J.G. Scott, S. Mailis, R.W. Eason, D.A. Scrymgeour, V. Gopalan, T. Jungk, E. Soergel, Appl. Phys. B **80**(3), 341 (2005)
753. S. Mailis, C.L. Sones, J.G. Scott, R.W. Eason, Appl. Surf. Sci. **247**(1–4), 497 (2005)
754. V.Y. Shur, in *Nucleation Theory and Applications*, ed. by J. Schmelzer (Wiley-VCH, 2005), pp. 178–214
755. A.J. Boyland, S. Mailis, I.E. Barry, R.W. Eason, M. Kaczmarek, Appl. Phys. Lett. **77**(18), 2792 (2000)
756. G.D. Miller, R.G. Batchko, M.M. Fejer, R.L. Byer, Proc. SPIE **2700**, 34 (1996)
757. M. Fujimura, K. Kintaka, T. Suhara, H. Nishihara, J. Lightwave Technol. **11**(8), 1360 (1993)
758. S. Kurimura, I. Shimoya, Y. Uesu, Jpn. J. Appl. Phys. Part 1 **35**, L31 (1996)
759. C. Restoin, S. Massy, C. Darraud-Taupiac, A. Barthelemy, Optical Materials **22**(3), 193 (2003)
760. Y. Glickman, E. Winebrand, A. Arie, G. Rosenman, Appl. Phys. Lett. **88**(1) (2006)
761. X. Li, K. Terabe, H. Hatano, K. Kitamura, J. Cryst. Growth **292**(2), 324 (2006)
762. X.J. Li, K. Terabe, H. Hatano, K. Kitamura, Jpn. J. Appl. Phys. Part 2 **45** (12–16), L399 (2006)

763. K. Mizuuchi, K. Yamamoto, Electronics Lett. **29**, 2064 (1993)
764. X.J. Li, K. Terabe, H. Hatano, H.R. Zeng, K. Kitamura, J. Appl. Phys. **100**(10), 106103 (2006)
765. J. He, S.H. Tang, Y.Q. Qin, P. Dong, H.Z. Zhang, C.H. Kang, W.X. Sun, Z.X. Shen, J. Appl. Phys. **93**(12), 9943 (2003)
766. K. Tanaka, Y. Cho, in *Abstracts of The 9-th International Symposium on Ferroic Domains and Micro- to Nanoscopic Structures, (ISFD-9), Dresden* (2006), p. P 1.29
767. A. Harada, Y. Nihei, Appl. Phys. Lett. **69**, 2629 (1996)
768. T. Andres, P. Haag, S. Zelt, J. Meyn, A. Borsutzky, R. Beigang, R. Wallenstein, Appl. Phys. B **76**, 241 (2003)
769. S.O. Fregatov, A.B. Sherman, Phys. Solid State **41**(3), 457 (1999)
770. M. Mohageg, D.V. Strekalov, A.A. Savchenkov, A.B. Matsko, V.S. Ilchenko, L. Maleki, Optics Express **13**(9), 3408 (2005)
771. A.L. Gruverman, J. Hatano, H. Tokumoto, Jpn. J. Appl. Phys. Part 1 **36**(4A), 2207 (1997)
772. L.M. Eng, M. Bammerlin, C. Loppacher, M. Guggisberg, R. Bennewitz, R. Luthi, E. Meyer, T. Huser, H. Heinzelmann, H.J. Güntherodt, Ferroelectrics **222**(1–4), 411 (1999)
773. R.V. Gainutdinov, T.R. Volk, A.L. Tolstikhina, L.I. Ivleva, Sov. Phys. JETP Lett. **86**(4), 268 (2007)
774. M.T.I. Molotskii, J. Appl. Phys. **97**(1) (2005)
775. K. Terabe, M. Nakamura, S. Takekawa, K. Kitamura, S. Higuchi, Y. Gotoh, Y. Cho, Appl. Phys. Lett. **82**(3), 433 (2003)
776. Y.S. Cho, K. Fujimoto, Y. Hiranaga, Y. Wagatsuma, A. Onoe, K. Terabe, K. Kitamura, Appl. Phys. Lett. **81**(23), 4401 (2002)
777. K. Fujimoto, Y.S. Cho, Appl. Phys. Lett. **83**(25), 5265 (2003)
778. Y. Cho, K. Fujimoto, Y. Hiranaga, Y. Wagatsuma, A. Onoe, K. Terabe, K. Kitamura, Ferroelectrics **292**, 51 (2003)
779. Y. Hiranaga, Y. Cho, K. Fujimoto, Y. Wagatsuma, A. Onoe, Jpn. J. Appl. Phys. Part 1 **42**(9B), 6050 (2003)
780. Y.S. Cho, Y. Hiranaga, K. Fujimoto, Y. Wagatsuma, A. Onoe, Integrated Ferroelectrics **61**, 77 (2004)
781. Y. Daimon, Y. Cho, in *Abstracts of The 9-th International Symposium on Ferroic Domains and Micro- to Nanoscopic Structures, (ISFD-9)* (2006), p. P 1.20
782. Y. Daimon, Y. Cho, Appl. Phys. Lett. **90**(19), 192906 (2007)
783. Y. Kan, X. Lu, H. Bo, F. Huang, X. Wu, J. Zhu, Appl. Phys. Lett. **91**(13), 132902 (2007)
784. G. Rosenman, P. Urenski, A. Agronin, Y. Rosenwaks, M. Molotskii, Appl. Phys. Lett. **82**(1), 103 (2003)
785. M. Molotskii, A. Agronin, P. Urenski, M. Shvebelman, G. Rosenman, Y. Rosenwaks, Phys. Rev. Lett. **90**(10) (2003)
786. A. Agronin, Y. Rosenwaks, G. Rosenman, Appl. Phys. Lett. **85**(3), 452 (2004)
787. A. Agronin, M. Molotskii, Y. Rosenwaks, G. Rosenman, B.J. Rodriguez, A.I. Kingon, A. Gruverman, J. Appl. Phys. **99**(10), 104102 (2006)
788. B.D. Terris, J.E. Stern, D. Rugar, H.J. Mamin, Phys. Rev. Lett. **63**(24), 2669 (1989)
789. G. van der Zwan, R.M. Mazo, J. Chem. Phys. **82**(7), 3344 (1985)

790. E.J. Mele, Am. J. Phys. **69**(5), 557 (2001)
791. M. Molotskii, J. Appl. Phys. **93**(10), 6234 (2003)
792. M.I. Molotskii, M.M. Shvebelman, Philos. Mag. **85**(15), 1637 (2005)
793. A.Y. Emelyanov, Phys. Rev. B **71**(13), 132102 (2005)
794. M. Molotskii, Y. Rosenwaks, G. Rosenman, Annu. Rev. Mater. Res. **37**, 271 (2007)
795. M. Houe, P.D. Townsend, J. Phys. D: Appl. Phys. **28**, 1747 (1995)
796. S. Zhu, Y.Y. Zhu, N.B. Ming, Science **278**(5339), 843 (1997)
797. D.A. Scrymgeour, A. Sharan, V. Gopalan, K.T. Gahagan, J.L. Casson, R. Sander, J.M. Robinson, F. Muhammad, P. Chandramani, F. Kiamilev, Appl. Phys. Lett. **81**(17), 3140 (2002)
798. R.S. Cudney, L.A. Rios, H.M. Escamilla, Optics Express **12**(23), 5783 (2004)
799. H. Ishizuki, I. Shoji, T. Taira, Appl. Phys. Lett. **82**(23), 4062 (2003)
800. S. Kurimura, Ferroelectrics **340**, 91 (2006)
801. T.W. Ren, J.L. He, C. Zhang, S.N. Zhu, Y.Y. Zhu, Y. Hang, J. Phys.: Condens. Matter **16**(18), 3289 (2004)
802. S. Saltiel, Y.S. Kivshar, Opt. Lett. **25**(16), 1204 (2000)
803. S. Saltiel, Y.S. Kivshar, Opt. Lett. **25**(21), 1612 (2000)
804. M. de Sterke, S.M. Saltiel, Y.S. Kivshar, Opt. Lett. **26**(8), 539 (2001)
805. A. Chowdhury, S.C. Hagness, L. McCaughan, Opt. Lett. **25**(11), 832 (2000)
806. A. Chowdhury, C. Staus, B.F. Boland, T.F. Kuech, L. McCaughan, Opt. Lett. **26**(17), 1353 (2001)

Index

Springer Series in
MATERIALS SCIENCE

Editors: R. Hull R. M. Osgood, Jr. J. Parisi H. Warlimont

50 **High-Resolution Imaging and Spectrometry of Materials**
Editors: F. Ernst and M. Rühle

51 **Point Defects in Semiconductors and Insulators**
Determination of Atomic and Electronic Structure from Paramagnetic Hyperfine Interactions
By J.-M. Spaeth and H. Overhof

52 **Polymer Films with Embedded Metal Nanoparticles**
By A. Heilmann

53 **Nanocrystalline Ceramics**
Synthesis and Structure
By M. Winterer

54 **Electronic Structure and Magnetism of Complex Materials**
Editors: D.J. Singh and D. A. Papaconstantopoulos

55 **Quasicrystals**
An Introduction to Structure, Physical Properties and Applications
Editors: J.-B. Suck, M. Schreiber, and P. Häussler

56 **SiO_2 in Si Microdevices**
By M. Itsumi

57 **Radiation Effects in Advanced Semiconductor Materials and Devices**
By C. Claeys and E. Simoen

58 **Functional Thin Films and Functional Materials**
New Concepts and Technologies
Editor: D. Shi

59 **Dielectric Properties of Porous Media**
By S.O. Gladkov

60 **Organic Photovoltaics**
Concepts and Realization
Editors: C. Brabec, V. Dyakonov, J. Parisi and N. Sariciftci

61 **Fatigue in Ferroelectric Ceramics and Related Issues**
By D.C. Lupascu

62 **Epitaxy**
Physical Principles and Technical Implementation
By M.A. Herman, W. Richter, and H. Sitter

63 **Fundamentals of Ion-Irradiated Polymers**
By D. Fink

64 **Morphology Control of Materials and Nanoparticles**
Advanced Materials Processing and Characterization
Editors: Y. Waseda and A. Muramatsu

65 **Transport Processes in Ion-Irradiated Polymers**
By D. Fink

66 **Multiphased Ceramic Materials**
Processing and Potential
Editors: W.-H. Tuan and J.-K. Guo

67 **Nondestructive Materials Characterization**
With Applications to Aerospace Materials
Editors: N.G.H. Meyendorf, P.B. Nagy, and S.I. Rokhlin

68 **Diffraction Analysis of the Microstructure of Materials**
Editors: E.J. Mittemeijer and P. Scardi

69 **Chemical–Mechanical Planarization of Semiconductor Materials**
Editor: M.R. Oliver

70 **Applications of the Isotopic Effect in Solids**
By V.G. Plekhanov

71 **Dissipative Phenomena in Condensed Matter**
Some Applications
By S. Dattagupta and S. Puri

72 **Predictive Simulation of Semiconductor Processing**
Status and Challenges
Editors: J. Dabrowski and E.R. Weber

73 **SiC Power Materials**
Devices and Applications
Editor: Z.C. Feng

Springer Series in
MATERIALS SCIENCE

Editors: R. Hull R. M. Osgood, Jr. J. Parisi H. Warlimont

74 **Plastic Deformation in Nanocrystalline Materials**
By M. Yu. Gutkin and I.A. Ovid'ko

75 **Wafer Bonding**
Applications and Technology
Editors: M. Alexe and U. Gösele

76 **Spirally Anisotropic Composites**
By G.E. Freger, V.N. Kestelman, and D.G. Freger

77 **Impurities Confined in Quantum Structures**
By P.O. Holtz and Q.X. Zhao

78 **Macromolecular Nanostructured Materials**
Editors: N. Ueyama and A. Harada

79 **Magnetism and Structure in Functional Materials**
Editors: A. Planes, L. Mañosa, and A. Saxena

80 **Micro- and Macro-Properties of Solids**
Thermal, Mechanical and Dielectric Properties
By D.B. Sirdeshmukh, L. Sirdeshmukh, and K.G. Subhadra

81 **Metallopolymer Nanocomposites**
By A.D. Pomogailo and V.N. Kestelman

82 **Plastics for Corrosion Inhibition**
By V.A. Goldade, L.S. Pinchuk, A.V. Makarevich and V.N. Kestelman

83 **Spectroscopic Properties of Rare Earths in Optical Materials**
Editors: G. Liu and B. Jacquier

84 **Hartree–Fock–Slater Method for Materials Science**
The DV–X Alpha Method for Design and Characterization of Materials
Editors: H. Adachi, T. Mukoyama, and J. Kawai

85 **Lifetime Spectroscopy**
A Method of Defect Characterization in Silicon for Photovoltaic Applications
By S. Rein

86 **Wide-Gap Chalcopyrites**
Editors: S. Siebentritt and U. Rau

87 **Micro- and Nanostructured Glasses**
By D. Hülsenberg and A. Harnisch

88 **Introduction to Wave Scattering, Localization and Mesoscopic Phenomena**
By P. Sheng

89 **Magneto-Science**
Magnetic Field Effects on Materials: Fundamentals and Applications
Editors: M. Yamaguchi and Y. Tanimoto

90 **Internal Friction in Metallic Materials**
A Handbook
By M.S. Blanter, I.S. Golovin, H. Neuhäuser, and H.-R. Sinning

91 **Time-dependent Mechanical Properties of Solid Bodies**
By W. Gräfe

92 **Solder Joint Technology**
Materials, Properties, and Reliability
By K.-N. Tu

93 **Materials for Tomorrow**
Theory, Experiments and Modelling
Editors: S. Gemming, M. Schreiber and J.-B. Suck

94 **Magnetic Nanostructures**
Editors: B. Aktas, L. Tagirov, and F. Mikailov

95 **Nanocrystals and Their Mesoscopic Organization**
By C.N.R. Rao, P.J. Thomas and G.U. Kulkarni

96 **GaN Electronics**
By R. Quay

97 **Multifunctional Barriers for Flexible Structure**
Textile, Leather and Paper
Editors: S. Duquesne, C. Magniez, and G. Camino

98 **Physics of Negative Refraction and Negative Index Materials**
Optical and Electronic Aspects and Diversified Approaches
Editors: C.M. Krowne and Y. Zhang

MIX
Papier aus verantwortungsvollen Quellen
Paper from responsible sources
FSC® C105338

If you have any concerns about our products,
you can contact us on
ProductSafety@springernature.com

In case Publisher is established outside the EU,
the EU authorized representative is:
Springer Nature Customer Service Center GmbH
Europaplatz 3, 69115 Heidelberg, Germany

Printed by Libri Plureos GmbH
in Hamburg, Germany